Snipers at War

Snipers at War

AN EQUIPMENT AND OPERATIONS HISTORY

John Walter

Foreword

Martin Pegler

GREENHILL BOOKS

NAVAL INSTITUTE PRESS
ANNAPOLIS, MARYLAND

To Jack the Dog (2005–17),
who was always first in line when a biscuit was to be had.

———————

Snipers at War
First published in 2017 by
Greenhill Books,
c/o Pen & Sword Books Ltd,
47 Church Street, Barnsley,
S. Yorkshire, S70 2AS

www.greenhillbooks.com
contact@greenhillbooks.com

Published and distributed in the United States of America and Canada by the
Naval Institute Press, 291 Wood Road, Annapolis, Maryland 21402-5043

www.nip.org

Greenhill Books ISBN: 978–1–78438–184-4
Naval Institute Press ISBN: 978–1–68247–266–8

CIP data records for this title are available from the British Library
Library of Congress Cataloging Number: 2017949271

Printed and bound in England by CPI Group (UK) Ltd, Croydon, CR0 4YY

Typeset in 10.5/13.3 pt Minion Pro

Contents

Foreword

The history of accurate shooting is as old as projectile weapons themselves. From the inception of the first hand-bows capable of lethality at ranges far in excess of any other weapon, men have striven to improve the accuracy and range of the projectiles they shot. There were of course, limits to this where the power of stringed weapons was concerned, but the introduction of the firearm into Europe in the middle of the fourteenth century revolutionised warfare. The early 'handgonnes' were crude, inaccurate and incapable of shooting at the distance of a good longbow but progress was inexorable, as they turned from miniature cannon to musket. Yet, for centuries, shooting was a black art, few gunmakers understanding the chemical processes that provided the explosive power of gunpowder, or the physics that enabled a bullet to travel even a moderate distance and hit its target. In this new book John Walter traces this complex quest for accuracy, a process that on the surface may appear as nothing more than the application of science, but no such 'science' as we understand it existed until at least the mid-eighteenthth century. As John Walter lucidly reveals in this book, what happened was a very complicated fusion of many disparate elements: metallurgy, chemistry, optics and physics, that coalesced into what we now call technology.

This process did not simply happen overnight and the story, which is both engrossing and complex, is looked at in great detail. Exactly how gunmakers, chemists, foundry-men and a host of other individuals managed to perfect the manufacture of high quality steel, rifled barrels, more efficient propellants and aerodynamic projectiles is lucidly explained. With these advances came a burgeoning demand from civilian shooters for greater precision as hunting and the sport of shooting became increasingly popular, but there was also a delineation between the requirements of the military and those of the sportsman. Linear warfare as practised from the seventeenth to the mid-nineteenth century required speed of shooting, not accuracy. It was weight of fire that won the day on the battlefield, not the ability of a soldier to pick off an individual at long range. Much of what was learned about accurate shooting was actually due to the requirements of game hunters and shooting societies, not the demands of the military.

Undoubtedly the most significant advance was the introduction of the rifled barrel. Whilst riflemen had been employed during the eighteenth century, they were used in small numbers and John Walter critically examines how the issue of rifles to sharpshooters in the Crimean War and the American Civil War heralded the first major changes in tactical use of the long-arm since the introduction of the military musket. In fact, the lessons of the Civil War were to prove the foundation upon which all later sniping was to be built and the methods devised are still applied today with modern sniper training. As the author shows, this was just the beginning and the evolution of the military sniper from these early days is the subject of the later chapters. Military sniping, which emerged from the shadows during the First World War, had, despite senior commanders' indifference, progressed into a highly specialised form of warfare by the Second World War. Snipers were employed in huge numbers in every theatre; the Eastern Front, in Western Europe and across the Pacific. How the art of sniping had to be re-discovered during this period is an engrossing tale that continued into the world of 'small wars' post-1945.

From the mid-twentieth century the role of the sniper was re-evaluated, as armies began to adopt highly accurate sporting rifles in a combat sniping role and, as a result, converted military rifles were largely abandoned in favour of dedicated sniping rifles based on the designs of the finest target rifles in production. This development has not ceased in the twenty-first century; indeed, if anything, the march of progress has increased in pace, and John Walter looks at not only the huge improvements in modern weapon design and ballistics but also at the incredible advances in optical science and electronics that, combined, have enabled the modern sniper to become the most valued and highly trained specialist on the modern battlefield.

I found the book totally engrossing and, above all, infinitely readable and I challenge any reader, however knowledgeable, not to find some fascinating revelation in its pages.

Martin Pegler

Preface and Acknowledgements

When I began work on *Snipers at War*, I had no clear vision of what the end-product would be. So much had been written about snipers and sniping, particularly in the last decade, and so many memoirs had been published, that I saw my task as simply to collate existing information in a way that would have broad appeal.

This plan lasted only until I was asked to prepare abstracts for the Frankfurt Book Fair. This meant submitting material which had to be convincingly accurate. But even my first attempts revealed that the story was not as straightforward as I had assumed: 'facts' sometimes rested on the flimsiest evidence, claims didn't always have much basis in reality, and credible stories had been overlooked. The more I searched, the more I found; and the more I found, the more I needed to explain why details were being challenged.

The ever-increasing accessibility of on-line resources also shaped the way in which *Snipers at War* evolved. Digitisation of patents and individual servicemen's records, fast-growing genealogical archives, and the ease with which out-of-print books can be accessed have all been helpful. It may be prudent to add, however, that personal experience of on-line family trees, wonderful repositories of data though they can be, suggests that many are wrong – and that the errors range from comparatively minor, to countless generations derived from baseless assumption.

In his Foreword, Martin Pegler has very generously suggested that there is something in *Snipers at War* for everyone. This is what I was trying to achieve, of course, but it is difficult to do so within the limitations of even a few hundred pages. Consequently, the book has a conscious pre-Vietnam historical bias largely because, though even yesterday is history, secrecy can still inhibit even-handed summary. The true story of snipers in Indonesia or Sri Lanka, and now in Syria, will be told only when unchallengeable information becomes available.

Nor is *Snipers at War* intended to be a catalogue of equipment, instead concentrating only on selected weapons to allow the inclusion of genealogical research; information can be extracted from some of the books listed in the Bibliography, though *The Sniper Encyclopaedia*, intended for publication early next year, will fill many of these gaps.

Of course, I could not have completed *Snipers at War* without assistance. I am particularly grateful to Michael Leventhal, who commissioned the project and turned a not-so-blind eye as it grew and grew ... and grew; to Martin Pegler, who not only reviewed what I had written with a diligently critical eye, but gave me access to his American Civil War Sharpshooters book before it had been published; and to editor Donald Sommerville, who found many things which had eluded everyone else.

Adrian Gilbert, author of *Sniper: One-on-One* and *Stalk and Kill: The Sniper Experience,* helped me overcome problems during the formative stages of the book; Leroy Thompson supplied advice, based on first-hand Special Forces experience, and details of Simo Häyhä's career; Charles Sasser generously allowed me to quote from *One Shot—One Kill*; Mark Spicer, sniper-trainer par excellence, let me divert questions to his authoritative *An Illustrated Manual of Sniper Skills*; and Leigh Neville shared information published in *Guns of the Special Forces 2001–2015*. Alla Begunova, author of *Angely Smerti: Zhenshchni-snaĭpery, 1941–1945,* answered many of my questions about Soviet sniping; Rhoda Maling kindly lent me a copy of her friend Denis Edwards's memoir, *The Devil's Own Luck*; David Foreman, translator of the Nikolaev and Pavlichenko memoirs, helped to overcome problems arising from my sketchy knowledge of the Russian language; and Bill Harriman provided some vital snippets. I am grateful to Heckler & Koch, among others, for information from which accuracy could be analysed; and Tom Irwin and Alice Bond of Accuracy International were always among the first to respond to requests for pictures and information.

Images have come from many sources, usually credited in individual captions, but there are a few whose origins remain uncertain; I have tried to acknowledge copyright wherever possible, and will gladly correct errors or omissions in any future edition. I owe a huge debt to Lisa Oakes of James D. Julia, Inc., auctioneers of Fairfield, Maine (www.jamesdjulia.com), for patiently dealing with many requests for help. *Snipers at War* would be much poorer without so many excellent photographs.

But all would be for nothing had not Alison, Adam and Nicky supported me through awkward times. Our grandchildren – Findlay, Georgia and Holly – provided much-appreciated and often hilarious distractions.

John Walter
2017

From the Bow to the Bullet

The idea of competition has existed since Stone Age hunters first fastened flint heads to their staves and competed with rival families to bring down a mammoth. Those who threw consistently accurately would have risen in status. Power might have been one thing, but technique was another.

Millennia passed, but the principles remained the same. If a weapon could be projected, a premium was placed on skill. Competitive throwing of the Greek javelin is said to have been included in the Olympic Games as part of the pentathlon, from 708 BCE until 393 CE, and Roman spear-throwers were often lauded for their performance in battle.

The introduction of the ox-horn bow is lost in antiquity, but probably occurred to different people at the same time. It was to be a great advance in technology. No more were long-distance strikes confined to spears, which could easily be returned by opponents; if arrows were available, archers could fire at will. Engagement range increased considerably.

Specialist bodies of archers were recruited, allowing battles to be won simply by aerial threat. The death of Achilles in the Trojan War, according to Greek myth, was due to an arrow fired by Paris striking him in an unprotected heel. And the Battle of Hastings (1066) was influenced by the arrow that immobilised the English king Harold by striking him in the eye or, perhaps, the forehead. Though Harold's wound may not have been immediately fatal, it was serious enough to change the course of the conflict by distracting the defenders at a crucial moment.[1]

Even from the beginning, Mongol horsemen used the bow to engage specific targets; yet there was a tendency in European warfare to loose arrows as a volley or wave, relying more on quantity than individual target-selection to subdue opponents.

The first step towards reliance on ultimate accuracy came with the crossbow, which has a surprisingly long pedigree. The first crossbows to appear in Europe were the work of the Greeks, diminutives of the siege engines used by Greek and then Roman armies.[2] At much the same time, the crossbow was introduced in China – where, for perhaps the first time, mass-production techniques created the bronze multi-part trigger mechanism fitted to the bows

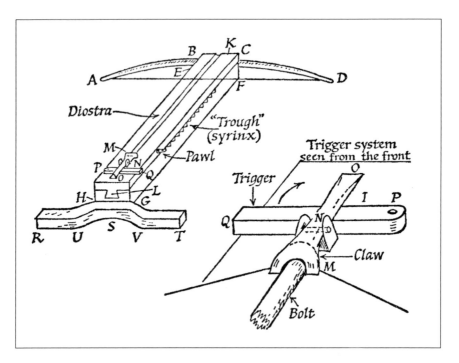

A bow-catapult or *gastraphētēs*, showing its similarity with some crossbows of the Middle Ages. The drawing is based on a description by Hero of Alexandria, writing in *Belopoeica* in the first century CE, from J. G. Landels, *Engineering in the Ancient World* (1997).

of the many crossbowmen of the Terracotta Army (and, presumably, those that were in use with the Chinese forces of the time).

In Europe, however, the heyday of the crossbow passed with the sack of Rome in 410 and the fall of the western Roman empire. The shortbow was already pre-eminent in Britannia, and similar weapons were favoured by the barbarian hordes that had raged across Europe.

By the time of the siege of the French city of Senlis in 947, however, the crossbow had reappeared in military service. It is said then to have come to England in 1066, with William of Normandy, 'the Conqueror', even though the Bayeux tapestry shows nothing but conventional bows.

Use of crossbows spread rapidly: perhaps because they were easy to shoot, if not necessarily to draw. Crossbows were used by Crusaders during the siege of Jerusalem in 1099, soon to be copied by the Saracens. Anna Komnene (1083–1153), the daughter of Emperor Alexios I of Byzantium, observed that the crossbow was a 'weapon of the barbarians, absolutely unknown to the Greeks ... a truly diabolical machine'.[3] Pope Innocent II clearly concurred with Anna Komnene. His predecessor Urban II had already objected to their use in 1096, and so, during the Second Lateran Council of 1139, Innocent banned not only crossbows but bows of all types used 'against Christians and Catholics'. This proscription had a temporary, but considerable effect in many European states.

Guillaume le Breton testified that the crossbow was 'unknown in France' until 1185, when Richard the Lionheart ordered the French to be instructed in its use. By the thirteenth century, crossbows were to be found in most military inventories. King John of England (reigned 1199–1216) had a company of mounted crossbowmen, as did Philippe Auguste of France and Friedrich II of the Holy Roman Empire.[4] The quantities involved could be surprisingly large. In 1295, for example, Philippe IV of France ('Philip the Fair') bought 2,000 crossbows from the *sénéchaussée* of Toulouse to arm men being raised to fight a war in Aquitaine. Philippe's agents also acquired in Bruges, among a variety of military stores, 1,885 crossbows and 666,258 quarrels to equip his fleet.[5]

Perhaps the best-known exponents of the medieval crossbow were the Genoese, whose state-supported bowmen were often hired to those who would pay for their services: the French and other Italian states. Yet the tactical deployment of crossbowmen was often handled badly by their masters. This was true of many individual campaigns in the Hundred Years War, including the great naval Battle of Sluys (24 June 1340), when English archers overcame the threat of crossbowmen aboard the French ships by loosing their arrows much more quickly and initially at a range the crossbow bolts could not match.

Crécy, Poitiers and Agincourt reinforced the supremacy of the longbow in the field. Crécy, fought on 26 August 1346, was a particular disaster for the Genoese, who had fled when faced with a 'veil of English arrows' – only to be ridden down by charging French knights, who paid little heed to the fate of men whom they regarded with contempt. The Genoese contingent, variously estimated as 2,000–6,000 strong, suffered terribly; their leader, Ottone Doria (or 'Odon Dorioa'), was just one among perhaps a thousand dead.

To counter charges of cowardice, the crossbowmen blamed a sudden downpour for soaking their bowstrings. The English, by contrast, had been able to detach their bowstrings instantly to keep them dry, but the design of crossbows prevented the Genoese from acting similarly. This explanation was disputed even by their French masters, who suspected that their crossbowmen had simply run away, but tests undertaken in the early 1890s by Ralph Payne-Gallwey provided support.

The laminated-bone bow-arms of many fourteenth-century crossbows were usually short and often straight. Consequently, the bowstring hung loose; inadequately waterproofed, if moistened by mist or rain it would slacken ever more greatly. The effects of moisture duly compromised the draw, considerably reducing the power that could be generated. Most later crossbows not only had the bow-arms bent so that tension kept the string taught, but lacquer and similar methods of waterproofing prevented loss of utility.

Crossbows have often been regarded as the 'poor relation' of the longbow – especially in England, where training with the longbow was still mandatory in the reign of Richard II (1377–99). A statute of 1389 had imposed on 'servants and labourers' an obligation to obtain bows and arrows with which to practise on Sundays and holy days at the expense of all other games.[6] Edward IV (king, 1461–83) ordered that all men should possess a bow of 'yew, wych, hazel or ash' of their own height, furnished with arrows the length of a man's arm. Henry VIII (1509–47) decreed not only that practice with anything other than a longbow was forbidden, but also that anyone found in possession of a crossbow would incur a fine of £10.

Henry also laid down in statutes the minimum distances at which men over twenty-four years of age were expected to practise: not less than 220 yards if 'flight arrows' were used, and not less than 140 yards with 'sheaf arrows'.

However, it was easier to fire a crossbow from cover or when prone. In addition, unlike the longbow, from which an arrow had to be loosed almost instantly, the interposition of a trigger in the crossbow mechanism allowed the bowman to wait patiently for the right moment to fire. The crossbow, therefore, was a far better stealth weapon.

Crossbows could be powerful, but had many serious disadvantages: they were comparatively complicated (becoming more so as the centuries passed) and notoriously difficult to draw. The lack of significant mechanical advantage in the short bow-arms compared with a longbow meant that the power needed to draw the bowstring was greatly increased. This in turn restricted the rate of fire, and also the engagement range.

Experiments reported by Payne-Gallwey[7] showed that a 14-inch bolt weighing about 3 oz, fired from a large siege bow, would fly through a ¾-inch deal board placed 60 yards away. Extreme range was reportedly 460 yards, but it is not clear how far the bow had been elevated to make the shot. Modern experiments suggest that the initial velocity of a medieval crossbow bolt was only about 175 ft/sec; yet the draw needed to cock Payne-Gallwey's bow was calculated to be about 1,200 lb – more than half a ton! – compared with only 120 lb for the most powerful longbow.

The crossbowman still presented a very real threat to the armoured knight, however, especially when bolts with a special head were used. Designed to minimise deflection, these had four small points on a squared tip. A Savoyard text of 1395 tells of armour that was certified to resist the impact of the bolt from a standard crossbow, and of a better grade that would resist a bolt fired from a more powerful siege bow (*arbalète à tour*).[8] However, at the Battle of Homildon Hill in 1402, Archibald Douglas, fourth Earl of Douglas, was severely wounded by five arrows which penetrated his mail-and-armour defences.

Trained longbowmen could fire as many as fourteen arrows in a minute, even if required to do so en masse; crossbowmen would do well to fire four quarrels or bolts in the same time, and then only if their bow was comparatively simple. Even in the earliest days, the bowman often lay on his back to place both feet on the bow-arms so that he could retract the bowstring until it was held on the 'tricker'. The introduction of foot-stirrups helped, as did technological advances such as the goat's foot lever and the *crannequin*[9] or windlass, the first of which appeared as the thirteenth century drew to a close. The problem was so extreme that crossbowmen were often provided with *pavises*, large man-height shields that were carried onto the battlefield by shieldsmen or 'pavisiers' to protect bowmen while they reloaded. In the Montaperti campaign of 1260, for example, the army of Florence deployed three hundred *pavesari* to protect a thousand crossbow-men.

But the luckless Genoese bowmen at Crécy lacked their pavises, which were with the laggardly baggage train when the battle began. There could be no protection from the English arrows that descended from the skies. It is still by no means clear if the retreat of the Genoese, who swore fealty not to France but only to themselves and their state, was panic-stricken or simply a tactical withdrawal. The French knights granted no quarter.

A depiction of the Battle of Crécy (1346), from a version of Froissart's *Chronicles* commissioned early in the fifteenth century by Flemish nobleman Louis de Gruuthuse, now in the collection of the Bibliothèque Nationale in Paris. Note the absence of pavises, and the crank-spanned bow in the left foreground.

A major advantage of the crossbow lay in its ability to engage individual targets at long range, which could be particularly beneficial if cover could be used advantageously. The defenders of castles and fortifications were, therefore, well placed to select targets at will. It had been recognised since the emergence of organised warfare that eliminating commanders was one of the best ways to win a battle.

To paraphrase Sun Tzu, it was necessary only to sever the head to kill a serpent. Roman legionaries had often gone to great lengths to protect the standards that customarily marked the position of their commanders, and such ideas persisted into the wars of the nineteenth century.

Many high-ranking officers fell victim to the crossbow bolt, deliberately assassinated, but at others times simply victims of combat. The former group may have included William II 'Rufus' of England, who was killed while hunting in the New Forest in August 1100 – an accidental ricochet, or perhaps a deliberate plot to place his brother on the throne.[10]

The latter group included Richard I ('The Lionheart', *Cœur de Lion*) of England, who was wounded by a bolt at the siege of Gaillon in 1192 and then fatally wounded seven years later at the investment of Châlus in the duchy of Aquitaine. Many, especially clergy, considered Richard's untimely death to be divine judgment: the inevitable consequence of accepting the use of the crossbow against fellow Christians, which had been proscribed sixty years previously by the Second Lateran Council.

Crossbows were still being used in combat in the sixteenth century, at a time before the gun had been developed sufficiently far to provide competition, and had had a lengthy target-shooting career. The earliest authenticated target-shooting competition of any type was held in 1286 in Schweidnitz, and a specially built target range in Görlitz had been constructed by 1377.

There were many who believed the bow to be a better weapon than the first guns. During the reign of Elizabeth I of England (1558–1603), Sir John Smith, one of her ablest generals, stated:

> I will never doubt to adventure my life, or many lives (if I had them), amongst 8,000 archers, complete, well-chosen and appointed, and therewithal provided and furnished with great store of sheaves of arrows, as also a good overplus of bows and bow-strings, against 20,000 of the best harquebusiers and musketeers there are in Christendom.[11]

A company of bowmen was raised to guard King Charles I as late as 1643, almost as soon as the Civil War began; and a competition held in the summer of 1792 in the Cumbrian village of Pacton Green, when twenty arrows and twenty musket-shots were fired at a target with a diameter of one clothyard (about forty inches) placed a hundred yards away, was won by the bowmen with sixteen hits. Their opponents had managed only twelve.

'Shooting the Popinjay' was promoted from the thirteenth century onward. Usually undertaken in Europe with crossbows, often of exquisite quality, this involved knocking a brightly coloured wooden parrot or 'Papingo' (derived from the medieval French *popinjay*) from the top of a pole or tree. In Prussia in 1354, the Grand Master had ordained the erection of a suitable target in each and every city in his dominion.[12]

The appearance of the first cannon on the battlefield, its provenance still hotly disputed, was to end the supremacy of the bow. At the beginning, however, this was not obvious; the first cannon were cumbersome, slow-firing, prone to explode, and could hurl projectiles only short distances. Only when steps had been taken to transform what would in modern parlance be called 'crew-served' into individual weapons was progress made.

This is not the place to give a detailed history of the earliest man-portable firearms, except to observe that once the crudity of the first matchlocks gave way to the surprisingly sophisticated wheel-lock – an obvious sign not only of technological progress but also that the status of the gun had risen – it was a short step to universal rearmament. The wheel-lock was often the plaything of the rich, but greater nobles such as the Elector of Saxony were sufficiently rich and powerful to arm corps of bodyguards with firearms.

The earliest wheel-lock guns were made by clockmakers in southern Germany and Bohemia, and the combination of reliable ignition, long barrels and comparatively small calibre often gave surprisingly good accuracy. This suited the firearm not only to target-shooting but also to an assassin. Several states attempted to ban guns, and especially short-barrelled pistols that could be easily concealed. However, armies saw in the arquebus and its successor, the musket, a way of equipping large numbers of ill-trained men with a weapon that was far easier to master than a bow.

Credit for introducing the first 'standard issue' firearm is usually given to Gustav II Adolf of Sweden ('Gustavus Adolphus') during the Thirty Years War of 1618–48. At this time, the matchlock was still the weapon of the infantry and wheel-lock pistols served the cavalry. When the wheel-lock, difficult and time-consuming to make, gave way to the simpler flinted locks in the seventeenth century, the idea of production in quantity became a reality.

This was not the mass-production method of later days, but merely ensured that the products of countless individual gunmakers conformed to basic standards: the calibre, for example, or the length of the barrel. Individual muskets still varied appreciably from each other, as the only way of obtaining large quantities was to recruit as many individual gunmakers as possible. Each man had his own ideas and techniques, and so each gun was effectively unique.

Yet performance steadily improved. Though most armies relied on the smashing power of volleys fired by serried ranks of well-drilled men, and though the accuracy of the smooth-bore infantry musket was very poor at any range beyond a hundred yards – compressing the battlefield into a narrow strip, filling the air with clouds of powder smoke sufficient to prevent individual targets being engaged – the value of rifling had been appreciated for many years. Though widely attributed to the English scientist Benjamin Robins and his contemporaries in the mid-eighteenth century, the idea of stabilising a projectile by spinning it in flight had been known to bowmen many centuries previously.

The earliest date on which rifling gun-barrels was practised has always been disputed, largely because no evidence has ever been produced to back any of the claims. It is possible that the earliest guns had straight grooves, intended to minimise the effects of fouling, and that the benefits of rifling arose as much by chance as design.

Predilection for decoration, generally with sculptural qualities, could have led an enterprising gunmaker to twist a barrel formed with flats – usually octagonal – to give a spiralling effect. This would have twisted the straight-cut grooves, of course, and the gun would have shot far better. Once established, the idea that twist-rifling improved accuracy would then have become commonplace within a few years.

August Kötter of Nuremberg, who died *c.* 1525, is often identified as the inventor of twist-rifling, even though it probably originated a generation previously. It seems likely that the straight grooves came first (there is no real confirmation), *c.* 1460, and that the twist-type followed ten to fifteen years later.[13] The oldest surviving rifle of this type, dated 1476, is owned by the Armeria Reale in Turin, and another with multiple grooves, *rayée à mèche* or 'rifled with stripes', once the property of the Emperor Maximilian, bears arms that can be dated precisely to 1486–93.[14]

Shooting competitions are mentioned as early as 1426, and by 1472 were being contested regularly in southern parts of Germany and throughout the cantons of Switzerland where crossbow-shooting was still popular. In 1563, a competition advertised in Berne classified guns with straight and twist-type rifling in separate categories. By the end of the sixteenth century, however, straight-groove rifling had been largely discarded.

A few guns were made by hammering strip-iron around a special mandrel and then, once the mandrel had been removed, twisting the entire barrel to give a bore with the cross-section of a heart or a star. It is possible to see in these an influence behind the Whitworth rifle and other designs with polygonal bores of the nineteenth century.

The rifle soon developed into a weapon of precision, by sophisticated multi-lever triggers and improvements in sights that included an occasional 'sighting tube' in which inspiration for the optical sight may be seen. Surprisingly sophisticated aperture sights also appeared in the sixteenth century, with windage and elevation controlled by screw-wheels. Improved performance

The renowned Dutch admiral Maarten Tromp was killed by an English marksman during the Battle of Scheveningen (1653). It is assumed that the sailor used a flintlock musket, probably of the so-called 'dog-lock' pattern common in England at the time. Tromp's death had an important effect on the course of the battle, which, though inconclusive, did enough damage to the Dutch fleet to ensure an English victory in the first Anglo-Dutch War.

encouraged some far-sighted noblemen to form corps of riflemen, beginning in Hesse in 1632. The riflemen were recruited from *Jäger* and *Förstleute*, hunters and foresters, who required little tuition in the arts of observing, tracking and shooting.

Where the Elector of Hesse led, others were to follow until, by the eighteenth century, almost all central European armies (even those of minor states and principalities) had corps of riflemen. In Britain, however, only the large-bore 'Brown Bess' musket was approved for rank-and-file until long after the Napoleonic Wars had ended. The British had, perhaps, missed a trick. Though the principal firearm of the English Civil War (1642–9) had been a musket, usually a matchlock, the first 'turn-off' barrels had appeared. Though these guns looked like muzzle loaders, they were charged by unscrewing the barrel, placing the ball and the powder in the chamber, and then screwing the barrel back against the frame. Made with early forms of flinted lock, guns of this class could be exceptionally accurate.

There is no real evidence that marksmanship was used deliberately to eliminate officers or artillerymen on battlefields such as Marston Moor, but the capability was there. On 31 July 1653, shortly after the Civil War had ended, Lieutenant-Admiral Maarten Harpertszoon Tromp, commander of the navy of the Republic of the United Netherlands, was killed during the Battle of Scheveningen by a sharpshooter stationed in the tops of the British ship *James*, commanded by Admiral Sir William Penn.

The global situation was changed by conflict in North America. The Seven Years War (1756–63) was a struggle for supremacy between the British and their allies on one side and the French and their allies on the other. The British had seized French territory in North America, and a series of bloody campaigns ensued. Eventually, General Wolfe defeated the forces of the French General Montcalm on the Plains of Abraham above Québec on 13 September 1759, and the Battle of Quiberon Bay (20 November 1759) proved to be similarly conclusive at sea.

Louis-Joseph de Montcalm-Gozon, Marquis de Saint-Veran, did not survive the decisive attack on Québec, dying a day later from a musket-ball in his back. Wolfe had been struck three times in the arm, shoulder and chest during the advance, though there is nothing to give credence to claims that any of these shots was the work of a sharpshooter. Those who attended the dying general attributed the hits to the fortunes of war and to Wolfe's fateful decision to lead his men by example.

The war in Canada and the northern part of what was to become the United States, where hostile terrain made conventional lines-of-battle impossible, promoted not only skirmishing but also irregular warfare of unbridled savagery in which Native Americans played a leading role. The British and the French both made good use of these tribes, whose hunting, tracking and observational skills were far better than those of European soldiers pitched into a campaign dominated by the spectacular geography of northern America.

Among the renowned irregulars were Rogers' Rangers, raised by the Massachusetts-born Robert Rogers (1731–95) initially for service in New Hampshire, who were destined to play an active part in the Seven Years War and are now hailed as the precursors of today's Special Forces. Dressed largely in green, Rogers's men gained a reputation for daring raids – admittedly, sometimes inflated by optimistic claims – and highly effective intelligence-gathering. Marksmanship was just one of their attributes.

British commanders who had seen the value of Rogers's sharpshooters at first hand, or had made good use of Iroquois scouts, soon demanded riflemen of their own. But few recruits were to be readily found in the British Isles, nor was British gunmaking organised in such a way that rifles could be obtained in quantity. The answer was to recruit from the German states that supported the British cause: Hanover, Brunswick–Wolfenbüttel, Hesse-Kassel and Schaumburg–Lippe. Virtually all of the guns were obtained in Germany and Bohemia, though William Grice and Robert Wilson copied guns sent home from the America colonies in the late 1750s.

This was at a time when the *Jäger* rifle had become regulation issue in the armies of Austria, Prussia and even Russia, though the earliest representatives, acquired from Zella St Blasii and Suhl, seem to have been issued in 1710–11 to sharpshooters in the pay of the king of Norway. Ironically, Charles XII of

Denmark, who had been wounded by a Russian marksman in 1709, was killed by one of the Norwegians in 1718.

Jäger rifles usually had short heavy barrels, spurred trigger guards, plain stocks with patch boxes in the butt, and furniture cast of iron or brass. The barrels were often 'swamped', increasing in diameter towards the muzzle, so that the additional weight helped to control the shot, and a set-trigger mechanism was commonly fitted. Most of the *Jägerbüchsen* had bars brazed to the side of the muzzle to receive a sword bayonet, though some state armies preferred to issue short swords.

Among individual *Jägerbüchsen* was the Austrian M1768, used by the élite marksmen of the frontier guards, which was double-barrelled and had a separate flintlock on each side of the breech. The top barrel was rifled; the lower barrel was a smooth-bore, and the gun, which weighed about 5 kg, was issued with a combination forked-rest and pike to steady the firer's aim.

The first Prussian *Jägerbüchse*, known eventually as *Alte Corpsbüchse*, dated from 1744, introduced by Frederick the Great for the *Feldjäger zu Fuss*, and last was the *Neue Corpsbüchse* of 1810. The Austrians issued their first rifle of this type in 1745, to the *Jäger-Trupp* raised from riflemen living in the Tirol. Each man was expected to provide his own firearm and, therefore, there was no real standardisation until the first regulation pattern was approved in 1759. Developed as part of the reforms instituted after the disastrous defeat of the Prussians by the forces of Napoleon Bonaparte, the *Neue Corpsbüchse* was the only one of the state's flintlock *Jäger* rifles to have a set-trigger.[15]

The Baker rifle could be seen as a logical consequence of the use by the British and their allies of *Jägerbüchsen* during the American wars. When the War of Independence began in 1776, the problems that had faced the British during the Seven Years War resurfaced. Soldiers serving in North America had been made well aware of the ability of colonist marksmen.

The deaths of Major-General Edward Braddock, shot through the chest during the Battle of Great Meadows on 9 July 1755, and of Brigadier-General George Augustus Howe, fatally wounded during a skirmish on 6 July 1758, were widely attributed to sharpshooters. Both deaths had important effects on the campaigns: Braddock had been the commander-in-chief of the British Army in North America, and Howe was second-in-command of British forces which were to be decisively defeated at the Battle of Carillon a day after his death.

Target-shooting was a popular if usually informal pastime, and those who lived in the wilderness had to shoot to live. This did not accord with British experience, as the average soldier had rarely shot anything other than a fowling piece before military service. When fighting began, the reputation of colonial riflemen was at its height. Indeed, the earliest battles of the War of Independence – some no more than extended skirmishes – were generally influenced by the high standing of these men, if only rarely by their performance. De Witt Bailey has suggested that:

> The idea that 'the rifle won the Revolution' originated in an era . . . when local incidents became accepted facts . . . and these generalities were expanded upon for well over a century by patriotic but hardly objective American writers. It is equally incorrect to imagine, in the time honoured conception, American armies composed of sharp-shooting riflemen. In actuality, the American armies were largely musket-armed and trained, and throughout the war attempts were being made to reduce the proportion of rifles in use in favour of the musket.
>
> All of the major battles fought during the American War were conducted in more or less normal European fashion. Indeed, until American troops learned to withstand the psychological effect of British bayonets the field inevitably went to the British. The stereotype impression of massed redcoats confronting scattered riflemen who shot from behind trees and fences – and thereby won the engagement – is almost pure fiction.[16]

Senior American commanders considered their riflemen to be vulnerable in battle, particularly if caught in the open or while reloading. Infantrymen would usually fire 'running ball' in such situations, sacrificing efficiency to speed, but the way in which their guns had to be loaded denied riflemen such an option. On 27 August 1776, during the Battle of Long Island, American riflemen had been routed by a bayonet charge.

The course of battles could also be transformed by individuals, among them the renowned sharpshooter Timothy Murphy. Unfortunately, the story of Murphy's marksmanship during the Battle of Bemis Heights (1777, also known as the Second Battle of Saratoga) has become so firmly entrenched in the mythology of the Revolutionary War that it is impossible to distinguish fact from fiction.

The date of Murphy's death is well established, but paucity of records from the mid-eighteenth century – a time when North America was still being torn apart by colonisers, principally the British and the French – hinders any attempt to define the place and date of his birth. Consequently, there are at least three entirely different versions of the story with claims to being 'factual'.

It seems most likely that Murphy was born in New Jersey, probably in 1751; he may have been christened on 23 March 1755 in Minisink, Sussex County; and it seems that he lived in New England for most of his life.

The most lurid of the alternatives suggests that Murphy was born 'William Baskins' in 1753, on an island in the Susquehanna river. His father was then killed during an Indian raid, and his mother and elder sister Margaret were taken captive. The boy remained with the tribe for several years, but was then released when a ransom was paid by a British officer, William Johnson, who had him re-baptised as 'Timothy Murphy' (perhaps the surname of his prospective adoptive parents). When war began, he joined the Continental Army and fought at Saratoga in 1777.

This story overlooks the probability that other 'Timothy Murphys' lived and fought through the American Wars. Yet it contains elements of truth: the *Pennsylvania Gazette* reported the raid in which the Baskins were taken on 29 July 1756, and there is evidence to show that daughter Margaret was eventually freed in 1764. But it also seems likely that the boy, 'Timothy Murphy' for the rest of his life, became a blacksmith in Fort Michilimackinac, Michigan, and died in 1840.[17]

The military career of Timothy Murphy is also contested. One account has him, 'accompanied by his brother John', enlisting with 'Captain John Lowdon's Company of the Northumberland County Riflemen' on 29 June 1776, then progressing to Morgan's Riflemen by way of the 12th Pennsylvania Regiment. But depositions placed in 1925 with the Connecticut Society of the Sons of the American Revolution by Alonzo and Ezra Cornell, great-great-grandsons of Rifleman Murphy, suggest that he had served with the 1st and 4th Pennsylvania Regiments, quoting *Military Records of Schoharie County* by George Warner as a source. However, Murphy was undoubtedly one of 500 marksmen picked by Colonel Daniel Morgan for service in Morgan's Riflemen; and there is equally no doubt that Murphy was on Bemis Heights on 7 October 1777.

The battle seemed to be swinging to the British when General Benedict Arnold, spotting the influence of an officer on a grey horse urging his men onward, said to Daniel Morgan: 'That gallant officer is General Fraser. I admire him, but it is necessary that he should die, do your duty' (a remark often attributed to Morgan himself). What happened next is disputed.

The most popular story has Murphy firing four rapid shots, using his double-barrelled flintlock rifle. The first shot cut through the horse's crupper (one account claims that it tore away part of the bridle); the second brushed the horse's mane 'behind the ears'; the third fatally wounded Fraser; and the fourth struck Sir Francis Clerke. Or, alternatively, the first shot missed; the second struck Fraser's horse; the third struck Fraser; and the fourth hit Clerke.

But Murphy would have had to reload both barrels of his rifle between two groups of shots. Even assuming the tale that he had climbed a tree is discounted in favour of merely using a fork in the tree as a rest, he would have lost time. So did he fire as part of a volley – a sensible precaution if the chances of a kill were to be as high as possible?

Incontrovertible evidence has yet to be produced to back claims that Timothy Murphy alone was responsible for the deaths of Fraser and Clerke. The oldest of supposed eye-witness testimonies, that of Revolutionary War army officer and sometime congressman Ebenezer Mattoon (1755–1843), published on 10 November 1835 in the *Saratoga Sentinel*, stated that an 'elderly man, with a long hunting gun . . . at that instant discharged his gun, and the general officer pitched forward on the neck of his horse'. Murphy, who was in his mid-twenties at the time of the battle, does not fit this description. However, a history of Schoharie county written in 1845 by Jeptha Simms suggests that Morgan had:

> Selected a few of his best marksmen . . . 'instructed to make Fraser their especial mark' and that 'Timothy Murphy . . . was one of those selected'. [Each man had] a chance to fire, and some of them more than once, before a favourable opportunity presented for Murphy; but when it did, the effect was soon manifest. The gallant general was riding upon a gallop when he received the fatal ball, and after a few bounds of his charger, fell mortally wounded.

This is probably the earliest published record to credit Timothy Murphy personally. Simms claimed to have been given the details by one of Murphy's sons; the sharpshooter had died in 1818, but had left several children who could have supplied information.[18] The story is partly supported by a letter written on 28 November 1781 (just four years after the event) by Joseph Graham, a British officer captured at Saratoga with Burgoyne. Graham quoted a conversation in which, without naming Murphy personally, Morgan had said:

> I saw they were led by an Officer on a grey horse – a devlish brave fellow; so, when we took the height a second time, says I to one of my best shots . . . you get up into that there tree, and single out him on the white [grey] horse. Dang it, 'twas no sooner said than done. On came the British again, with the grey horseman leading; but his career was short enough this time. I jist tuck my eyes off him for a moment, when I turned them the place where he had been – pooh! he was gone.[19]

The balance of probability is that Murphy did fire the decisive shots, but there must still be an element of doubt. Ranges up to 800 yards have been claimed for the fatal hits, but a test undertaken on the battlefield by John Plaster, using a laser rangefinder, gave 330 yards. This is well within the capabilities of a long rifle, but would also testify to the accuracy of Murphy's shooting.

The victims were Brigadier-General Simon Fraser of the 24th Foot, born in Balnain in Inverness-shire in 1729, who had been appointed to his command in Québec in 1776 by Sir John Burgoyne himself; and Sir Francis Carr Clerke, 'seventh Baronet of Hitcham in Buckinghamshire', adjutant of the 3rd Regiment of Foot Guards and aide-de-camp to Burgoyne, whose dispatch Clerke was unlucky enough to be delivering to Fraser.

Contrary to widely expressed belief, Clerke was not a high-ranking officer, and neither his nor Fraser's wounds were immediately fatal.[20] Yet the death of

General Fraser had a dramatic influence on the campaign fought in and around Saratoga, as it led directly to British defeat and the surrender of Burgoyne. It scarcely matters who fired the fatal shot.

However, there was one occasion in the war when the British would have benefited had the marksman chosen to fire. The shooter was Captain Patrick Ferguson (1744–80), the inventor of the breech-loading rifle that could, with better luck, have made its name in combat. An opportunity was lost when the Briton declined to shoot a man in the back. On 11 September 1777, during the Battle of Brandywine some weeks after the brief encounter, Ferguson was hit in the right arm by a musket ball. While recuperating in hospital, he learned from a surgeon that some injured Patriots had confirmed that Washington was most probably the soldier Ferguson had seen. The Briton was unrepentant.[21]

The encounter with Washington was once questioned, particularly as one of the two men reconnoitring troop dispositions was dressed 'as a hussar' – a type of cavalryman the Continental Army did not then deploy. But the second man is now usually identified as Polish-born Count Casimir Pulaski (Kazimierz Michał Władysław Wiktor Pułaski, 1745–79), commander of Pulaski's Legion and regarded as one of the 'fathers of the US Cavalry'. Pulaski is known to have favoured the uniform of a French hussar.

A good-quality Ferguson sporting rifle by Barber of Newark, with the breech open. The rapid-thread plug is lowered simply by rotating the trigger guard. The ball, wad and powder can then be dropped into the chamber and the plug raised behind them.
Courtesy of James D. Julia, auctioneers (www.jamesdjulia.com)

It also seems probable that Ferguson, from an aristocratic background,[22] had seen English sporting guns made on the 'Chaumette principle'. Sealing the breech with a detachable plug had been practised for many years, dating back at least to the middle of the seventeenth century as wheel-locks dating from 1650–65 still survive. The chamber could be exposed simply by screwing a vertical plug at the breech up or down, depending on individual construction. A ball was pushed into the chamber with a finger, a wad could follow optionally, and then the remaining space was filled with black powder. The plug was wound back into place, and the gun could be fired.

The first to turn the plug with a lever doubling as the trigger guard may have been Frenchman Isaac de la Chaumette, who is said to have demonstrated a carbine of this type to the King of France in 1704 and to have received an English patent in 1721.[23] Chaumette's system was stronger than most other breech-loaders of its day and guns were made in some numbers by Samuel Bidet, a fugitive Huguenot gunmaker who had settled in London early in the eighteenth century. They co-existed with an older, more traditional design (smooth-bore or rifled) which still needed a separate spanner to rotate the plug.

Many English drop-plug sporting guns made in 1740–70, by gunmakers such as Wogdon and Heylin, combined the plug with the trigger guard. Several turns were usually required to expose the chamber, and so one of the principal advances claimed by Ferguson in his English patent, No. 1,139 of 2 December 1776, was the use of a rapid-pitch thread to drop the plug simply by a three-quarter turn of the trigger guard. In addition, deep grooves were cut in the breech-plug – crossing the multiple screw-threads – to carry grease.

Drop-plug guns were prone to foul, particularly if laid aside without cleaning after a few shots had been fired, and dried fouling was often bad enough to lock the plugs so tightly that tools were required to remove them. Ferguson elected to use grease in the hope that fouling would be minimised, and modern trials have tended to support his views. The rifle demonstrated to British officials, and then to King George III, was fired continuously without (it is said) a single hitch.

There is little doubt that, if minor faults were overcome, the Ferguson could have been an outstanding weapon, far better than the Baker rifle that was to become standard issue. One of the greatest strengths was the speed with which it could be loaded, without compromising performance, and the ease with which it could be used behind cover. Accuracy was also far better than any smooth-bore musket. The trials had been undertaken at distances as great as 200 yards without missing the target, and a claim that the Ferguson retains sufficient accuracy to pose a threat at 300 yards seems reasonable. This is at least four times the effective engagement range of a Long Land Pattern 'Brown Bess' musket.

The death of Patrick Ferguson at the Battle of King's Mountain, engraving by Charles Jeens, after Alonzo Chappel. *Anne S. K. Brown Military Collection*

Patrick Ferguson had been granted permission to raise a corps of riflemen, 200 strong, mainly from the 6th and 14th Regiments of Foot, but the sudden outbreak of war in America forced a reduction in their numbers to just a hundred, largely because rapid training had become paramount. Ferguson and his men then sailed for America, and were soon performing well enough to attract the approval of General Howe. Unfortunately, after Ferguson's serious injury at Brandywine, the rifles were placed in store – from which some or all of them may never have reappeared – and the men rejoined the line infantry.

On his return to service in May 1778, Ferguson served in New Jersey under Sir Henry Clinton. He was commissioned on 25 October 1779 as a major in the 71st Foot, and became Inspector of Militia in the Carolinas on 22 May 1780. Patrick Ferguson was to recruit Loyalists to fight for the British cause, which was deeply unpopular with local Patriots, but his renowned even-handedness soon gave way to frustration and then the ill-treatment of those who declined to follow him.

Loyalist troops were signally defeated by Patriot riflemen at the Battle of Musgrove's Mill, and Ferguson – with one of his mistresses! – fell on 7 October 1780 during the Battle of King's Mountain. He and his rifle were soon consigned to history.

The form of the rifles used by the colonists has been almost universally mis-represented as the 'long rifle' of later days: long and elegant, of comparatively small bore, well-made with decorative brasswork. However, guns of this type were not being made in the 1770s; they belong to a later era, appropriate more to the War of 1812 than the War of Independence.

The first rifles to be used in North America were generally brought to the country by immigrants from central Europe, and the first generations of American gunsmiths merely refined the designs with which they were most familiar. Short-barrelled *Jäger* rifles were not ideally suited to North American conditions, where game was often shot 'for the pot'. The need to shoot birds and small mammals favoured a smooth-bore, which could fire a single ball or small-diameter shot with equal facility, and the need to conserve weight while simultaneously lengthening the barrel led to generally refined contours.

Calibre was universally reduced to conserve lead, which was always in short supply, and the thickness of the barrel-walls was measurably reduced. Very little metal-bearing ore was being mined in North America in 1775, and the processed iron was generally acknowledged to be poor. Barrel-makers preferred to use old horseshoe nails re-formed into soft iron strips, which offered better quality and allowed thickness to be reduced without compromising strength.

Gradually, the stubby *Jäger* rifle gave way to the long-barrelled 'Kentucky Rifle' (something of a misnomer, as comparatively few of the guns were actually made in Kentucky). The 'Long Rifle' of 1775 usually had a bore of .50–.55, rifled with six or eight deep grooves, needed to minimise the effects of fouling from poor-quality powder. Rifling customarily made a turn in about four feet, but there was no consensus and a great many variations will be found.

The guns made in Pennsylvania and neighbouring states at the time of the War of Independence were usually plain-stocked of maple, with comparatively little drop at the butt-heel and a notably thick wrist. The attenuation of the fore-end draws attention to the lock, which is generally cruder and proportionately larger than those of nineteenth-century guns.

Furniture, invariably brass, was confined to a small nosecap (or 'stock cap'); a trigger guard, often spurred; a butt plate, sometimes separated into heel and toe plates; and a patch-box cover, occasionally with decorative side plates. Very few of the guns have set-triggers, a single trigger with a backward-curling tip being more common. The lock generally had a single bridle, a flat base plate with two lines scribed across the tail, a flat-body cock, and an angular pan.

The so-called 'Southern Rifle', made and used in the southernmost colonies, was a cruder affair stocked in whatever wood was available – some are walnut, others are ash – and with furniture reduced to a minimum. Most guns have nothing but a trigger guard, though nosecaps and butt-plates will occasionally be found. And the patch-box so characteristic of Pennsylvania rifles was usually absent; a large-diameter hole bored in the butt was often deemed sufficient to retain enough grease or candle-wax for the patches.

Rifles issued to the British Army after 1776 came, once again, largely from Hanover, Hesse and other allied German states. Most of them differed from the guns used in the Seven Years War only in detail. William Grice made about 700 of a .54-calibre adaptation of the German guns in 1776, which were sent to North America to arm light infantrymen deployed as skirmishers. These rifles were similar to the *Jägerbüchsen*, but had 36¾-inch swamped barrels and locks of typically British military pattern. They had plain walnut stocks with a sliding wooden patch-box cover on the right side of the butt, and a spurred, cast brass trigger guard.

This precise design was not perpetuated, but elements of it could be seen in the short-barrelled Baker rifle of the Napoleonic Wars (1793–1815). The first of its type to be standardised for issue in the British Army, developed by London gunmaker Ezekiel Baker, the regulation rifle could easily out-perform the New Land and India Pattern smooth-bore muskets, posing a very real threat at a range of 200 yards.

There was little novelty in Baker's design, which showed Dutch influences and had simply amalgamated several desirable features. The lock was initially a reduced-scale version of the New Land Pattern, though, after *c.* 1806, a flat ring-neck cock, a raised pan, and a safety bolt let into the tail of the lock-plate were added. Rifles were made prior to 1803 in .750 'Musket' and .625 'Carbine' bores, but the latter was preferred thereafter. The lighter carbine ball, which attained higher muzzle velocity, proved to be more accurate at long range.

The perfected .625 rifles were about 46 inches long, with 30-inch barrels, and weighed 9 lb. A bayonet bar was brazed to the right side of the muzzle, and the standing-block back sight – unmarked, but suitable for shots taken at distances up to 200 yards – had a small folding leaf set for 300 yards. A cheek piece usually lay on the left side of the butt, a brass-lidded patch-box was cut into the right side of the butt, and the slightest suggestion of a pistol grip was accentuated by the spurred cast-brass trigger guard.

The butt was generally angled so that the shoulder plate was vertical and the comb sloped up to the heel. This 'straight line' construction directed recoil into the firer's shoulder in such a way that, or so it was hoped, muzzle jump was minimised.

The 'Experimental Corps of Rifle Men' was raised in the spring of 1800 from fourteen regiments of foot, with an establishment of eight companies under the command of Colonel Coote Manningham. It was recognised in October as the '95th Regiment' (though seniority was back-dated to 25 August 1800 to mark the first combat engagement at Ferrol in the Peninsula). Eight hundred Baker rifles had been ordered from several London gunmakers, and it seems that most had been delivered into store by mid-1801. The riflemen of the 95th wore green and black instead of red and white, and were intended to act as skirmishers and sharpshooters whenever appropriate.

Marksmanship was not only practised but openly encouraged, and many a memoir testifies to the effectiveness of the riflemen on the battlefield. The French came to fear the prowess of what had become the 95th (Rifle) Regiment in 1802 and was to be renamed as The Rifle Brigade at the end of the Napoleonic Wars.

In his book *With British Snipers to the Reich* (1948), Clifford Shore drew attention to the performance of the 60th Regiment in the Peninsular War when, quoting Marshal Soult, 'A battalion of ten companies . . . is armed with a short rifle; the men are selected for their marksmanship; they perform the duties of scouts, and in action are expressly ordered to pick off officers, especially Field or General officers.'

The 60th subsequently gained many battle honours in the Peninsula as a 'Regiment of Foot': a status retained until 1824, when it was elevated to 'Rifle Corps'; the prefix 'King's' was granted only in 1830. However, a 5th Battalion of the 60th had been raised at Cowes on the Isle of Wight, apparently in 1797, to act as skirmishers. These men wore green jackets with red facings, but claims that they were armed with the Baker rifle are misleading; the first weapons to be issued seem to have been light-infantry fusils, similar to Land Pattern muskets, but of smaller calibre and fitted with shorter barrels. Baker rifles do not seem to have been substituted for these smooth-bores until *c.* 1805, once the merits of rifling had been demonstrated by the 95th Rifles and wider issue had been sanctioned.[24]

Thomas Plunkett of the 95th Regiment kills General Colbert with a long-range 'back position' shot from his Baker Rifle. Artist Harry Payne has drawn the lock of the Baker too far forward, but has otherwise captured the scene convincingly. *Author's collection*

Surprisingly large quantities of Baker rifles were made. By the end of 1805, output is said to have totalled 2,757 and enough barrels and locks had been supplied to 'set-up' another 3,000. By the end of the Napoleonic Wars in 1815, nearly 50,000 Bakers had been made by the gun trade in London and Birmingham. Several thousand guns had even been sent to arm Portuguese *caçadores*.

On 3 January 1808, during the Battle of Cacabelos, on the road leading back to Corunna, Rifleman Thomas Plunkett of the 95th fired the famous long-range shots that fatally wounded French Général de Brigade Auguste-François-Marie de Colbert-Chabanais and then killed one of Colbert's aides (sometimes said to have been the 'Trumpet-Major') who had rushed to the stricken general's side. The distance has been claimed to have been anywhere between 200 yards and 800 yards, but would undoubtedly have been much closer to the former than the latter. The implication is that Colbert judged himself to be safe as he would have been outside musket range.

> Seeing Colbert charging ahead of his men, distinctive because of his uniform and grey horse ... [Plunkett] raced ahead of the line and onto the bridge. Throwing himself onto his back and resting his Baker Rifle on his crossed feet with the butt under his right shoulder in the approved manner, Plunkett fired at and killed Colbert.[25]

Plunkett apparently used the popular 'back position', resting the rifle on his right calf which had been bent across his left knee. Captain Henry Beauffroy, writing in *Scloppeteria* later that year, described this as 'not only awkward but painful', though it became very popular in nineteenth century target-shooting circles and a variation was still being recommended during the Second World War if the topography of the shooting-site was favourable.

The French had had an opportunity to field riflemen of their own, as a suitable rifled carbine had been adopted in 1793. However, the commander of the French troops in the Peninsula and many of his senior officers saw riflemen as undisciplined wasters of ammunition who posed a threat to the discipline of the firing line by deliberately targeting enemy officers. This the French officers considered to be ungentlemanly, but events soon proved their views to be a grave tactical error.

At sea, marksmen in the mast-tops of men-o'-war, regardless of their nationality, were charged with eliminating as many officers as could be seen on the deck of an opposing vessel. This was particularly important as ships often lay alongside in the heat of battle, each attempting to take the other by boarding, and even a momentary loss of control could be the difference between success and failure.

Consequently, there were many high-ranking casualties. Probably the best known is Horatio Nelson, struck down on 21 October 1805 by a ball fired from the maintop of the French battleship *Rédoutable*. The marine who claimed to have fired the shot, Robert Guillemard, recalled that:

> The two decks were covered with dead bodies . . . I perceived Captain [Jean-Jacques] Lucas motionless at his post, and several wounded officers still giving orders. On the poop of the English vessel there was an officer covered with orders, and with only one arm. From what I had heard of Nelson, I knew it was he. He was surrounded by several officers, to whom he seemed to be giving orders. I saw him quite exposed and close to me. I could even have taken aim at the man I saw, but I fired at hazard among the groups of sailors and officers. All at once I saw great confusion on board the *Victory*; the men crowded round the officer whom I had taken to be Nelson . . .[26]

Nelson's death occurred when honours were largely won and had little effect on the course of battle. This was not always so, as sharpshooters were able to influence several naval engagements during the American War of Independence. These included the struggle off Flamborough Head on 23 September 1779 between the USS *Bonhomme Richard*, captained by Scots-born John Paul Jones, and HMS *Serapis* (the command of Richard Pearson). Larger and more powerful – at least on paper – the American ship was an old French East Indiaman which had been loaned to the US Navy; the British ship, a 'fifth rate', was merely months old.

When the first broadsides were fired, at least three of the six 18-pounder guns aboard *Bonhomme Richard* burst in quick succession, killing not only their crews and many other sailors on the gun deck, but also severely damaging the structure of the ship. The battle raged until the frigates became locked together, at which point boarding was contemplated. The British tried and were repulsed, and the Americans tried with, initially, the same result.

Serapis' guns continued to rake *Bonhomme Richard* between decks at point-blank range, doing terrible damage, but the accurate musketry of American marines in the mast-tops drove the British sailors below decks. At this point, though the British were able to kill more than fifty of the riflemen, they were unable to resist another boarding.

Captain Pearson struck *Serapis*' colours and John Paul Jones gained an important victory. Both ships were wrecked. *Bonhomme Richard*, holed below the waterline, was taking on so much water that Jones transferred his crew and his flag to the British frigate and set sail for the Dutch United Provinces. *Serapis* was subsequently handed to the French Navy, but was wrecked by fire and explosion off Madagascar two years afterward.

The early stages of the Napoleonic Wars had seen the issue, albeit briefly, of one of the most remarkable weapons of its era: the military air-rifle. Reservoir-type airguns had been made in small numbers for many years, but only for sporting use until gunsmith Bartolomeo Girandoni[27] of Vienna demonstrated a repeating flintlock and a butt-reservoir airgun to Emperor Josef II on 1 March 1779. The emperor was impressed by both submissions, ordering his army to test their qualities in the field.

The trials were encouraging, and so 1,000 repeating muskets and 500 airguns were ordered. Girandoni and his workshop were hastily transferred to Vienna, and production began painfully slowly; by the end of November 1784, only 111 repeaters and about 290 airguns had been completed. Seven hundred more airguns had been delivered by the autumn of 1787.

The Girandoni *Repetier-Windbüchse* M1780 was about 50 in. long, and weighed a little over 9 lb. Calibre was about .51, and the bore was rifled with twelve shallow grooves spiralling to the right to make one turn in the 30 in. barrel. The mounts and frame were brass, the barrel was browned, the fore-end was usually walnut, and the iron butt-reservoir was covered in leather.

A ball could be transferred from the tube magazine on the right side of the breech each time the spring-loaded breech slide was pushed sufficiently far laterally. An external hammer was cocked, then released by pressing the trigger to strike and thereby momentarily open the air valve. This allowed a pulse of the air compressed into the hollow butt to propel the bullet along the bore.

Muzzle velocity was low, about 975 ft/sec for the first ten shots, but then declined appreciably; by the thirtieth shot, it was only 550 ft/sec. However, the discharge was silent and the bullet, which had about the same muzzle energy as a modern 9 mm Parabellum, was lethal enough at ranges up to about 120 yards. This was little more than that of a flintlock musket, of course, but was useful if the marksman could approach his target stealthily enough in the knowledge that silent discharge would not disclose his position. In addition, being a breech-loader, the airgun was much more accurate than a musket would have been.

The biggest problem lay in charging. Not only were 2,000 pump-strokes required to charge each reservoir to working pressure (with many explosions during the process) but the emperor ordered that each gun was to be accompanied by three reservoirs. This presented a problem for the field army, particularly if the bulky pumps were within range of enemy artillery.

The only documented use of the Girandoni airguns occurred during the Russo-Turkish War of 1787–92, when 200 were sent to Austrian troops guarding the Hungarian border. When Josef II died, his successor, Leopold II, formed the Tiroler Scharfschützen ('Tirolese sharpshooters') to whom some of the Girandoni airguns were issued; others remained in the hands of the Jäger-Corps. Airguns were undoubtedly used against the French. However, no specific successes have been identified even though, in its day, the *Windbüchse* was an ideal close-range sniper's weapon.

It has been estimated that only about 1,540 M1780 airguns were ever made. In 1799, about a hundred of them were still serving the Tiroler Scharfschützen, but 308 had been 'lost' and more than a thousand were languishing in store. Finally, in 1801, the commander of the Scharfschützen asked that his airguns should be replaced with the flintlock *Jägerbüchse* M1795, with which he hoped to double the maximum range of engagement. The *Windbüchsen* were duly withdrawn in March 1801 and put back into store. Survivors may have been reissued in 1805 for use against the French troops of Napoleon Bonaparte, but were withdrawn again in 1815 and scrapped sometime in the 1820s. Very few survivors have ever been identified.

The War of 1812 was fought between the USA and Britain while the Napoleonic Wars raged in Europe. The use of snipers and marksmen by the US forces tended to repeat that of the early years of the War of Independence. At sea, marines in the mast-tops were vital to the success of USS *Constitution* in ship-to-ship engagements with HMS *Java* (formerly the French 'Pallas'-class frigate *Rénomée*) and then with HMS *Guerrière*, another ex-French prize.

In the earlier struggle, on 20 May 1812, US Marine sergeant Adrian Peterson struck down *Java*'s captain, Henry Lambert.[28] In the later engagement, on 19 August 1812, the American topmen wounded the most senior officers on *Guerrière*'s quarterdeck, including Captain James Dacres; and a British marksman shot and killed Lieutenant William Bush, the first US Marine officer to die in battle. The American victories both owed much to loss of command in the British ships at vital times.

In the land war, individuals, often working independently and armed largely with long rifles (or something generally comparable), were able to strike down many high-ranking officers. Among the victims was Major-General Robert Ross, commander of the British expeditionary force to North America, the much-hated 'Man who burned the White House' and the victor of the Battle of Bladensberg (28 August 1814). Born in 1766, Ross had risen to his high command by way of a successful career in the Peninsular War under Wellington.

The War of 1812 had effectively run its course – the Battle of New Orleans was an afternote – when, seeking to attack Baltimore, Ross was struck down on 12 September 1814 by two teenaged American marksmen. Daniel Wells and Henry McComas had stayed to cover the withdrawal of their colleagues during what came to be called the Battle of North Point.

Faced with a British advance, Wells and McComas elected to risk firing at the officer on a horse who was obviously organising his men. Both fired, and Ross fell, fatally hit by a bullet that passed through his right arm into his chest. His death was avenged by his men, who caught and killed the snipers almost as soon as Ross had dropped from his horse.

There were many similar events during the American Wars, and several chapters of John Plaster's *History of Snipers and Sharpshooting* (2008) tell this fascinating story in great detail. However, the absence of incontrovertible evidence often calls the claims of individual marksmen into question.

Identifying fall-of-shot satisfactorily on a battlefield obscured by the 'fog of war', especially at long range without the benefit of telescopes, is exceptionally difficult. Many high-ranking officers were struck down by random bullets simply because they were in the vanguard, but these 'kills' are sometimes ascribed to marksmen who had fired in the general direction at much the same time.

'A rifleman's bullet' has been hailed as the killer of General Sir Edward Pakenham during the Battle of New Orleans on 8 January 1815, but testimony of Major Duncan MacDougall of the 85th Foot, the aide-de-camp into whose arms Pakenham fell, suggested that only one of three wounds (and not the fatal hit to the spine) was due to a bullet. The two others MacDougall attributed to grape-shot fired by the US artillery, which also apparently accounted for the fatal wound suffered by Major-General Sir Samuel Gibbs as he attempted to rally the British after Pakenham's fall.[29]

The end of the Napoleonic Wars was also the end of an era in which massed-ranks' fire of large-calibre infantry muskets ruled the battlefield. This was partly due to the supersession of the flinted lock by percussion ignition, though more than twenty years would pass before the first steps taken with the Forsyth scent-bottle lock progressed to the self-contained copper cap.

This is not the place to discuss the development of the cap in detail, as there were many dead-ends. Nor is the paternity of the cap entirely clear, though it is customarily credited in Anglo-American circles to one-time artist Joshua Shaw, and by the French to François Prélat amongst others. But the introduction of the cap to military service, beginning in Britain with the Brunswick rifle of 1837, had one important effect: trials suggested that it reduced the number of misfires to a seventh of the total expected with the standard flintlock.

Combining rifling and the percussion cap, simultaneously improving range and certainty of fire, encouraged the first attempts to provide rifles which were capable of placing shots accurately at ranges far beyond the 200 yards expected of a Baker or a Brunswick.

The sub-diameter chamber of Henri-Gustave Delvigne (1800–76) and the pillar-breech of Louis-Étienne de Thouvenin (1791–1882) were moves in the right direction, though performance was degraded by forcing the bullet down onto the shoulder of the chamber or over the tip of the pillar. However, when Charles-Claude-Étienne 'Meunier *dit* Minié' (1804–79) introduced the self-expanding bullet – once again, a novelty claimed by others[30] – a major advance was made. The subject of French patent no. 7,978, sought on 16 April 1849, Minié's bullet allowed effectual engagement at 500 yards or more. Long-range sniping became a viable proposition; war would never be the same again.

Notes to Prologue

1. Harold, gravely wounded by the arrow, may have been killed by Norman swords or lances during the final mêlée around the Dragon of Wessex standard.
2. J. Landels, *Engineering in the Ancient World* . . . (1997), p. 101.
3. E. R. A Sewter, *The Alexiad of Anna Komnena* (1969), quoted by Philippe Contamine, *War in the Middle Ages* (1984), p. 71.
4. E. Audouin, *Essai sur l'armée royale du temps de Philippe Auguste* (1913), p. 67.
5. Contamine, *War in the Middle Ages*, p. 189, n. 38
6. In 1277, the Gascons had sent 120 crossbowman to assist Edward I in his campaigns in Scotland; crossbows were used by the English during the Battle of Falkirk (22 July 1298); and, as late as 25 October 1415 (the year of Agincourt), Rymer's Muster Roll of the army of Henry V still listed thirty-eight crossbowmen.
7. Ralph [William Frankland-]Payne-Gallwey, *The Crossbow* (1893).
8. Contamine, *War in the Middle Ages*, p. 188.
9. The word *crannequin* appears to be a diminutive of *crancstæf*, an Old English weaver's implement with a shaft bent at a right angle.
10. The bolt was said to have been fired by Sir Walter Tyrrell (or 'Tirell'), possibly at a buck. Conspiracy theorists – even at the time – suggested that it was a deliberate act to favour William's younger brother, who ascended the throne as Henry I.
11. W. W. Greener, *The Gun and Its Development* , 9th edition (1910), p. 12.
12. Competitions are still held in many places, including the Scottish burgh of Kilwinning, Ayrshire, where 'Ding Doon the Doo' ('Knock-down the Dove') has been held annually each summer since 1483 or earlier by the Ancient Society of Kilwinning Archers.
13. J.-R. Clergeau, 'Histoire de la rayure et des canons rayés: 1. partie', in *Gazette des Armes*, July 1977, pp. 67–70.
14. The Arms are most probably those of the 'King of Rome', to which title he had been elected on 16 February 1486. Maximilian subsequently became Holy Roman Emperor when his father, Friedrich III, died on 19 August 1493; the Arms of the Empire are easily distinguishable from those of the King of Rome.
15. Surviving rifles were altered to cap-lock ('M1810/35') and then to accept a Thouvenin-style pillar breech ('M1810/35/48'). Transformed into *Zündnadel-Defensionsbüchsen*, some lasted in second-line service until the Franco-Prussian War of 1870–1.
16. De Witt Bailey II, 'The Rifle in the American War 1775–1783', in *Guns Review*, Vol. 8, No. 12 (December 1968), pp. 8–11.
17. Published, with information about other sources, in *The Covington Leader* on 1 October 1986. Hugh T. Harrington, 'The Myth of Rifleman Timothy Murphy', in *Journal of the American Revolution* (https://allthingsliberty.com/), 25 March 2013, summarised not only the evidence but also the validity of underlying assumptions about Murphy's early life.
18. On 1 October 1780, Timothy Murphy married Margaretha Feeck (or 'Feck', 1763–1807, of Dutch/German descent), and subsequently had five sons and four daughters; on 22 April 1810, he married Mary Robertson (1783–1861) and had four sons. It seems that all but two of the children reached maturity.
19. Joseph Graham, 'A Recollection of the American Revolutionary War', *Virginia Historical Register*, Vol. 6 (1883), pp. 209–11.
20. Simon Fraser was carried from the battlefield to the nearby house of Baroness Frederika Rierdersel, wife of the commander of the Braunschweiger-Jäger serving in the British Army and of all the German troops at Saratoga, but died the same evening. Clerke was

captured when Burgoyne surrendered. Clerke died on 13 October. Born on 24 October 1748, he had entered the Foot Guards in 1770 and was promoted lieutenant on 26 July 1775. He still held this junior rank when he died.

21. The story has been taken from the Clan Ferguson Society of North America, http://cfsna.net/personalities/overseas/major-patrick-ferguson.

22. Ferguson was the second son of James Ferguson of Pitfour and Anne Murray, a sister of the fifth Lord Elibank. The elder Ferguson, an advocate, took the title 'Lord Pitfour' after rising to judge's rank as a senator of the College of Justice. Patrick Ferguson had entered the Army in 1759, to serve with the Royal Scots Greys, owing to the patronage of his maternal uncle General James Murray.

23. Surprisingly little is known about Isaac de la Chaumette, who may have been born c. 1670 in Rochechouart, near Poitiers, into a Huguenot family. He remained in France until at least 1715, so it is assumed he rejected his religion after the Edict of Nantes (1684) had caused so many of his relatives to flee. It is not entirely clear if he had settled in England when the 1721 patent was sought, and the date and place of his death remain a mystery.

24. The 60th Regiment had been raised in North America in 1755 as the 62nd (Royal American) Regiment of Foot and almost immediately renumbered '60th', the previous unit of the same designation being disbanded in 1748. A second 60th rifle battalion was formed in the Peninsula, and another was raised to serve in the War of 1812.

25. Sir Charles Oman stated in *A History of the Peninsular War* (1996 reprint, Vol. 1, pp. 568–9) that the shot had been taken 'from a range that seemed extraordinary to the riflemen of that day'. R. Rutherford-Moore in the article 'Plunkett's Shot', in *First Empire* 24 (1998), concluded that the range could have been 'anywhere between 200 and 600 metres'. But no records were kept and there is an absence of reliable eye-witness testimony. However, William Surtees stated in *Twenty-Five Years in the Rifle Brigade* (1996 reprint, p. 90) that Thomas Plunkett 'got sufficiently nigh to make sure of his mark': if the rifleman regarded the range as normal, it is highly unlikely to have been beyond 300 yards. A one-shot kill at this distance with a Baker rifle would be possible but fortunate, as the probability of a hit would be low. Two kills in two shots tests belief to its limits. The 'leg-over-leg' position with the left hand on top of the butt, can give a very stable platform for a long-range shot. It avoids the left hand gripping the fore-end, which can interfere with the natural harmonics set up in the barrel on firing.

26. Robert Harvey, *Mavericks: The Maverick Genius of Great Military Leaders* (2008), p. 123.

27. Girandoni was born in the village of Cortina d'Ampezzo on 30 March 1744, and seems to have trained initially as a watchmaker. He died in Pensing, near Vienna, on 21 March 1799, but work was continued by his one-time assistant Francesco Colli into the 1820s.

28. A British marksman's bullet severely wounded the commander of the American ship. Commodore William Bainbridge.

29. However, the terrible losses among British officers during the battle suggest that the American sharpshooters had been firing most effectively: three generals, eight colonels and lieutenant-colonels; eight majors; eighteen captains; fifty-four lieutenants and other subalterns. These deaths were sufficient in themselves to destabilise the command structure of Pakenham's army.

30. Including William Greener and, more plausibly, Irish-born John Newton Norton (1784/5– 1867), 'late 34th Foot', whose bullet, said to have been developed in 1832, was inspired by the way in which the elastic pith of the lotus plant expanded to seal the bore of the blowpipes used by tribesmen of southern India.

CHAPTER ONE

Rifles and Sights: What Does a Sniper Need?

One of the most important factors in the success of any military weapon is its power. Each group has its own defining characteristics, even though performance data from the pre-1800 era – when testing techniques were primitive – are either unreliable or non-existent. The situation improved dramatically only when the electric chronograph was perfected.

The progression of individual precision-shooting weapons, from the crossbow to the cap-lock rifle, has been sketched in the previous chapter. But the employment of the sniper, in the modern sense of the word at least, had to await the perfection of the optical sight.

Though embryonic designs dated back to the 1830s, if not earlier, the telescope sight was not used in any numbers until the US Civil War. The Confederate marksmen made good use of Davidson short-tube sights imported from Britain with Kerr, Whitworth and similar small-bore cap-lock rifles, and the Federal sharpshooters made use of Amadon, Malcolm and other types of long-tube sights.

Rifles and ammunition

There has been a slow but steady move from the slow-moving projectiles of the black-powder age to the high-velocity smokeless small-bore rounds of today, but all weapons suffer loss of power as range increases.

Assuming bullet weights are broadly comparable, high muzzle velocity is commensurate with flat trajectory at appropriately limited range: the projectile, as it flies towards the target, neither rises significantly above nor falls greatly below the line of sight for a surprising distance. This has great advantages in combat, as the gun will shoot where it is aimed almost without regard to the sights.[1] Conversely, a slow-moving heavy bullet loses much of its impetus at long range, has a looping trajectory, and will, therefore, drop much more rapidly than its fast-moving rival.

From a sniper's perspective, a fast-moving bullet with a comparatively flat trajectory can still hit the target even if the range has been misjudged, but slower-moving projectiles such as the 8 mm Austro-Hungarian Mannlicher or the French Lebel put a premium on accurate range-gauging.

It is vitally important that the bullet reaches the target orientated as near horizontally as possible. Otherwise, it could fail to pass down (for example) a pipe being used to protect the sniper. Of course, the bullet might still bounce down the pipe, but the chances of a kill are clearly somewhat reduced by uncertainty that the ricochet will strike the intended target squarely and by the loss in kinetic energy with each individual graze.

Quantifying in-flight velocity loss and how it affects bullet strike has occupied mathematical and military minds for many years. But though computing results requires knowledge and considerable interpretational skill, there are several key factors that can be readily understood.

A cylindrical projectile retains velocity much better than a ball of the same calibre largely because, even the though cross-sectional area is identical, it is much heavier. The difference can be expressed in *sectional density*, which is simply the weight of the projectile divided by the cross-sectional area. The longer the bullet, the greater the weight; sectional density improves until an over-long bullet becomes unstable and may even tumble in flight.

The shape of the bullet is also crucial to efficiency. Though a flat-nose cylinder will outperform the ball, it encounters too much air resistance to be useful – a bullet of this type will hit very effectively at short range, but its extreme range will be greatly curtailed. Rounding the nose was such an obvious answer that 'conoidal' projectiles of this type, influenced by the work of men such as Delvigne, Norton, and Minié, were fired by all the rifled muskets introduced in the middle of the nineteenth century.

This was a great step forward technologically. Musket balls had several dis-advantages, including restricted weight, a tendency to bounce down the bore unless carefully patched, and a marked lack of directional constancy at the muzzle. Consequently, they lost velocity rapidly and were inaccurate enough – even at short ranges – to support the wry observation that 'one went high, one went low, but where on earth did the third shot go?'

The cylindro-conoidal Minié bullet, with a hollow base to expand the skirt into the rifling, was a revelation. Though much heavier than the ball, greatly reducing velocity with a powder charge of the same weight as that of the musket, performance at long range was a revelation to men whose battlefield had been compressed so that volley-fire could have its devastating effect. The Minié bullet extended the range of engagement at least fivefold, even though commanders were often reluctant to depart from 'whites of their eyes' close-range combat. Trials undertaken with regulation muskets and rifled muskets in the USA, as early as 1843, showed the old view to be entirely mistaken.

The smooth-bore and the rifled Minié adaptation were both .69 calibre, but there the similarity ended. The musket, firing a 412-grain ball, registered a muzzle velocity of 1,500 ft/sec compared with only 954 ft/sec for the rifled gun and its 740-grain bullet. Muzzle energies were 2,059 ft-lb and 1,496 ft-lb respectively, a theoretical advantage to the smooth-bore of 38 per cent but both clearly lethal. At 400 yards, however, the ball had lost so much velocity that kinetic energy had reduced to 199 ft-lb; the Minié bullet was still offering 797 ft-lb owing to much smaller loss of velocity. By 1,000 yards, effectively the limit of the smooth-bore's range, energy figures had dropped to 32 ft-lb compared with 369 ft-lb for the Minié. The .69 ball was no more threat than a .22 LR rimfire bullet would be at fifty yards, whereas the .69 bullet was storing more kinetic energy than a .45 ACP bullet does at a hundred yards.

The ever-widening disparity was due almost entirely to the far better ballistic shape of the bullet. In addition, the ball ceased to have inherent accuracy after about 200 yards (the most optimistic estimate!) while the Minié bullet was capable of striking a man-size target at twice this distance.

Compared with smooth-bore muskets, the accuracy and range of the rifle had increased significantly. Reducing calibre usually improved performance, largely because loss of projectile weight was more than counter-balanced by an increase in sectional density. In addition, smaller-diameter projectiles usually had a noticeably finer nose-form to reduce air resistance.

The .69-calibre Minié was replaced in US service in 1855 by the .58 version, which was broadly comparable with the British .577 Pattern 1853 rifle-musket ('Enfield'), which itself had replaced the .702 Pattern 1851 ('Minié'). Trials at 1,000 yards had shown the old .69 ball to be travelling at 187 ft/sec, giving an energy of 32 ft-lb, and to be descending rapidly at an angle of 56 degrees; flight time was 9.8 seconds. For the .58 bullet, the figures were 395 ft/sec, 257 ft-lb, 13 degrees and 4.7 seconds respectively. The vast improvement was clear to all, and the move away from large-calibre projectiles, with a 'smashing effect' valuable only at short range, became inevitable.

As the Crimean and US Civil Wars approached, therefore, armies had an infantry weapon which was capable of killing an opponent at prodigious distances. Yet it was still an externally primed muzzle-loader, vulnerable to misfortunes ranging from misfiring to unintentional multiple loading in the heat (more specifically, noise) of battle. The breech-loader was to be the ideal solution, but many observers in 1861 remained unconvinced. It was to take a war between the US states, with considerably more than a million casualties, to show that the days of the cap-lock were over.

As the standard military rifle progressed from the rifle-musket of the 1850s (calibre usually between 0.55 and 0.60) through the first generation of purpose-designed breechloaders of the 1870s (0.40–0.45) to the small-bore magazine rifles of the 1890s (0.236–0.315), so the bullets not only flew increasingly far but also extended the effective range of engagement. The Lee-Metford and Lee-Enfield rifles had long-range sights of doubtful utility, developed in the hope that volley fire could be useful. Tests had even been undertaken in South Africa at ranges up to 3,000 yards.

One of the key indicators of performance is the *ballistic coefficient*, the universally accepted expression of the ability of a projectile to retain velocity. Ballistic coefficients combine basic elements of sectional density with a 'form factor' derived from the geometry of the projectile nose, which will have been shaped to minimise the effects of air resistance.

Knowing the ballistic coefficient allows specially prepared tables to be used to reveal the residual velocity at intermediate points in the projectile's travel. This in turn allows the way in which the bullet drops as it flies to be calculated, allowing back-sights to be graduated appropriately and telling individual firers – assuming they have not set their sights – how much 'hold-over' to use.

Prior to 1900, virtually all military cartridges were loaded with round-nose bullets. These were overtaken with the adoption on 25 March 1903 of the German *S.-Patrone*, loaded with an ogival ('pointed') bullet patented in Germany by DWM in 1904. US Patent 841,861 of 22 January 1907 names the inventor as Arthur Gleinich, a 'physicist and ballistician of König-Wüsterhausen'. Soon, every army had followed the lead, excepting the Austro-Hungarians and the Italians. Ogival bullets, sharper-nosed than their predecessors, had better drag coefficients. Consequently, assuming all other criteria remained unchanged, they could fly farther and retain velocity better than the roundheads.

Most infantry rifles of the First World War were issued with cartridges which were broadly comparable, whatever their nationality. Bullets ranged from 148 grains or 9.60 gm for the Russian M1908 'Type L' ball round to 244 grains or 15.81 gm for the M1890 Austro-Hungarian Mannlicher; and velocity at the muzzle varied between 1,985 ft/sec (605 m/sec) for the Russian M91 ball round to 2,855 ft/sec (870 m/sec) for the S.-Patrone.

The calibre of pre-1918 rifle cartridges fell into a comparatively narrow band. The USA and Russia used .30 (7.62 mm) ammunition, the Turks used 7.65 mm (.301), the British used .303 (7.7 mm), and the Austro-Hungarians, the French and the Germans all used 8 mm (.315). There are small dimensional differences that the calibre designations did not recognise during the First World War, and the German 8 mm round was subsequently classified as '7.9 mm' or (in Britain) '7.92 mm'. Only the Italians – and the Japanese, involved only peripherally in the Far East – departed greatly from this near-consensus, by accepting 6.5 mm (.256). But there is little doubt that the Germans outgunned the British in the First World War, as the striking energy of the S.-Patrone at 100 yards was about 20 per cent greater than that of the Mk VII .303 ball round.

The drag coefficient of a blunt-ended cylindrical projectile travelling at Mach Two, 2,235 ft/sec, much the same as many First World War rifle bullets, is usually reckoned to be about 1.69. If the edges of the head are rounded, depending on the precise dimensions of the radiusing, the coefficient drops to 1.20–1.35; hemispherical and conical noses reduce values to about 0.93 and 0.83 respectively. An entirely conical projectile could offer a drag coefficient as low as 0.3, but would be impossible to stabilise in the bore. However, these figures do show that there was much to be gained by adopting ogival-nose bullets and that a 10–15 per cent increase in performance would be possible. Almost all modern military-service cartridges, therefore, contain ogival bullets.

Excepting the French Balle Mle 86 D, all of the rifle-calibre bullets of the First World War had bases that were simply squared-off to the axis of the bore. The French *Balle D* was exceptionally long, solid (not jacketed) and had what is now called a 'boat tail', rounded towards the heel to improve the drag coefficient by reducing turbulence in the bullet wake.

Though placing a premium on the manufacturing standards needed to ensure that the shaping of head and heel were consistent, and that the section of the bullet that engraved in the rifling was long enough to assure longitudinal stability, the gains were appreciable: better performance at long range than the flat-based Balle M it replaced, and, so the French claimed, better accuracy.

Trajectory and flight time present problems of their own. The bullet is subject to the effects of gravity as soon as it emerges from the muzzle, requiring the barrel to be elevated to pre-determined angles to hit marks placed at the particular distances marked on the back-sight leaves. The lower the muzzle velocity, the greater the drop will be. A ball fired by a flintlock musket dropped an unbelievable amount as it neared the end of its trajectory, and even the most modern high-velocity cartridges can only minimise the problem.

The .69-calibre French Mle 1763 'Charleville' musket, which could be regarded as a high-point of mid-eighteenth-century infantry weaponry, fired a .627-diameter ball of approximately 370 grains at a muzzle velocity of 1,400 ft/sec, giving a muzzle energy of 1,605 ft-lb. If the muzzle of the musket was elevated at an angle of one degree, the ball travelled about 933 feet before dropping back to a line drawn horizontally from the muzzle; at this point, it was travelling at 563 ft/sec and kinetic energy had dropped to 260 ft-lb – still more than enough to inflict a fatal wound. The ball had reached a maximum height of about 5.49 feet at a distance of 530 feet. At about 3,700 feet, however, velocity had declined to a point where the ball would be falling almost vertically.

The 5.56 mm SS109D round fired in the M16A1 and most comparable modern ultra-small-calibre rifles has a muzzle velocity of about 946 m/sec, declining to 480 m/sec at 500 metres and 259 m/sec at 1,000 metres. Maximum range is about 3,000 metres, but the bullet has ceased to have any real power at this distance.

For a set-range of 2,000 metres, requiring about 8.5 seconds for the projectile to reach the target, the M16A1 muzzle must be elevated at about 6° 18′, but the angle of descent when the bullet reaches the target is 19° 51'. The trajectory vertex, only 6.9 metres above the horizontal aim-line when the sights are set for 1,000 metres, is, rather staggeringly, 100 metres high at 2,000 metres.

Figures taken from promotional literature accompanying the Walther WA2000 sniper rifle – an abortive project which did much to destroy the company's profitability – illustrate how far bullets will drop if the gun is fired when the bore is horizontal. The 'control weapons' were Kar. 98k rechambered to fire .300 Winchester Magnum ('.300 Win Mag') and .308 Winchester (7.62 × 51 NATO) rounds. The rifles were sighted for 100 metres and the projectile drops were taken at 100-metre intervals. The figures are surprising, particularly as the .300 Win Mag, loaded with a 9.7 gm bullet, had a muzzle velocity of 944 m/sec: bullet-drop measured 6.7 cm at 200 metres, 17 cm at 300 metres, 45.7 cm at 400 metres, 92.6 cm at 500 metres, and 213.8 cm at 600 metres. Comparable results for the .308 round, firing a 9.4 gm projectile at 866 m/sec, were drops of 8.5, 33.5, 85.2, 138.1 and 297.7 cm respectively.

One obvious problem concerns the use of a variety of cartridges in a single magazine, which was often practised. Bullets loaded into otherwise similar cartridge cases can vary in construction: armour-piercing bullets will contain a hardened core, tracers contain a bright-burning pellet to mark their path, and incendiary bullets usually contain phosphorus. Their weights can differ considerably.

This was particularly true of the German rounds, as the heavy ball or *schwere Spitzgeschosse* (sS) weighed 12.83 gm, whereas the iron-cored bullets, *Spitzgeschosse mit Eisenkern* (SmE) and *mit Kern* (SmK), weighed 11.5–11.6 gm; the *Spitzgeschosse mit Kern und Leuchtspur* (SmK L'spur, tracer) weighed only 10.17 gm. The Russian L-type ball weighed 9.6 gm, compared with 11.8 gm for the heavyweight boat-tail 'D'. The BZT armour-piercing incendiary/tracer bullet weighed 9.2 gm. The bullets of the British Mk VII and VIIz ball cartridges weighed 174 grains, the same as that of the armour-piercing W. Mk Iz, but the G. Mk II tracer bullet weighed only 154.

The differences in projectile weight obviously had an effect on muzzle velocity: the Russian D-type ball reached only 815 m/sec compared with 875 m/sec for the BZT. Trials show that accuracy is acceptable as long as the firer uses only one of these cartridge types, but can deteriorate greatly if two are intermixed. The difference in velocity gives excessive vertical dispersion, but can also affect lateral dispersion by increasing the 'cone of flight'.

The Germans and the Russians, fighting with great brutality, used 'explosive bullets' to heighten the well-established fear of snipers. However, many reports understandably confused the tendency of long-body bullets to tumble when they struck a target (and the German 7.9 mm bullet was particularly prone to do so) with the truly explosive or incendiary projectiles containing material which detonated on impact or spontaneously combusted when exposed to air.

The German *Beobachtungs-Patrone* (B.-Patr., observation cartridge) was an explosive-incendiary round loaded with a special 10.85 gm bullet – chromium tipped, later black-bodied with a bright-metal tip – containing a fuze, a striker and a detonator of lead styphnate, barium peroxide and calcium silicide. A 0.4 gm phosphorus charge yielded a cloud of bluish-white smoke when the projectile struck the target.

In January 1945, concerned by the ever-increasing reports of Russian snipers using explosive bullets, the head of the German Army equipment office circulated a suggestion that similar ammunition should be issued to German snipers. Though the Luftwaffe and possibly also the Army had been using B.-Patronen in rifles for some time, this was strictly unofficial; Hitler had actually forbidden the use of explosive rounds in anti-personnel roles until such time as he decided it was appropriate.

On 15 February 1945, Hitler approved the issue of 'B.-Cartridges for use only in the East', and on 29 February the Army quartermaster-general was instructed not only to issue cartridges to sniper-training companies but also to ensure that each qualified sniper received twenty B.-Patr. to supplement the ninety ball rounds he would take to the front.[2]

Drift of a bullet in flight can be due to bullet rotation, to crosswinds, or to combinations of the two. Rifled barrels impart a surprising rotational velocity to the bullets as they travel down the bore to the muzzle. When a 7.62×51 projectile emerges, it can be spinning at 45,000 rpm and the bullet tends to wander off the line of sight in the direction of spin. However, there are many contradictory factors. The April 1909 edition of the handbook for the Springfield rifle, *Description and Rules for the Management of the U. S. Magazine Rifle Model of 1903, Caliber .30*, stated that:

> The rifle has a right hand twist, and the drift proper is therefore to the right. There is, however, a slight lateral jump [on firing] to the left, and the total horizontal deviation of the bullet, excluding wind, is the algebraic sum of the drift and the lateral jump. The trajectory is found to be very slightly to the left of the central or uncorrected line of sight up to a range of 500 yards, and beyond that range to the right . . .

Spin-induced bullet drift was corrected by the back-sight leaf to 500 yards, past which the firer had to make additional adjustments manually. At 1,000 yards, the bullet drifted 13 inches to the right of the bore-line, but about 6.3 inches were automatically corrected by the sight-leaf. Yet this still left the firer to add a deflection of 6.7 inches to his point of aim; at 1,500 yards, drift to the right was 61.75 inches, only 26.75 of which were corrected by the back sight.

Rates of spin can be calculated to be optimal for particular bullets; no single rifling pitch can suit every type of cartridge, as the characteristics of bullets vary enormously. Today's manufacturers rarely make mistakes, but a change in rifling can have a considerable effect. When M16A1 rifles were revised to fire the Belgian SS109 bullet instead of the original US M193, altering the pitch of the rifling appropriately, a comparatively unstable projectile capable of inflicting horrendous wounds was replaced by something which was more stable and (to many minds) less effective on the battlefield.

Depending on direction, crosswinds can reinforce or oppose spin-drift. It is often suggested that the wind can only have a tiny effect on something as small and as rapidly moving as a bullet. However, the effects can be significant; at 100 metres, the Swiss 7.5 × 55 round chambered in the StGw. 57, firing a bullet weighing 11.25 gm at a muzzle velocity of about 880 m/sec, deflects 15 mm in a 90-degree crosswind of 2 m/sec and 69 mm in a wind of 6 m/sec. At 500 metres, the deflections are 47 mm and 2,190 mm respectively. This clearly illustrates how, even for a well-trained sniper, hitting a target at long range still relies greatly on accurate observation and good fortune.

Not all cartridges of the same general dimensions perform similarly, owing to variations in the shape and weight of their bullets. For example, in a 2.5 m/sec crosswind, a 6.5 × 68 sporting round loaded with a 6 gm bullet (attaining 1,150 m/sec at the muzzle) and a similar round loaded with a 9.1 gm bullet (910 m/sec) give deflections of 160 mm and 120 mm respectively at 300 metres. The better sectional density of the heavier bullet is the greatest contributor to the reduction in drift.

Another illustration can be seen in cartridges generating similar velocities with bullets of radically differing weight: .222 Remington (3.25 gm bullet, 970 m/sec at the muzzle) drifts about 310 mm at 300 m in a 2.5 m/sec crosswind, whereas 8 × 68S (11.7 gm bullet, 990 m/sec) drifts merely 140 mm. Once again, sectional density is the greatest limiting factor.

But a word of caution: in reality, crosswinds are not necessarily consistent throughout the flight of the bullet and are rarely tangential to the flight-path. This complicates the process of estimating the difference between the point of aim and the point of strike. In addition, gusts or the proximity of buildings, trees, walls or even undulations of terrain can all affect bullet-drift sufficiently to cause a shot to miss the target.

Weather conditions can have measurable effects on projectiles. Heavy rain can be an inhibitor to predictable flight, as can atmospheric pressure and ambient temperature. Performance tables for the US M1903 rifle assumed that the M1906 cartridge fired its bullet at 2,700 ± 20 ft/sec on a 'normal summer day', 70° Fahrenheit. However, the expected variation of velocity per degree was assessed to be 1.5 ft/sec. Though this was insignificant if ambient temperatures remained within normal bounds, in sub-zero or tropical heat the impact on a long-distance target could be significantly altered by loss or gain in velocity.

This factor would be particularly important if the rifle was being fired in an Alsakan winter or in Siberia, where temperatures can plummet to fifty degrees below zero. It also shows the undesirability of firing a rifle on a cold day, unless it has been allowed to adjust to its surroundings and 'sighter shots' have been fired to warm the barrel and the breech mechanism.

Muzzle velocity could be reduced by as much as 125–150 ft/sec in Arctic conditions, and so the target-strike would be considerably beneath the expected point of impact. In addition, special lubricants were needed to ensure the rifle cycled satisfactorily. Even the Red Army encountered problems of this type during the Winter War with Finland (1939–40); and the lessons had not been learned by the Germans when 'General Winter' slowed their advance into the USSR in 1941–2.[3]

Target movement is another factor to consider. A man moving at walking pace can make surprising progress in the several seconds a bullet can take to reach his range. Running will make an even greater difference, especially if the prospective target is wise enough to change speed and direction irregularly. This presents the firer with the problem of assessing the amount of 'lead' to allow, which, without the help of modern rangefinders or similar aid, is almost impossible to gauge without experience.

The telescope sight

With rare exceptions such as Simo Häyhä, Billy Sing, Bert Kemp and one or two others who relied exclusively on rifles with open sights, no sniper could operate effectively without the assistance of an optical sight.

The first telescope was made at the beginning of the seventeenth century by two Dutchmen, then improved by Galileo Galilei and eventually by Johannes Kepler. The *Galilean* form (which reappeared briefly during the First World War) relied on two lenses to enlarge the image, which was focused behind the ocular lens; the Kepler form, though relying on the same lens pair, allowed the image to focus in front of the ocular lens. This had a major advantage, in that eye relief (the distance between the lens and the firer's pupil) could be substantially increased, but the image in the ocular lens was inverted. Consequently *Keplerian* refractors require an additional lens to re-erect the image.

Basic theory behind the telescope was understood by the eighteenth century, and many instruments had been made for military and naval use. It was simply a matter of time before the ability of a telescope to magnify (and hence locate a mark at long range) was married to the ability of a gun to hit the mark.

Precisely when this occurred is still disputed, and credit is usually given to experimenters working in the second quarter of the nineteenth century. Yet, given the technological progress that was being made a hundred years earlier, the use of alidades in surveying tools, and tube-like gun sights dating back to the arquebus, it is difficult to accept that the applications to small firearms – however primitive they may have been – did not pre-date 1800.

One of the most seemingly intractable problems of early lenses (and, indeed, glassware) was irregular dispersion of light. It had been identified by Isaac Newton, but no progress was made until, in 1733, Englishman Chester Moor Hall made the first truly achromatic lens by combining individual lenses of flint and crown glass to minimise chromatic aberration. However, though Hall is now usually credited with the invention, the first patent was granted in 1758 to London optician John Dollond and the mathematical explanation, by Samuel Klingenstierna of Uppsala, was published in 1760. Unfortunately, the manufacture of achromatic lenses was hindered by a lack of suitable good-quality glass; costs were high, and the potential rewards were perceived as small.

The breakthrough was made by one-time carpenter and bell-founder Pierre-Louis Guinand, born of a French father and a Swiss mother in La Corbatière, Neuchâtel, in 1748. Guinand trained as a clock-case maker until, after examining an English-made telescope, he became fascinated with optics and determined to make better lenses than those he had seen. Eventually, the crucial discovery was made: stirring molten flint glass in a crucible largely eliminated image-distortion by distributing the major components – silica, lead oxide, potash – much more evenly. Heavyweight lead, in particular, had been prone to separate when moulded.

By 1805, after thirty years of hard work, the Guinand process had been perfected. It allowed optical glass to be made in large quantities, and, therefore, reduced unit costs appreciably. Among those who recognised the potential was a commercial counsellor named Utzschneider, who was an adviser to the king of Bavaria. Guinand moved to Bavaria in 1807 to collaborate with Joseph von Frauenhofer (1787–1826), born in Straubing, who had soon developed polishing machines to maintain curvature, a spherometer, and a heliometer to measure light transmission. Though Guinand and Frauenhofer sought to improve microscope lenses, the telescope, and ultimately the telescope sight, were also to benefit.

Guinand returned to Switzerland in 1813, then went to Paris to open a workshop with his son Aimé. Achromatic lenses made by the Guinands were supplied to two French opticians, Robert-Aglaë Cauchoux and Noël-Jean Lerebours, who received a gold medal from La Société d'Encouragement pour l'Industrie Nationale for their lunettes in 1824, the year of Pierre-Louis Guinand's death. Aimé Guinand continued to make lenses in Choisy-le-Roi in partnership with George Bontemps, but the stirring process was to remain a closely guarded secret. Finally, in June 1838, Henri Guinand – eldest son of Pierre-Louis – divulged the secrets of the stirring process to two members of the Académie des Sciences to establish its pedigree. In 1840, the contributions of Guinand and Bontemps to lens-making were finally recognised by the Société d'Encouragement.

The development of achromatic lenses was vital to the development of the optical sight, which, in its modern form, is widely credited to civil engineer John Chapman on the basis of comments made in 1844 in *Instructions to Young Marksmen*.[4] However, the claim of David Davidson, inventor of 'Colonel Davidson's Rifle Telescope', takes priority.

Davidson was born in Haddington, Midlothian, Scotland, on 18 August 1811[5] to Henry Davidson, 'writer' and future Sheriff Clerk of East Lothian, and Martha Mary Chisholm 'his spowse'. In 1830, he joined the East India Company as a cadet and was then commissioned into the Bombay Army with the rank of lieutenant. An accomplished sportsman, Davidson was soon experimenting with guns and ammunition in the area around Fort Asseereegurh (now Asirgarh), where he was stationed.

According to Davidson's autobiography, he had purchased a flintlock sporting rifle made in London by Samuel Staudenmeyer. As an experiment, a draw-type telescope was then fixed to it. The date of this trial is no longer known, but was undoubtedly some time prior to 1835. Davidson continued his experiments, ordering a cap-lock sporting rifle from James Purdey in 1839, fitted with a telescope sight which, when it arrived in India, was not to his liking and was almost immediately modified. There had also been an optically sighted air cane, and at least two cap-lock pistols.

Rifled Cannon, a paper written in 1839 by Davidson (then still serving in India), had been presented to the Royal Society in Edinburgh by Charles Piazzi Smyth,[6] and it is believed that substantial numbers of rifle telescopes had been made by the time Davidson retired his army commission and returned to Scotland in 1848. Several instruments made by optician John Robertson of Haddington were exhibited at the Great Exhibition of 1851, when Prince Albert reportedly admired them greatly.

Piazzi Smyth gave a lecture to the British Association in 1850, entitled 'On the Application of Telescope Sights to Rifles'. In it, he remarked that there were four principal objections to open sights:

> 1st. That there are three objects to be brought in a line, the sight at the breech, that at the muzzle, and the object aimed at; and that these three things being at very unequal distances from the eye, cannot be seen equally distinct at the same time. 2nd. Unless the barrel is of inordinate length, there is not sufficient radial length between the sights to give the opportunity of pointing accurately. 3rd. As only one of the breech-vanes ['sight leaves' in current parlance] can be raised at a time, there are no means of making allowances for distances intermediate between

those for which the vanes are calculated. 4th. In sunshine there is a phase of the muzzle-sight which is very prejudicial to correct aiming.

Piazzi Smyth recommended 'applying to the barrel a small telescope with cross wires', an opinion which was to conflict with that of Hans Busk, chronicler of the Volunteer movement, who promoted the merits of the full length American-style sight. Experience proved Piazzi Smyth to be right, and Busk to be wrong. Full-length sights had no place in combat. Piazzi Smyth acknowledged that though his experiments with telescope sights had begun in the mid-1840s, the work of Davidson pre-empted him by many years.

David Davidson patented his detached collimator sight, 'Improved Apparatus for Pointing Ordnance and Restoring the Aim of the Piece ...' in Britain on 17 January 1855. This was featured in the 20 October 1855 issue of *Mechanics' Magazine.* Davidson then commissioned gunmakers John Dickson & Sons and opticians Alexander Adie & Sons, both of Edinburgh, to develop and promote his sight. The basic design remained largely unchanged – optically, the rifle telescope was no more than standard – but the mounting system was gradually refined.

The perfected version, 'Telescopes for Fire-Arms', patented in Britain on 19 December 1862, No. 3,399/62, relied on the body pivoting on a special stud in the stock and a front mount (equipped with a Vernier scale) to which the sight-body could be locked. An internal windage and elevation adjuster was also fitted. In a paper read before the Royal United Services Institution in March 1865, 'Lieutenant-Colonel D. Davidson, 1st City [of] Edinburgh Rifle Volunteers, and late H. M. Bombay Army' described how:

> A sportsman, in searching for deer, finds a telescope ... indispensable. Military riflemen, too, when engaged at long ranges, avail themselves of its aid. Of this we have a striking example in an incident which occurred in the rifle pits before Sebastopol. One soldier was observed lying with his rifle carefully pointed at a distant embrasure, and with his finger on the trigger ready to pull, while by his side lay another with a telescope directed at the same object. He, with the telescope, was anxiously watching the moment when a gunner should show himself, in order that he might give the signal to the other to fire.
>
> Though the present application, which renders the telescope available for the longest ranges, is new, the principle is no new thing, but has long been tested both in India and America. I cannot say how long the telescope has been used as a sight for rifles in America, but thirty years ago, which I rather think was anterior to its use in the New World, I introduced it into India, and used it, along with many other sportsmen, to whose rifles I applied it, with singular success against the antelope on the plains of the Deccan ...

Dickson-made sights sometimes have graticles to estimate range and, if necessary, hold-over. This was a surprisingly modern idea which only found favour in the second half of the twentieth century and undermines the claim of the Kahles company to have pioneered the 'drop compensator'.

A long life ended in Newington, Edinburgh, on 18 May 1900, when Sir David Davidson, 'Retired Officer Indian Army, Knight Commander of the Bath', died of old age. His achievements largely passed with him, only to be fully rediscovered and publicised in 2010 by Stephen Roberts.[7]

The life of John Ratcliffe Chapman, born on 9 January 1815 in Revesby, Lincolnshire, is difficult to disentangle; even genealogy still fails to provide convincing descent. His father is sometimes said to have been mechanical engineer William Chapman (1749–1832), but baptismal records name John Ratcliffe's parents as John Chapman and Isabella Lyon.[8] Chapman's telescope sight does not seem to have been exploited until after he had settled in North America in 1842, and its chronology is still unclear. Claims have been made that simple telescope sights were already in use in the eastern USA in 1835, but few (if any) of them seem to be backed by fact. In his book, which is said to have been written in 1844 though published only in 1848, John Ratcliffe Chapman indicates that he had been:

> Aware that telescopes have been in use for some time, but to the best of my knowledge they never perform as well as the globe sight . . . I shall now proceed to describe such improvements in sighting which I have lately been able to effect by the introduction of a telescopic sight . . . The tube in which the lenses are fixed is three feet and one inch long, 5/8th of an inch diameter . . . [and weighs] 10 ozs. It can be made very good and true out of sheet iron. To the front end a saddle of steel is fitted and brazed . . . one object to be attained being stiffness, for when fixed on the rifle a discharge has the tendency to pitch it forward and break out the dovetail. A carriage is made to slide through the bead sight dovetail . . . through which two screws pass into the saddle serving as axis or pivots of elevation and depression . . .

Chapman seems to have supplied sights to Morgan James (c. 1815–78) of Utica, New York State, who was renowned among the finest gunmakers of his day. Among the guns made by James were heavyweight target cap-lock rifles, with sights – crude though they undoubtedly were – which facilitated exceptionally precise shooting.

James was attracted to the way in which the telescope sights were fitted. He experimented with methods not only of ensuring that the lenses stayed in place but also that they could survive the shock of recoil. This became of ever-growing importance as the heavyweight target rifles (in which great weight minimised recoil energy) gradually gave way to lighter sporting guns in which the ease with which fast-moving targets could be engaged was a prerequisite.

Morgan James seems to have bought his sights from several gunmakers who doubled as opticians. They include George H. Ferris, born on 8 January 1820 in Utica, New York State. Ferris was listed as a 'gun smith' in the 1850 US Federal census (married with four children), the 1860 Federal census, and in the 1875 New York State census. The 1880 Federal census gives his occupation as 'gun maker' but he died *c*. 1885. Few of Ferris's telescope sights survive; the rifles on which they will be found are customarily marked 'James & Ferris'.

One of the first major improvements in the design of telescope sights was made by William Malcolm of Syracuse, New York State. Born in Sullivan on 13 October 1823, the year his Scots-born father opened a shop in nearby Syracuse, Malcolm was listed in the censuses as a 'gunmaker' (1860 Federal) or 'optician' (1865 New York State, 1870 and 1880 Federal). He had benefited from a scientific education, and had experimented with lenses for many years before perfecting his manufacturing techniques.

By 1855, Malcolm realised that existing lenses had serious faults – a corona or 'halo' of colours at the edges of unclear images, lack of unified focus across the lens surface, and minimal adaptability. Inspired by scientific developments in Europe (notably in France), he produced an achromatic lens by carefully sandwiching two or more lens-discs, each diffracting light at slightly differing angles. If these compensated for the difference in wavelength of the components of 'white' light (which diffracts as a rainbow of seven colours), the coronae disappeared.

William Malcolm worked tirelessly to develop what would now be called an 'orthoscopic' lens, ground sufficiently precisely for the image to appear sharply focused across the entire lens surface. These advances were revelations even to those who had had experience of the Chapman-type sights. They allowed Malcolm to make sights on a commercial scale, possibly with the involvement of William A. Sweet of Syracuse.

Malcolm telescopes survive not only as 'free of mounting', but also on sporting guns, usually with windage and elevation adjusters built into the back mount. Magnification can vary from 3× to as much as 20×. The sights offered good quality, and were usually specially adjusted to suit the eyesight of the purchaser. This made them expensive – even though they performed particularly well – and limited their re-sale value, unless the purchaser's eyesight duplicated that of the original owner. Consequently, though sales were enough to make Malcolm a good living (the 1870 census gives the value of his real estate as $30,000), they were unable to compete with the sights that were introduced by large-scale manufacturers with a better grasp of marketing.

By the time William Malcolm died in Syracuse on 12 July 1890, survived by his wife and two of his four children, his sights were no longer commercially viable.[9] However, he had helped to lay the foundations for what was to become a worldwide industry.

*   *   *

Brief details of more modern sighting equipment – optical, infra-red, image-intensifying, laser gauging, etc. – will be found in the appropriate chapters. Among the key performance arbiters are:

Magnification, which is simply the relationship between the size of the object seen through the sight compared with that seen with the naked eye. The true magnification can sometimes be gauged approximately by keeping the second eye open and allowing the brain to compare the two images.

Alternatively, a piece of thin paper can be placed over the eyepiece of a sight pointed at a bright light. As the paper is drawn backwards, the circle of light diminishes until it reaches a minimum and then begins to grow. This establishes the eye-point – the optimum distance at which the eye should be placed – and the sight aperture, which is the diameter of the circle of light at its smallest point. Magnification can be calculated by dividing the sight aperture into the diameter of the objective lens (though this is prone to give a false figure unless undertaken carefully).

The limited ability of the human eye ensures that all sights will perform adequately as long as the diameter of the objective lens exceeds 12 mm. Large-diameter objectives make no difference to resolution of detail, assuming magnification remains unchanged, though they do offer greatly improved transmission of light.

To determine *relative brightness*, the result of dividing the effective diameter of the objective lens by the sight aperture is simply squared. If a 6× telescope sight has an objective lens diameter of 40 mm and a sight aperture of 5 mm, therefore, its relative brightness is 64.

Transmission of light is limited by the iris in the eye, which adjusts automatically to ambient light. But iris diameter rarely exceeds 3 mm in daylight and, therefore, any relative brightness greater than 9–10 is wasted: at dusk, conversely, the iris can expand to a little over 5 mm for an optimal relative brightness of 25–30. Large-objective sights can provide relative brightnesses as great as 100, allowing the eye to see detail in conditions where ambient light is insufficient to satisfy a fully opened iris.

Resolution of any lens in a small-diameter telescope is limited by the performance of the human eye. A normal eye can resolve detail as fine as a minute of arc, one sixtieth of one degree. To find the potential acuity of the eye when presented with a magnified image, therefore, the minute of arc is simply divided by the true magnification: a 4× sight would improve the resolution to a quarter of a minute of arc (15 seconds of arc).

Field of view is defined by dividing the angle of view by magnification. An enormous objective lens can contribute little to performance unless the

OPTICAL SIGHT PROBLEMS

The greatest obstruction to the use of the earliest telescope sights lay in the manufacture of lenses good enough to allow the smallest of marks to be taken at ranges as great as 500 yards. It was virtually impossible in the mid-nineteenth century to grind a single-element lens to avoid distortion. Other shortcomings of optical sights include:

Chromatic aberration, rising from the inability of a lens to focus light rays at a single point.[10]

Spherical aberration, caused by light rays from the outer margins of the lens focusing ahead of those from the centre, blurring the image. Though this can be minimised by accurate grinding, the work must not disturb essential colour corrections. 'Aplanatic' lenses have been corrected for spherical aberration if all the differing parts of the image focus simultaneously when adjustments are made.

Image distortion, which occurs when the lenses have been ground so poorly that parts of the image may seem twisted or bent. Focusing on a brick wall or similar grid-like pattern in the middle distance will show whether horizontal and vertical lines are straight, and whether they focus sharply across the entire field of view simultaneously. 'Orthoscopic' sights will have been corrected for image distortion, coma and field-curvature.

Coma, occurring when a lens, unable to focus light passing through it obliquely, 'smears' an image outward towards the edges

Field curvature, arising when elements in the lens system, particularly the erector components, produce an image with blurred edges and a pin-sharp centre – or, conversely, a blurred centre with crystal clear margins.

eye-piece lens is also enlarged. Too restricted a view makes target acquisition difficult; this is unimportant if the target is static, but the same cannot be said of hunting and it is vital to find a compromise. At 100 yards, assuming the size and power of the lenses remains unchanged, field drops from 41 ft at 3 × to only 15 ft at 9 ×.

The Guinand method of making achromatic lenses had been introduced to Britain in 1848, when Bontemps fled the French revolution of that year to work with Chance Brothers of Smethwick, Birmingham. A manufactory was built to make the hard crown and dense flint glass needed for the lenses of telescopes, and exports were soon being made to Austria, Switzerland and the German states. The popularity of achromatic lenses grew rapidly, but quality was constrained by manufacturing limitations. The French, to whom much of the development should be credited, could not match inventiveness with output, and even in Britain, cradle of industrialisation, glassmaking remained primitive. This was still evident in 1917:

The Ballard was introduced during the Civil War, but success awaited the involvement of Marlin. This No. 5 Pacific Model, dating *c.* 1880, has a full-length Malcolm sight with double crosshairs and a silver eye piece. Rifles of this type kept sharpshooting alive in an era of military apathy. *Courtesy of James D. Julia, auctioneers (www.jamesdjulia.com)*

> Accustomed as we are . . . to ready supplies of excellent optical glass, it's easy to forget how exacting a material it is . . . Military optical glass needed to be both chemically and physically uniform – free from striae [lines due to chemical variations trapped when glass melt cooled], bubbles [volatiles trapped during cooling] and inclusions [then called 'stones'] of foreign material in the melt. The glass also needed to be free from internal stress that could cause variations in optical properties. It needed to maintain a correct, uniform refractive index and dispersion ratio (Abbe number) throughout. It required high transparency, and had to be free of colours and stains introduced by chemical impurities. And it needed to be durable and stable.[11]

Calculating performance

Every projectile-weapon, be it crossbow or the latest .50-calibre long-range sniper rifle, has inherent limitations. That these have changed over many centuries of development is undeniable; but it is equally true that they still constrain performance within pre-determined boundaries. These boundaries fall into two classes: internal, arising from the characteristics of the gun and the ammunition it fires, and external, when influences such as distance, bullet-drift, crosswind and movement of the target are brought to bear.

Many theories and counter-theories have been put forward to explain why bullets rarely arrive at the same place on the target from shot to shot.

Propellant-gas pressure in the cartridge case rises steadily from the moment the striker ignites the primer and the priming flash ignites the main charge, peaking at about the moment when the thrust on the base of the projectile is sufficient to start the latter into the rifling. A period of near-equilibrium follows, when the potential increase in chamber pressure (which would have occurred had the bullet failed to move) is negated by the gradually declining compression ratio caused by the bullet moving up the barrel. The greater the bore volume that lies behind the projectile – which is clearly increasing as the bullet moves forward – then the lower the pressure must be.

If the barrel is too short, the projectile can still be accelerating as it leaves the muzzle. If the bore is of ideal length, the bullet achieves constant velocity shortly before it emerges; and if the barrel is extended beyond this point, friction between the bore wall and the bullet-material may be sufficient to overcome the steadily declining propulsive force, so that the projectile actually loses velocity before emerging.

Consequently, barrel length can have an important influence on performance. It is possible, for example, that the barrel of Germany's WWII Karabiner (Kar.) 98k is too short to make best use of the power of the standard German 7.9 mm cartridge. This could explain why accuracy tests of the Zielfernrohr (Zf.) Kar. 98 were generally uninspiring – even the Germans eventually admitted that their guns could not match the performance of Obrazets 91/30 goda (Obr. 91/30g) Mosin-Nagants – and the long-barrelled Gewehr (Gew.) 98 of the First World War proved to be a better (if undeniably cumbersome) sniper rifle.

Many factors control the efficiency with which a gun will shoot, even assuming it has the perfect bullets. If we know what happens when we pull the trigger, then what happens to the bullet between the time it emerges at the muzzle and when it reaches the target? There is undoubtedly a correlation between consistent generation of velocity and performance at the target, and, therefore, a gun whose muzzle velocity fluctuates greatly will invariably shoot badly.

However, it is quite possible for bullets to pass consistently down a bore to the muzzle and be chronographed as they emerge with very little fluctuation, but equally possible that the rifling is unsuitable. The chronograph will detect a bullet which has stripped in the rifling, but cannot detect, for example, that the rifling pitch is too 'slow' – the grooves do not turn sufficiently quickly – to give acceptable gyro-stability in the projectile's flight.

Alternatively, the bullet may rotate too quickly and wander from its intended path; in neither case is the accuracy likely to be noteworthy. Yet consistent power generation is at least a departure point from which a prediction of accuracy can be attempted, as most gunmakers, well-practised with types and pitches of rifling, rarely make serious mistakes.

A reasonable way of measuring accuracy potential of guns, even after all these limiting factors have been recognised, has still to be suggested. Test targets supplied by manufacturers are often shot from a machine-rest and are seldom duplicated consistently in the field, even though better results can occasionally be obtained by chance.

These 'test-targets' often feature only a few shots pasted into a hole cut in the centre of a target, and are commonly confined to no more than five shots. Ten shots would obviously be more desirable, as the chances of getting three or even five good shots from any single sequence is high: and, therefore, a bad guide to quality if the remaining shots are poor and have been discarded from the 'paste-up' target.

Deciding how many shots represents a sequence is vitally important. A good five-shot group is fine, and a gun that gives five consecutive tight groups would also be desirable; but what if one, two or even all five of the groups landed in different places? There is rarely any indication from published multiple-group tests that the centres of each group align; they may be displaced from one another by several inches.

Consequently, I have almost always used sequences of 25 shots; not too many to be impossible to shoot – yet just enough to ensure mathematical probity, minimise the worst excesses of an occasional displaced shot, and escape the worst effects of chamber heating during the short test period.

The next steps are best illustrated by means of actual groups obtained at 300 yards with a standard Russian Obr. 91/30 rifle, firing ball cartridges chosen deliberately from a single batch of 1950s Soviet ammunition. Using five 5-shot groups often removes the difficulty of measuring shot-fall in one large ragged 25-shot hole (which it could be at short range), but raises an additional problem of locating the aim point as each target was changed. For the purposes of experiment, though by no means an unimpeachable solution, the crosshairs of the optical sight were simply held on aiming marks drawn on each target.

Having obtained the five 5-shot groups, a numerical expression of their size (and therefore, quality) must be deduced. The easiest way is simply to measure the diameter of each group individually, which provides a rough-and-ready guide but makes no concession to the relative distances of individual shots from the centre. This can be very misleading: simple measurement is not necessarily a fair arbiter and it is often essential to look more closely at an overall picture.

Many systems of accuracy-determination have been mooted in the past and, indeed, the British Army once used a system known as 'Figure of Merit' – explained in Appendix IX of Christopher Roads's textbook, *The British Soldier's Firearm, 1850–1864*, where its strengths and weaknesses are highlighted.[12]

Figure of Merit, like most essentially similar methods, was based on the radial displacement of each shot from the group centre – a principle accepted for the purposes of this chapter, but with significant modifications.

TEST TARGET
Mosin-Nagant 91/30, 7.62 mm cal., 300 yd
RECT: 9.2 × 10.9 in., MØ 6.98 in., MOA
2.3, FS 6.15 in.
SNIPER SCORE: 73

How to Use the Diagrams

Each trial assesses an individual gun, highlighting what is – and sometimes is not – possible by plotting the strike-point of twenty-five individual shots. Targets measure 6 ft × 6 ft, with an 8-inch 'head size' bull's-eye (individual shots are enlarged on the diagram for clarity, but maintain their true centres). The size of the *shot-rectangle* RECT, containing all the shots, is given in inches, height × breadth; this can show how the 'mean-strike' is displaced from the aim-point. The *mean diameter* MØ, in inches, is obtained from the 25-shot group by calculation; MOA is a *minute-of-angle* assessment; and FS is the distance the *farthest shot* lies from the true group centre. If all the shots strike within the 8-inch bull (including any 'cutting shots') the gun is reckoned to be capable of 'one shot, one kill' and is given a *sniper score* or SS of 100 – or more if the shots are tightly grouped – at the specified test distance.

Accepting radial-displacement theory requires precise measurement and time-consuming calculations, as the strike-point of 25 shots must be plotted individually. It will not be obvious from the five separate groups where the centre of the cumulative sequence or 'big group' lies. This is where the aim-point mark has its secondary use.

Taking the first of the shot-groups, the distance each individual shot lies from the aim point must be measured. These details should be plotted on the horizontal ('x') and vertical ('y') axes. Once all the shots of each of the five groups – or the single 25-shot one – have been traced, the table of dimensions should resemble the one given below.

Shot number	A	C	E	B	D	F	G
1	−2.52	−4.54	20.58	−4.44	−1.98	3.93	4.95
2	+3.78	+1.76	3.11	−0.36	+2.10	4.40	2.74
3	+0.90	−1.12	1.25	−0.96	+1.50	2.24	1.87
4	+0.72	−1.30	1.68	−8.40	−5.94	35.31	6.08
5	+5.46	+3.44	11.86	−2.46	0.00	0.00	3.44
6	+2.46	+0.44	0.20	+2.46	+4.92	24.18	4.94
7	+6.72	+4.70	22.13	+0.12	+2.58	6.64	5.36
8	−1.50	−3.52	12.36	−7.50	−5.04	25.43	6.15
9	+3.24	+1.22	1.50	−3.12	−0.66	0.44	1.39
10	+3.12	+1.10	1.22	−0.84	+1.62	2.62	1.96
11	+1.86	−0.16	0.02	−5.22	−2.76	7.63	2.77
12	+3.72	+1.70	2.90	−3.30	−0.84	0.71	1.90
13	+2.10	+0.08	0.01	+2.52	+4.98	24.78	4.98
14	+5.76	+3.74	14.02	−5.52	−3.06	9.38	4.84
15	+4.92	+2.90	8.43	−7.44	−4.98	24.82	5.77
16	−0.18	−2.20	4.82	−1.80	+0.66	0.43	2.29
17	−0.48	−2.50	6.23	−3.72	−1.26	1.59	2.80
18	+2.10	+0.08	0.01	−3.24	−0.78	0.61	0.79
19	+2.04	+0.02	0.00	+1.74	+4.20	17.62	4.20
20	+1.80	−0.22	0.05	−1.14	+1.32	1.74	1.34
21	+2.46	+0.44	0.20	−5.94	−3.48	12.13	3.51
22	−1.56	−3.58	12.79	−0.18	+2.28	5.19	4.24
23	−1.08	−3.10	9.59	+0.66	+3.12	9.72	4.39
24	+0.42	−1.60	2.55	−0.66	+1.80	3.23	2.40
25	+4.14	+2.12	4.51	−2.70	−0.24	0.06	2.14
	(19 shots) +57.8			(5 shots) +7.5			
	(6 shots) −7.4			(19 shots) −68.9			
Totals	+50.40			−61.44			87.23

The distances each shot lies from the aim-point mark, horizontally and vertically, have been tabulated. The table states these distances as positive values if the shots lie to the right of the aim-point horizontally, or negative values if they lie to the left; similarly, a shot that strikes above the aim-point is positive vertically, but is negative if it lies below. These 'plus or minus' elements have an important effect on the calculation of the actual centre of the 25-shot group.

Column A contains measurements of all the shots considered horizontally. Nineteen of them struck to the right of the mark; the remaining six flew to the left. Therefore, 19 have values ranging from +0.4 to +6.7 inches, and the others lie between −0.2 and −2.5.

If the values of the 'positive' shots are added together, a total of +57.8 is obtained while the 'negatives' amount to –7.4. The sum of these two is +50.4, which indicates that the centre of the group lies to the right of the aim-point. As 25 shots were fired, the actual displacement of the centre is found simply by dividing the total (50.4) by the number of shots (25), giving a value of 2.02.

One co-ordinate of the group centre has now been fixed, and we must repeat the process in the vertical plane (column B) to discover if the group centre lies above or below the aim-point. Five of the 25 shots have positive values (which means that they lie above the aim-point), +0.1 to +2.5, and, with one having no value, 19 are negatives –0.2 to –8.4. It is clear that the group centre must lie beneath the aim-point as the total of the 'negatives' is –68.9 and that of the 'positive shots' is merely +7.5.

The sum of these is –61.4, therefore, and as 25 shots contributed to the total, the actual displacement (total divided by shots) is –2.46 in. As this result has a negative value, the actual 'big group' centre lies a short distance below the aim-point.

Locating the group centre allows the distance each shot lies from this centre to be calculated. Columns A and B show how far each shot lies from the aim-point, as they have all been measured individually. But it has also been discovered that the group centre lies 2.02 in. to the right and 2.46 in. below the aim-point. Consequently, an allowance must be made before progress can be made and, effectively, the value of '+2' must be added to all figures in column A and the value of '–2.5' subtracted from the figures in column B. Subtracting a minus quantity, of course, is the same as adding 2.5 to the figures in B.

This creates columns C and D. Note that applying corrections can mean that shots previously allocated positive values horizontally can now be negative, and that some which were negative vertically are now positive. This simply indicates that though they lie to the right or below the aim-point, they are actually to the left or above the group centre (which, as has been seen, is not the same thing).

The shot whose centre lay precisely on the vertical axis-line (no. 5) obviously contributes nothing to the total, but must nonetheless be allocated 'zero value' and included in the total number of shots fired.

The next steps are best understood from a typical group. The diagram overleaf shows the aim-point, which was already established, and the relative position of the combined group centre: not the centre of this particular group but rather that resulting from all 25 shots in the sequence. The distance from the centre of shot 'B' to the centre of the big-group 'G' is obtained by simple geometry using Pythagoras' Theorem.

The square of displacement distance 'BG' is, therefore, equal to the square of the length 'AB' added to the square of the length 'AG'; 'AB' is the horizontal displacement, the corrected column C on our first table, while 'AG' is the vertical displacement taken from column D. For the first shot, the values are –4.54 for AB and –1.98 for AG, as the shot struck to the right and above the calculated group centre. Using $BG^2 = AB^2 + AG^2$ gives $BG^2 = 4.54^2 + 1.98^2$ or 20.58 plus 3.93 (a total of 24.51); columns E and F on Table 1 display the squared values of C and D respectively.

Remembering that the value obtained, 24.51, is actually the square of the displacement distance BG, we have to use a calculator to obtain the actual value: $\sqrt{24.51}$ is 4.95, and so the first shot lies about five inches from the centre of the 'big group'. The process is repeated for all 25 shots, until it is discovered (column G) that the nearest shot to the group centre is the eighteenth, only 0.79 in. away, and that the farthest is number eight – 6.15 in. distant. Finally, values in the last column are added together to give 87.23 in.: the total displacement of all 25 shots considered individually without regard to the direction in which they lie from the group centre.

This 'string total' could be used as a final result (still often used in bench-rest shooting), but today's shooters usually talk in terms of 'group diameter' and steps must be taken to fall in line with their preferences.

The average radial displacement (the total divided by the number of shots) of the 87.2 in. test string is 3.49 in. Of course, diameter is twice the value of radius; consequently, 6.98 in. should be considered as a reasonable 'mean group diameter' ('MGD') assessment of the trial. This is lower than the average diameter of the five groups and considerably smaller than the rectangle containing all the shots, but has been obtained in a way which minimises the effect of a few strays or fliers. Had there been a substantial number of odd shots or one of the supposedly good groups was displaced in relation to the remainder, MGD would have been far worse.

A 30-shot trial undertaken at 100 metres with a 5.56 mm Heckler & Koch G41 Zf., supported with a wrist rest, gave a mean radius of 10.3 mm (0.406 in)

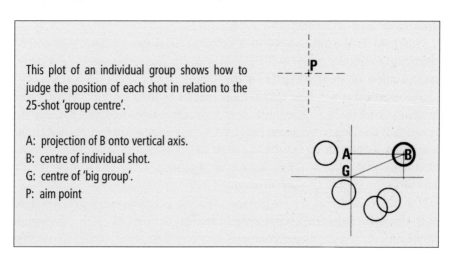

This plot of an individual group shows how to judge the position of each shot in relation to the 25-shot 'group centre'.

A: projection of B onto vertical axis.
B: centre of individual shot.
G: centre of 'big group'.
P: aim point

or a mean diameter of 20.6 mm (0.812 in). A solitary 'flier' lay 30.3 mm (1.19 in) from the centre, so the grouping potential of the gun could have been claimed as 47 mm, the average diameter of the three ten-shot groups, or as the 60.6 mm-diameter circle that contained all thirty rounds. The MGD system, however, gives a fairer prediction of accuracy by allowing an inevitable (if occasional) 'flier'. A result of 0.206 cm or 0.7 MOA accords well with practice.

It is possible to derive an acceptable theoretical prediction of the performance of an individual gun/ammunition combination on virtually any occasion. Though this involves the use of standard deviation and confidence limits to make the process mathematically acceptable, the system can nonetheless make realistic predictions even though results obtained under differing conditions clearly cannot be compared directly.

Predictions are made by deducing the difference d of each individual shot from the average, squaring d to provide d^2, then totalling the d^2 figures and dividing the resulting total by the number of shots n. Square-rooting $d^2 \div n$ gives the standard deviation, s.

Use of confidence limits (C) then allows a theoretical assessment to be made of the performance that can be expected on every occasion (assuming the same ammunition is used) – not just that of the individual trial-series from which the figures were obtained.

The value of C can be set appropriately. The first versions of the analysis system worked with 98 per cent probability ($C = 2.33$) but experience has since suggested that 95 per cent ($C = 1.95$) is more realistic, as it allows for one shot in every 25-shot sequence to be rogue.

An explanation of the mathematical theory behind the derivation of the final expression δ can be found in the fourth edition of *The Airgun Book* (1987),[13] but δ is effectively the mean group diameter qualified by limits placed equally above and below the mean. These limits are found by multiplying C and s, then dividing the result by the square-root of n (as there are 25 shots, $\sqrt{n}$ will always be 5 in the trials discussed here).

The trial of the Obr. 91/30 Mosin-Nagant gives qualifying limits of 1.36. This suggests, therefore, that there is 95 per cent probability that any shot (or combination of shots) fired from the same gun/ammunition combination, under identical conditions, will strike the target within a group of mean radius 3.49 ± 1.36 inches. The lower limit, '$\delta-$', can be disregarded, as shots within the mean radius will clearly be effective; only those that lie outside the mean radius can degrade the 'sniper score'.

As diameter is twice radius, 9.70 inches (6.98 + 2.72), '$\delta+$', is the vital result. It can be used to predict how well the rifle is likely to perform on a 'one shot, one kill' scale. A glance at the target (and even the five individual groups) shows that the shots are widely dispersed, instead of clustering in the centre with a few outliers. The tighter the clustering, the better the score; the wider the spacing within the same mean radius, the greater the value of δ+ becomes and the gun will perform noticeably more poorly.

It has been suggested that the ratio between the areas of the circles bounded by the mean diameter and δ+ can help to predict the chance of the sniper's bullet striking the target. The mathematics of this are tenuous, and must be tested far more greatly than simply relying on a handful of experiments. The mean diameter of the Mosin-Nagant trial is 51 per cent of δ+, but the eight-inch bull's-eye must be taken into account.

As the mean diameter circle of this particular experiment lies within the bull (even though δ+ is outside), so the 'sniper score' rises to 73. If both mean diameter and δ+ circles lie within the bull's-eye, the rifle will score considerably more than the 100 that would theoretically signify 'one shot, one kill'.

Results can be related to variations in range by a constant such as a 'minute of angle' – one sixtieth part of a degree, generally taken to be equivalent to a one-inch group at 100 yards, though the true conversion is 1.047 inches.[14]

One problem with MOA is that it cannot handle metric and imperial measurements with equal facility: a one-inch group at 100 yards is approximately the same as a 29.07 mm group at 100 metres, but metric conversions are not simple-ratio transformations.

In addition, MOA is a straightforward statement of actual group diameter, whereas the MGD system (actually known as MAP or 'mean accuracy projection') includes a special element to allow a reliable prediction to be made.

Degradation of accuracy with range is notoriously difficult to predict. It has even been claimed that a gun shooting 1 MOA will give groups of an inch at 100 yards, 2 inches at 200 yards, 5 inches at 500 yards, and 10 inches at 1,000 yards. This is illusory; loss of velocity and spin-induced bullet drift are just two of many factors undermining the validity of a linear relationship.

Accuracy undoubtedly deteriorates as range increases, and, however much the effects are ameliorated by high velocity or good bullet design, group size steadily increases beyond the 'straight-line' predictions. The problem is simply assessing magnitude.

The mean radii of 25 shots from a 5.56 mm AR18 rifle were 2.0 in. at 100 yards, 5.4 in. at 200 yards, 8.2 in. at 300 yards, 12.4 in. at 400 yards and 16.4 in. at 500 yards. Even this simple experiment shows that a gun shooting exactly 2 MOA at 100 yards does not retain this accuracy as range increases, degrading to 2.7 MOA at 200 yards and 3.28 MOA at 500 yards.

The 5.56 mm SS109D bullet, with a muzzle velocity of 946 m/sec, is travelling at 831 m/sec at 100 metres, 734 m/sec at 200 metres, and 480 m/sec at 500 metres. The velocity has been roughly halved (actually to 54 percent) after the bullet has travelled 500 metres, and accuracy, while not half as good, has worsened to 64 per cent of the original value.

A similar experiment undertaken with a 7.62 mm L1A1 rifle gave MOA as 2.2 at 100 yards, 2.6 at 200 yards and 2.8 at 500 yards – not the same magnitude of change that the AR18 seemed to show. However, the 7.62 mm SS77 boat-tailed bullet retains velocity far better than the comparatively lightweight 5.56 mm SS109D: 840 m/sec at the muzzle declining to about 515 m/sec at 500 metres. Muzzle velocity declines by 39 per cent, therefore, but accuracy has degraded only by 27 per cent.

Deducing mathematically justifiable correlations between the changes in velocity with accuracy clearly cannot be simply a matter of introducing 'loss of velocity' to the equation. However, trials suggest that degrading accuracy by linking it to velocity-loss can give a broad basis for comparison if the cartridge is shared by the participating guns.

Actual results

It is very difficult to find acceptable trial results for the period prior to 1850, when the effects of industrialisation led to the realisation that accurate measurement was key to all engineering progress. Too many attempts to gauge accuracy had been based on systems such as Figure of Merit (explained previously) and those in which the shots that missed the target not only went unrecorded but were rarely allocated a value in the subsequent assessments.

'Projectile weapons' fall broadly into groups:

1. **Primitive weapons** such as crossbows, arquebuses and early muskets. It has been all but impossible to find reliable medieval or even post-medieval records that would allow a justifiable accuracy assessment to be attempted. Surviving bows, with bow-arms of horn and sinew, cannot be risked in any test; and, of course, would probably break under the strain. Nor can the engravings of shooting competitions be trusted, because they inevitably foreshorten range and distort perspective.

The best-known of all crossbowmen is William Tell, from the Swiss canton of Uri. Today's legend – originating with the *Weisses Buch von Sarnen, c.* 1474 – tells how the Habsburg emperor, seeking to incorporate Uri in his domains, appointed Gessler as *Landvogt* (bailiff) of Altdorf. The *Landvogt* hung his hat on a pole in the village square, demanding a bow of loyalty before it. On 18 November 1307, however, Tell refused to bow and was arrested. Gessler told Tell that his life (and that of his young son) could be spared only by shooting an apple off his son's head. Tell, of course, duly split the apple with his first shot.

Though the Tell legend has medieval roots and may well be based on fact, it is generally assumed to be apocryphal. The range would have to have been very short, perhaps no more than fifteen yards, but a hit at this distance is by no means improbable.

William Tell shoots the apple from his son's head (at comparatively short range), not only saving both their lives but also ultimately inspiring the creation of the Swiss Confederation. From a wood engraving published in *Cosmographica* in 1554.

However, the story tells little about the ultimate performance of a crossbow. Perhaps the best that can be achieved is to record that 'an ell at a hundred paces' could be an acceptable standard. The ell, a measure of cloth (also known as the 'clothyard', the length of an English arrow), is 40–45 inches; paces are usually reckoned to be about twenty-eight inches.[15] The typical medieval archer would have been appreciably shorter than today's average man and his pace may have been 25–26 inches, but this is insignificant in the context of largely unknown shooting results.

Placing 25 bolts into a 40-inch circle at 75 yards would equate to an MOA equivalent of 53! Clearly, much more research needs to be done before any valid conclusions can be drawn, though hunters of today do not reckon the maximum effective engagement range of a crossbow to lie beyond 40 yards.

The performance of smooth-bore muskets is also difficult to establish with certainty; though trials were conducted by the British and the French, among others, they were concerned more with the probability of hitting a man-size target in volley fire than defining the individual shot-strike. The Brown Bess was a notoriously poor shooter owing to excessive windage between the ball and the bore, and it seems that a 30-inch group at 100 yards would verge on miraculous (MOA 30). A trial undertaken in Woolwich in 1835 with a New Land Pattern musket gave appalling results: dispersion at 100 yards was 56 inches vertically and 72 inches horizontally. At 200 yards, shooting was too bad to be recorded.

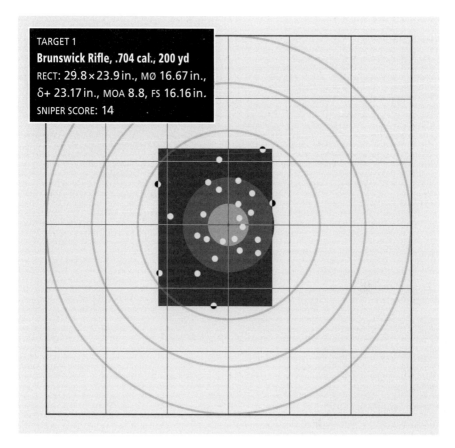

TARGET 1
Brunswick Rifle, .704 cal., 200 yd
RECT: 29.8 × 23.9 in., MØ 16.67 in.,
δ+ 23.17 in., MOA 8.8, FS 16.16 in.
SNIPER SCORE: 14

2. 'Primitive-precision' weapons such as the American Long Rifle, the standardised British flintlock rifle (Baker) and early cap-lock rifles such as the Brunswick. This category contains some surprisingly good performers. The diagram obtained from the Brunswick rifle at 200 yards proved to be far better than had been anticipated; indeed, a test undertaken in the 1830s gave results equivalent to a 'sniper score' of 89 at 100 yards, giving a very good chance of a 'head-hit'.

3. Cap-lock rifle-muskets such as the British P/53 .577 Enfield, the Austrian .58 M/54 Lorenz, the Russian '6-line' Obr. 1856g., the French 17.8 mm Fusil d'Infanterie Mle 1857, and the US .58 M1855 and M1861 Springfields.

This group contains the range of essentially similar rifle-muskets that enjoyed a brief heyday in the 1850s and 1860s, before being swept away by metallic-cartridge breechloaders. At their best, they were capable of surprisingly good shooting at ranges several times that of the musket, but manufacture of the self-expanding balls was not always consistent at a time when the value of strictly enforced dimensional tolerances was not understood.

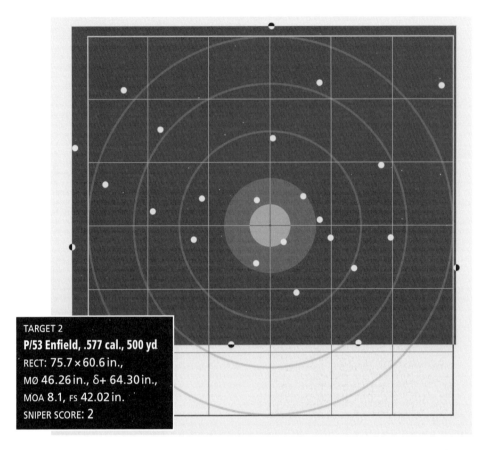

TARGET 2
P/53 Enfield, .577 cal., 500 yd
RECT: 75.7 × 60.6 in.,
MØ 46.26 in., δ+ 64.30 in.,
MOA 8.1, FS 42.02 in.
SNIPER SCORE: 2

The Whitworth rifle, especially the perfected .450-calibre pattern firing mechanically fitting hexagonal-section bullets, was the great exception. Results published in *The Times* in 1857 drew attention to the great disparity between results obtained from the .577 P/53 Enfield and those from the then-experimental .450 Whitworth. The diagrams shown here (Targets 2 and 3) are adapted from the 500-yard government-trials targets, simply by adding five shots taken at random from otherwise comparable trials to allow an 'analysis of 25'.[16]

The P/53 performed exceptionally badly on this occasion, which was probably due to poor-quality ammunition.[17] But the Whitworth gave such good results at 500 yards that it would have out-performed many of the rifles of the Second World War at the same distance. The key to this success lay in Whitworth's fierce determination to keep manufacturing tolerances to an absolute minimum. For example, the bore diameter was held to 0.450 + 0.001, compared with 0.303 ± 0.002 for some of the No. 4 Mk I rifles made in Britain during the Second World War.

The Whitworth of the 1850s, therefore, was allowed only *one quarter* of the tolerance-latitude of its 1940s successor. As the projectiles were also made to

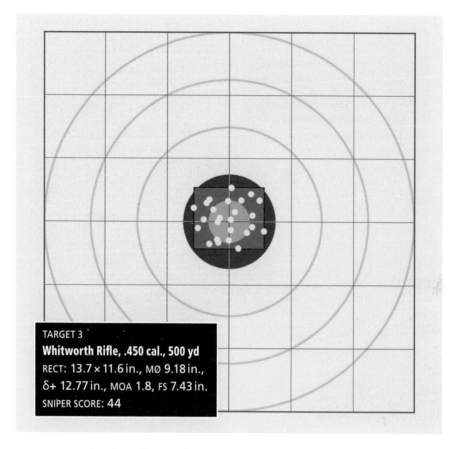

TARGET 3
Whitworth Rifle, .450 cal., 500 yd
RECT: 13.7 × 11.6 in., MØ 9.18 in.,
δ+ 12.77 in., MOA 1.8, FS 7.43 in.
SNIPER SCORE: 44

exacting standards, Whitworths could shoot so unbelievably well that their performance in the hands of Confederate sharpshooters (which has often been questioned) was exemplary . . . as long as caveats were observed: the rifles were extremely susceptible to the degrading effects of propellant fouling and to extremes of heat and humidity, and so the bore needed to be regularly (and thoroughly) cleaned. Joseph Whitworth had produced a rifle that met his manufacturing ideals, but not the needs of battle.

4. Cap-lock breech-loaders. The advent of breech-loading improved accuracy compared with muzzle-loading smooth-bore muskets, but not necessarily in comparison with long barrelled rifle-muskets firing self-expanding bullets. Target 4, from an 1859-type Sharps US Army rifle, gives a clue to the performance of Berdan's sharpshooters: good, but not necessarily on a par with a Whitworth cap-lock (Target 3).

This category could also be considered to include Dreyse *Zündnadelgewehre*, though the cap was buried in the papier-mâché sabot in which the sub-diameter oviform bullet was carried. The accuracy of the Dreyse – even the *Jägerbüchse* – was very poor, and range was also surprisingly short. The French

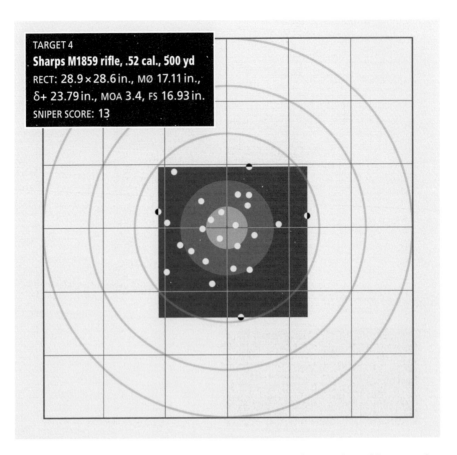

TARGET 4
Sharps M1859 rifle, .52 cal., 500 yd
RECT: 28.9 × 28.6 in., MØ 17.11 in.,
δ+ 23.79 in., MOA 3.4, FS 16.93 in.
SNIPER SCORE: 13

Mle. 66 Chassepot was similar to the Dreyse, but the combustible cartridge was a better design in which the bullet (not the sabot) engaged the rifling. The 11 mm-calibre Chassepot could out-shoot the Dreyse by a considerable margin and, as far as accuracy was concerned, was also a match for most of the first-generation cartridge breechloaders.

5. First generation metallic-cartridge breech-loaders. These enjoyed only a very short period in vogue. They were usually conversions of rifle-muskets to fire self-contained metal-case cartridges, and so their calibre duplicated those of the cap-locks. The British Snider rifle was converted from the P/53 Enfield rifle-musket, and the first 'Trapdoor Springfields' issued to the US Army were .58 rimfire conversions of the Model 1855 and 1861 rifle-muskets. Accuracy, as the Snider target shows, was slightly better than the rifle-muskets had usually achieved largely because the self-contained cartridges shot more consistently. However, the comparatively poor ballistic shape of the bullets prevented wholesale improvement.

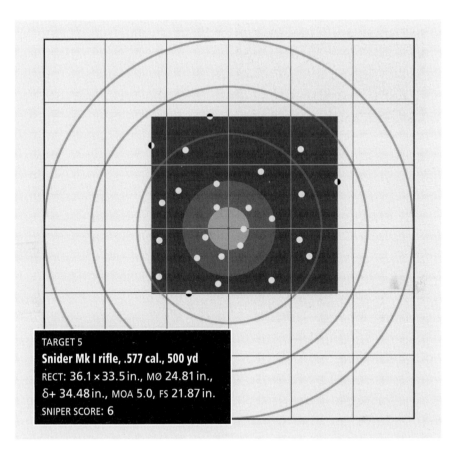

TARGET 5
Snider Mk I rifle, .577 cal., 500 yd
RECT: 36.1 × 33.5 in., MØ 24.81 in.,
δ+ 34.48 in., MOA 5.0, FS 21.87 in.
SNIPER SCORE: 6

6. Second-generation breech-loaders. Converting large calibre cap-lock rifle-muskets proved to be little more than a temporary measure, as virtually all military authorities agreed that a reduction in calibre was beneficial: not only did it reduce the weight of individual cartridges, but the bullets were better ballistically. Accuracy, it was hoped, would improve commensurately. Calibres usually lay between 0.4 and 0.5 inches, with 11 mm (.433) and .45 predominating. Tests undertaken in Britain in the early 1870s proved that the .450/577 Martini-Henry out-performed the Chassepot needle rifle, but the necked Martini-Henry cartridge, especially in the earliest days of production, did not always shoot consistently. The straight-case .45–70 chambered in the US Army's 'Trapdoor Springfield' was a better design in many ways, and accuracy results from these guns could be very good. Target 6 shows a typical result obtained at 500 yards with hand-loaded cartridges.

The advent of reliable metal-case cartridges increased interest in target shooting, and the records of national and international matches often make fascinating reading, especially as they were undertaken at ranges up to (and sometimes in excess of) 1,000 yards. Cap-locks initially gave the metallic-

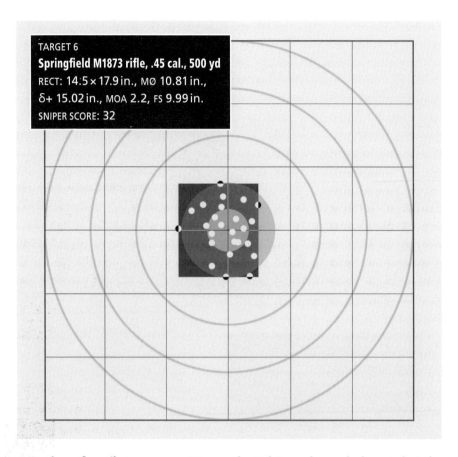

TARGET 6
Springfield M1873 rifle, .45 cal., 500 yd
RECT: 14.5 × 17.9 in., MØ 10.81 in.,
δ+ 15.02 in., MOA 2.2, FS 9.99 in.
SNIPER SCORE: 32

cartridge rifles effective competition – the Whitworth regularly out-shot the Snider – but uncertainty of ignition in damp conditions was an insuperable weakness of any cap-lock. Second-generation breechloaders could be capable of very good shooting. Though this was not generally reflected on the battlefield, where the concept of sniping had effectively ended with the American Civil War, buffalo hunters and long-range target shooters kept techniques alive, helped by increasingly sophisticated sights and by the gradual acceptance of the telescope. Gunsmith John Lower of Denver, Colorado, fired fifty shots 'offhand' from a Sharps-Borchardt rifle into a 23-inch circle at 200 yards, and a Martini-Henry barrel clamped in a rest placed 24 shots in a 20-inch circle at 1,000 yards.

7. **Small-calibre rifles firing smokeless ammunition**. The 1909 handbook of the .30-calibre US M1903 rifle gives the accuracy, 'as determined by firings to date', as mean absolute radii of 1.1 in. at 100 yards, 2.3 in. at 200 yards, 5.9 in. at 500 yards, and 13.3 inches at 1,000 yards. Assuming these to be equivalent to groups of 2.2–26.6 in., the MOA performance of the rifle is approximately 2.2 at 100 yards, 2.3 at 200 yards, 2.4 at 500 yards and 2.7 at 1,000 yards.

Taken in typical Afghanistan terrain, hot, dry and dusty, this sniper is armed with a 7.62 mm M21 rifle. Modern optical and electro-optical sights make the marksman's job far easier than it was in the world wars, as the range is often calculated by laser. *US Army photograph*

Though the .30–06 round performs better at long range than some military cartridges of generally comparable dimensions, these figures give a good idea of the performance of the pre-1945 sniper rifle. Tests of Obr. 91/30 Mosin-Nagants have returned about the same degree of accuracy to 800 metres, which the Russians themselves apparently regarded as the maximum effective range. Results shown on page 47 seem to support this conclusion, with an MOA return of 2.3 at 300 yards, but the Sniper Score of just 73 – for a rifle generally considered to be capable of very good 'head-kill' performance at this range – is disappointing. However, it was not a specially selected *snayperskaya vintovka* (merely an infantry rifle with an 'after-market' optical sight), and the unexpected dispersion of shots suggests that the ammunition may have been far from ideal. Consequently, only 13 of 25 are rated as 'head kills' even though the most distant shot lies merely six inches from the group centre-point.

British No. 4 Mk I rifles were expected to group four out of five shots in a 3×4 in. rectangle at 100 yards to be deemed acceptable, but individual guns could often place all their shots in a group half this size. These would be attaining 1.5–1.8 MOA. The very best of the sniper rifles, which had to pass additional tests (see page 223), could be capable of 1 MOA.

Tests with a Soviet-era AKM and a Czech vz. 58, essentially similar in size and performance, gave mean-diameter groups at 100, 200 and 400 yards of 2.4, 6.4 and 14.8 in. for the AKM and 2.0, 5.6 and 13.0 in. for the vz. 58. The velocity of the Soviet 7.62 × 39 mm M43 bullet, common to both guns, drops (in the AKM at least) from 2,325 ft/sec to about 1,050 ft/sec at 400 yards. The velocity loss is 55 per cent, therefore, and the degradation in accuracy resolves to 54 per

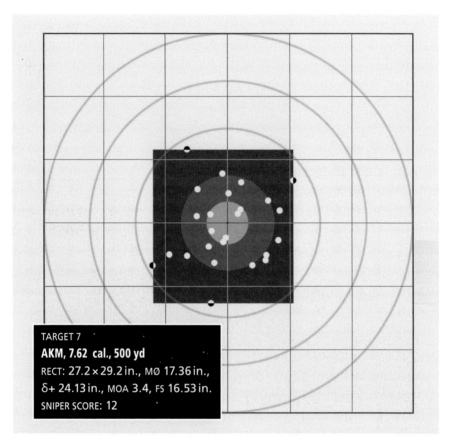

TARGET 7

AKM, 7.62 cal., 500 yd

RECT: 27.2 × 29.2 in., MØ 17.36 in.,

δ+ 24.13 in., MOA 3.4, FS 16.53 in.

SNIPER SCORE: 12

cent for the AKM and 63 per cent for the vz. 58, even though the latter proved to be more accurate at longer ranges.

This group also includes semi-automatic rifles such as the 7.62 mm SAFN and FN FAL, the Heckler & Koch G3 and the US M14. Target 8 shows the results of a trial with the 7.62 mm ArmaLite AR-10, the 'big brother' of the AR-15/M16 series; this particular combination of rifle and ammunition was not shooting well, as the MOA of 3.0 shows, yet every one of the 25 shots lies within the 'head kill' zone (inset). Consequently, the Sniper Score of 391 not only emphasises a 'one shot, one kill' prediction, but also how the individual shots would cluster towards the centre of the target if the sights were altered to centre the group on the bull's eye.

8. Ultra-small-calibre rifles. The .223 (5.56 × 45 mm) cartridge entered US Army service in the early 1960s, with the M16 rifle, in an attempt to provide lightweight ammunition which could offer the high velocity that would minimise range-gauging errors at short and medium ranges. Unfortunately, though accuracy was very good out to about 300 yards, the original US M193 bullet proved to be unstable and was generally replaced by the more conventional

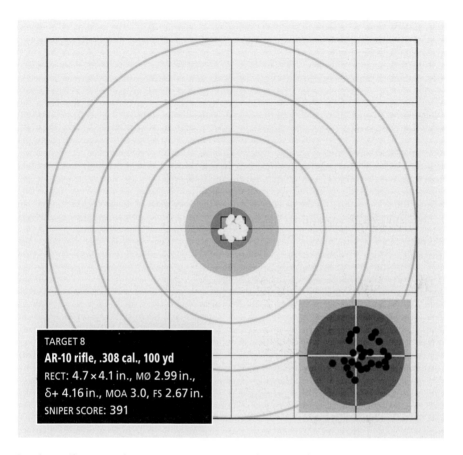

TARGET 8
AR-10 rifle, .308 cal., 100 yd
RECT: 4.7 × 4.1 in., MØ 2.99 in.,
δ+ 4.16 in., MOA 3.0, FS 2.67 in.
SNIPER SCORE: 391

but less effective Belgian SS109. Except in a few specific circumstances, snipers rarely use .223 rifles, as the bullet does not have the range of the 'full-power' .308 Winchester (7.62 × 51) or .30–06.

9. Precision-shooting rifles, even though they are based on standard designs, can be capable of exceptionally accurate shooting. For example, the US M24 SMS and M110 SASS rifles, firing M118LR .308 rounds (175-grain bullet, muzzle velocity 2,600 ft/sec) can give seven-inch groups at 800 yards (MOA 0.875).

The trend towards ever more powerful ammunition has extended both engagement range and potential accuracy: the SOCOM MK 13 Model 5 rifle, chambered for the .300 Winchester Magnum cartridge (7.62 × 67) firing a 190-grain bullet with a muzzle velocity of 2,900 ft/sec, will often give a group of eleven inches at 1,200 yards (0.92 MOA); and the AI L115A3, chambered for the .338 Lapua Magnum round (8.59 × 67, 300 grains, 2,800 ft/sec), claims to be capable of the same grouping at 1,500 yards (0.73 MOA), virtually guaranteeing a hit (if not necessarily a 'kill') at long range.

Claims have been made that the Russian SVD can perform similarly to the M24 SMS, but the methods of analysis are highly questionable.

10. **Large-calibre anti-matériel/long-range sniping rifles** have increased greatly in popularity in recent years, though they have a pedigree stretching back to the PTRD (Degtyarev) and PTRS (Simonov) anti-tank rifles of the early 1940s, which, made obsolete by ever-thickening tank armour, found a niche as long-range anti-personnel weapons. Attempts have also occasionally been made to use the .50 Browning heavy machine gun as a sniping weapon. Unfortunately, the ammunition for these guns was not designed primarily for accuracy. Though great strides have been made in recent years to update the .50 Browning cartridge, these still fall short of 'match-grade accuracy'. Published figures reveal that the US Army LRRS firing MK 211 cartridges (671-grain bullet, V_0 2,900 ft/sec) is currently expected only to give groups of 25 inches at 1,000 yards and 37 inches at 1,500 yards, each about 2.5 MOA. This suggests that most ultra-long-range kills owe as much to good fortune as a sniper's skill.

Notes to Chapter One

1. The term 'point-blank' is usually used to qualify the distance at which the lowest sight-setting and the strike point coincide. It can be set at anything from 25 yards for a handgun to 200 yards for a high velocity rifle. The 1909 handbook for the US Army's M1903 Springfield rifle gives the point-blank 'danger space' using the battle sight – a muzzle elevation of 0° 17' 57" – as 719 yards if the firer is standing (assumed bore height of 56 in), 629 yards kneeling (30 in) and 590 yards prone (12 in); if the bore is truly horizontal, these distances reduce to 400, 307 and 205 yards respectively. However, as even fixed 'iron' sights are set some distance above the bore, a bullet striking at ranges below the point-blank setting will usually hit low. This is rarely enough to make much difference in normal combat, but can be an important factor if an optical sight is set well above the bore compared with fixed sights.
2. Adapted from Richard D. Law, *Sniper Variations of the German K98k Rifle* (1996), quoting *Die deutsche Scharfschütze* ('The German Sharpshooter'), date unknown.
3. 'KF' or 'Kf.' marks on German optical equipment signify *Kältefest* – 'cold resistant'. However, the sights proved to be difficult to adjust in temperatures below −4 °C; as −50 °C could be registered in a Russian winter, German snipers were placed at a disadvantage. Though lubricants were improved, the problem was never entirely overcome.
4. *Instructions to Young Marksmen in all that relates to the General Construction, Practical Manipulation, etc., etc., as exhibited in the Improved American Rifle*, eventually published by D. Appleton & Company of New York in 1848
5. His birth is sometimes recorded as 9 September 1811, but this was his baptism.
6. Charles Piazzi Smyth, born in Naples on 3 January 1811, was Astronomer Royal for Scotland from 1846 until resigning his office in 1888 in protest against lack of funding. He died in February 1900 and was buried in St John's Church, Sharow.
7. Steven Roberts, 'Davidson's Telescope Rifle Sight' (©2010), accessed from www.springfieldarsenal.files.wordpress.com but linked to a now inaccessible source..
8. Isabella may have been born 'Isabella Showler', the name given to the bride in the marriage of John and Isabella in Revesby on 5 July 1814. This would have meant that

she was pregnant with John Ratcliffe at the time. John Chapman had probably died by 1841, as the census of that year records Isabella Chapman as head of household. The misidentification of John Ratcliffe Chapman's background has probably arisen from confusion with John Chapman (1801–54), preacher, radical politician, textile-machinery maker and 'Founder and late Manager of the Great Indian Peninsular Railway'. John Ratcliffe Chapman is listed in all the US Federal and New York State censuses from 1850 until his death as a 'farmer'.

9. Malcolm lost his right eye in 1869, when a cartridge exploded in a gun he was testing, and suffered from heart and respiratory problems for the last few years of his life.

10. Ordinary 'white' light is composed of a rainbow in which the wavelength (and hence focal length) of each component differs; if the lens is badly corrected, a blurred image with a coloured corona or 'halo' may be visible. Sandwiches of thin glass are used to unify focus, each diffracting or 'bending' a portion of the light at a different angle.

11. Stewart Wills, 'How the Great War Changed the Optics Industry', in *Optics & Photonics News*, January 2016.

12. In 1990, on the basis of 100 trials, I proposed a factor of 4.518 to enable Figures of Merit to be converted to group diameter. One of the problems of F.o.M. was that misses were allocated a value equivalent only to half the diagonal measurement of the target, even though they could have missed by far greater distances.

13. The system was developed in the 1980s to classify the fluctuating power of airguns, but is easily adapted to handle any projectile weapon.

14. A difference of less than 5 per cent has no significance even at the distances to which .50 Browning will carry. At 1,500 yards, the error is less than three-quarters of an inch.

15. The 'paces' used on rifle sights included the Russian *arshin*, equivalent to 27.99 inches, and the Austrian *Schritt* of 28.53 inches.

16. A random-number generator, working within extremes of the trial data, was used to create the five additional shots necessary for an 'analysis of 25' from the 20-round string. This solution is arguable, but, as the tests cited here are not intended to provide direct comparisons of individual guns, small errors can be overlooked.

17. Other P/53 trials gave better results, though they are rarely exceptional. One undertaken in 1867, suitably adjusted, gave MØ as 35.09 in., δ+ as 48.78 in., and MOA as 7.0. The farthest shot lay 32.18 in. away from the centre of the group, and the 'sniper score' was only 3.

CHAPTER TWO

Civil and Colonial Wars: The Sniper is Born

The Crimean War, fought from October 1853 until 30 March 1856, was basically a protracted and extremely costly struggle between the Russians and the Ottoman Empire, supported by Britain and France (and a few regiments raised in 1855 by the kingdom of Sardinia). The ostensible causes included perceived persecution by the Muslim Turks of the Christian minority but there were undercurrents of mutual suspicion. The British and the French rightly suspected that the Russians would take advantage of the parlous state of the Ottoman Empire, crumbling slowly away, to seize territory and threaten Anglo-French domination of the Mediterranean. The Russians rightly suspected that the British and the French would deliberately obstruct any expansion of Russian territory at the expense of the Ottoman Empire.

Casualties in the war were horrendous. The British lost more than 20,000 dead, but only 2,755 were killed in action; 2,019 died of wounds, and 16,323 of disease. The French had lost 10,240 killed in action, another 20,000 who died of their wounds, and a staggering 60,000 dead of disease. The Ottoman Empire lost 95,000–170,000 (estimates vary wildly) and Russian casualties have been estimated as approximately 400,000, 143,000 of whom died.[1]

The Crimean war is popularly remembered in Britain for the siege of Sevastopol, the battles of the Alma, Inkerman and Balaklava, the Charge of the Light Brigade, and huge losses of life arising from poor sanitation and ineffectual treatment of wounds and disease. But the conflict was much wider-ranging: though the primary campaign was centred in the Crimea, efforts were also made to subdue Russian interests in the Far East.

The war was the first in which the telegraph and railways played any note-worthy part, and the first in which unfettered journalism – and battlefield photography – helped to shape public opinion. It was also the proving ground for the perfected rifle-musket, when, in the British Army, the large-calibre P/51 Minié slowly gave way to the P/53 Enfield (though many infantrymen were still carrying muskets when the fighting ended).

The Highland Brigade at the Battle of the Alma, 20 September 1854, the first major
confrontation of the Crimean War. The British and French defeated the Russians.
From the Illustrated London News, *courtesy of Philip J. Haythornthwaite*

The regulation British cap-lock musket of 1850 was still the smooth-bored
P/42, a cap-lock derivative of the India Pattern Brown Bess of the Napoleonic
Wars, firing a large-diameter ball which was accurate only to 75–80 yards
and had lost most of its effectiveness at 200 yards unless volleys were ordered.
The introduction of the P/51, with its self-expanding bullet, raised effective
engagement range tenfold, albeit at the expense of a particularly vicious
recoil.

The advent of the rifle-musket encouraged marksmanship. In January
1853, the Brigade of Guards received their first P/51 rifle-muskets on a scale
of twenty-five per company, to be issued, according to instructions given by
commander-in-chief Lord Harding, only to the most proficient marksmen.
When the Guards were sent to the Crimea, only the 4th Division was still
carrying P/42 smooth-bores.

The accuracy of the P/51 was such that a practised marksman could make
occasional shots count at ranges up to 900 yards, the limit of the back-sight
graduations. Modern estimates have even claimed that a man-size target could
be hit 75 per cent of the time at 400 yards; 60 per cent at 600 yards; and 25
per cent at 800 yards.[2] Brevet Major Somerset Gough Calthorpe (1831–1912),
an aide-de-camp to Lord Raglan, reported that Rifleman Herbert of the Rifle
Brigade had shot a Cossack officer from his horse at a range judged to have

been about a thousand yards, as Herbert was deliberately 'holding over' the 900-yard position in his back sight.

On 18 October 1854, an order given by the commander of the 1st Guards Division created a section of sharpshooters in each battalion, comprising ten men and an NCO – all accredited shots – each section to be commanded by a captain and a lieutenant. The officers were to be appointed by roster, not by selection, as it was confidently assumed that any Guards officer would be competent to oversee operations. The men, it was believed, would be drawn from those within the Guards who had a rural background: the rabbit hunters, the poachers and the gamekeepers, all of whom had well-honed tracking and shooting skills.

Writing in *The Coldstream Guards in the Crimea*, published in 1897, Lt.-Col. Sir John Ross of Bladensberg – whose grandfather Major-General Robert Ross, 'The Man who burned the White House', had been killed by sharpshooters during the War of 1812 – suggested that the October 1854 order stated:

> The sharp-shooters will . . . endeavour to improve their cover behind any obstacle by scraping out a hollow for themselves in the ground, and they will carry . . . provisions so that they will be enabled to remain, being once under cover, for many hours . . . without relief.
>
> Whilst so established they will endeavour to pick off the enemy's artillerymen in the embrasures.
>
> The approach of the sharp-shooters to the spot they must occupy must be rapid, in a scattered order; each man acting for himself, and exercising his intelligence to the utmost of his ability. Each man will select the spot which suits him best, and be guided only in that choice by the cover he may find.

Suitably but only slightly amended, these directives applied as readily to the snipers of Stalingrad – not too distant from the Crimea – less than a hundred years later.

The first officer appointed to command the Guards Sharpshooters was Gerald Littlehales Goodlake. Born in Faringdon on 14 May 1832, son of Thomas Mills Goodlake and Emilia-Maria, daughter of baronet Sir Edward Barker, Gerald had enjoyed the privileges of his class: five years at Eton College, where he excelled in sport, and then, on 18 June 1850, a commission 'by purchase' as a second lieutenant in the 23rd Foot (one source says 21st). Goodlake transferred to the Coldstream Guards on 27 June 1851, as an 'Ensign and Lieutenant by Purchase'.[3]

The Guards left England for the Crimea on 18 January 1854. Shortly after arriving, Goodlake was promoted 'Lieutenant and Captain without Purchase'. He was to command the 'Sharpshooters of the Guards Brigade' only from 17 October to 27 November 1854, but this was long enough not only to see

Gerald Littlehales Goodlake, commander of the Guards Sharpshooters in the Crimea.

constant action but for him to win the Victoria Cross for gallantry and leadership shown in repulsing a Russian sortie on 26 October, during a battle dubbed 'Little Inkerman'.

Goodlake, who was not defensively minded, believed that attack was the best way to deal with the Russian rifle-pits from which constant sniping had taken a steady toll of British soldiers in general and artillerymen in particular. The Russian army included many men recruited into the *zastrelshchniky*, 'skirmishers', and the *shtutserniky* or 'riflemen'. Often raised in the snows of Finland (a Russian duchy until 1917), the wastes of Siberia or the steppes of central Asia, these experienced hunters proved to be very effective snipers.

The Russian Army was armed primarily with cap-locks: some smooth-bored, others which had been rifled. The first Minié type rifle-musket, the 'Seven-Line' Obr. 1854g., was only just beginning to reach the front line in 1855. Sharpshooters carried the so-called *Littichesky Shtutser, Obr. 1843g.*, which was little more than a copy of the British Brunswick Rifle of the 1830s.[4] However, these men posed sufficient danger to persuade the British and the French to take every risk to eliminate the rifle pits.[5]

Gerald Goodlake lived through the battles of the Alma, Balaklava and Chernaya, as well as the lengthy siege of Sevastopol, to rank as lieutenant-general at the time of his death on 5 April 1890. However his sharpshooters, for whom movement was an essential part of their effectiveness, were disbanded after the Battle of Inkerman. In February 1855, Goodlake recorded that of his original thirty men, thirteen were dead, five injured too badly to fight, and one had been commissioned. The problem was simply that there were too few effectual replacements, even from the rank-and-file of the Guards Brigade.

Writing in 1865, in the seventh volume of his *Invasion of the Crimea*, A. W. Kinglake suggested that there was:

> No doubt that [the officers of the Guards Sharpshooters[6]] and the men acting under them knew well that by hanging close to the enemy . . . they gained opportunities of doing really good services; but they would hardly deny, I would believe, that one motive at least, if not the main one, for engaging in these enterprises, was a love of adventure and sport.

Use was made, at least until the summer of 1855, of 'static snipers', in rifle pits – sometimes captured from the Russians – which could be several hundred yards ahead of the British lines. Thereafter, recognition of the many benefits of sniping faded perceptibly. Yet the Crimean War was undoubtedly the proving-ground of the rifle-musket as a sniper's weapon. It may also have been the first conflict in which Davidson telescope sights were used, but, unfortunately, there is as yet no incontrovertible evidence.

The American Civil War

When war began in the USA in the spring of 1861, the US Army was equipped largely with muzzle-loading rifle-muskets, the perfected 'Springfield' having found approval in 1855. Of course, substantial quantities of older cap-locks remained in use, and there were even some .54-calibre rifles.

Once the fighting had begun in earnest, with losses mounting so rapidly that conscription had to be introduced to swell the dwindling ranks of regular soldiers and enthusiastic volunteers, it became clear that demand for fire-arms was far outstripping manufacturing capacity. The government-owned Springfield and Harper's Ferry manufactories and the leading privately owned contractors such as Colt and Remington were stretched to their limits, and a vast array of suppliers became involved. Some were contracted by the government, but many simply speculated, often mistakenly, on the efficacy of their products.

The 1850s had seen a rapid growth in enthusiasm for breech-loading (generally stoutly resisted by the military authorities) and a wide variety of proprietary designs entered service in numbers ranging from a few hundred to tens of thousands.

Though by no means obvious at the time, this gave an unparalleled chance to assess the merits not only of breech-loading, but also of the metallic cartridge and the repeating rifle. Some guns made their inventors' fortunes, many fell by the wayside; and it is certainly true that enthusiasm waned perceptibly as soon as the war had run its course and outstanding contracts for firearms had been peremptorily cancelled.

One of the most important advantages of the breechloader was the ease with which it could be reloaded. This was important to the soldier on the battlefield, who needed to be certain that his gun had fired and that what he thought was the act of firing was not just a 'flash in the pan'. This could result in charge upon charge being loaded into the barrel of what had become a useless weapon; indeed, one rifle-musket retrieved from the field of Gettysburg had been loaded 23 times, and several thousand had been loaded 3–10 times.[7]

Snipers faced different problems, but the advantages of a breechloader were clear. The painting of a Federal sharpshooter by Winslow Homer, which was published in the form of an engraving in *Harper's Weekly* in 1862, illustrates the difficulties of loading an ultra-heavy muzzle-loader in a tree. The time taken to withdraw the rammer from under the fore-end, insert a new ball in the barrel, ram it down and then return the rammer could cost a marksman his life . . . even assuming that the smoke of discharge had not already betrayed his position.

Statistically, by far the largest number of Civil War breech-loaders were carbines. However, these were not usually acquired for inherent accuracy (though many would outshoot a rifle-musket at battle ranges); handiness and, in the Spencer's case, a magazine were much more desirable. But the same could not be said of the 35,000 breech-loading rifles acquired by the Federal government in 1861–5.[8]

The Sharps was the most popular single-shot breech loader purchased by the Federal government during the Civil War and the first breech-loader to be adopted by any corps of specialist sharpshooters. Patented by Christian Sharps on 12 September 1848, the dropping-block system had been under development for thirteen years when the fighting began. Consequently, the worst of the teething troubles had been overcome and enough guns were being made to satisfy the burgeoning commercial market.

The breech-block slid downward within a sturdy iron frame when the lever forming the trigger-guard was pressed down and forward. A combustible cartridge was pushed into the chamber, the block was raised, and the gun could be fired by pressing the trigger to release the side hammer to strike a cap.

The first guns were made by Daniel Nippes of Mill Creek, Pennsylvania, from April 1849. They had obliquely moving breech-blocks instead of the vertical pattern shown in the drawings accompanying the 1848 patent, and a Sharps-patent primer system was fitted into the frame ahead of the breech-block. A detachable primer wheel lay beneath a hinged cover. A second contract for a modified '1849-type' rifle had an 1848-patent Maynard tape primer.

These iron-mounted guns had a special platinum alloy sealing-ring in the breech-block face which, acting with an adjustable chamber bushing, was intended to prevent gas leaks. Nippes-made rifles had simple back-action locks and the breech lever was a separate forging which fitted snugly around the trigger guard.

The Sharps Rifle Manufacturing Company was formed in the autumn of 1851, though production was entrusted to Robbins & Lawrence of Windsor, Vermont. The design was gradually simplified, as the trigger guard was forged integrally with the breech lever and the hammer was mounted inside the back-action lock plate. Sharps carbines had been tested extensively by the US Army and US Navy in the autumn of 1850, initially with great success, but prolonged trials showed that the breech could leak gas alarmingly. Two hundred 1852-type carbines, delivered to the Army in 1853, included a disc priming system patented by Christian Sharps in October 1852.[9] The brass-mounted guns retained the platinum ring gas check and the adjustable chamber bushing of their 1851-type predecessors, but had a conventional external hammer.

Made by Robbins & Lawrence from 1853 until 1855, and then by the Sharps Rifle Company (1855–8), 204 1853-pattern rifles were purchased by the US Navy and the Marine Corps: half stocked, they had brass furniture, a single barrel band, and a tenon beneath the muzzle to receive a sabre bayonet. Four hundred essentially similar carbines were purchased by the Army in 1853–6.[10]

The 1855 pattern was developed for trials with the British, who eventually placed an order for 6,000 .56 carbines in January 1856. Meanwhile, in April 1855, 400 .52 carbines of the same type had been ordered by the US Army. These had the tape-primer on the right side of the body and a straight-neck hammer. A short fore-end was held by a single band; mounts were almost always brass, though a few were iron. However, only about twenty-five 1855-pattern rifles were purchased by the Army. They had full-length stocks, three barrel bands, and provision for a socket bayonet at the muzzle; 200 otherwise-similar half-stocked rifles were acquired for the US Navy in 1856–7.

Extensive trials showed Sharps' platinum ring to be an ineffectual seal, especially if wear developed in the contacting surfaces. An expandable gas-check ring let into the breech-block face was patented on 1 April 1856 by Hezekiah Conant of Hartford, Connecticut, but had been introduced to production several months prior to the grant.[11] The Sharps Rifle Company is said to have paid Conant $80,000 for rights to his patent, but experience showed that the seal was still not as effective as its promoters had hoped. Leaks were not minimised until Richard Lawrence patented an improved gas-check in December 1859,[12] but it had been used for at least a year before the New Model was introduced,

The M1859 Sharps rifle, also known as the 'New Model', had a stirrup and swivel between the main-spring and the tumbler, and the breech-block moved vertically; the back sight was new, and the addition of a cut-off in the pellet-primer mechanism allowed standard caps to be used.

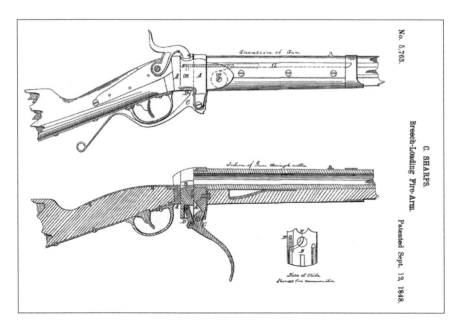

Drawings from the first patent granted to Christian Sharps. Note how the body is placed well ahead of the trigger, and the separate operating lever. Only the earliest guns took this form. *US Patent Office, Washington, DC*

The US Navy ordered 900 .56 rifles directly from Sharps on 9 September 1859, followed by 1,500 .52 calibre guns from John Mitchell of Washington, DC, in June 1861. Their barrels were held in the full-length stock by two brass bands, and a bar-and-tenon at the muzzle accepted a sword bayonet.

The first US Army orders for New Model rifles were placed in June 1861, when 109 .52 'Sharps Long Range Rifles with bayonets' were ordered from C. C. Bean of New York. By the end of June 1865, 9,350 rifles had been delivered into Federal stores. Excepting 600 with 36-inch barrels, Army rifles were generally similar to the Navy pattern though their 30-inch barrels were held by three iron bands instead of two. They had case-hardened lock plates, a patch-box set into the right side of the butt, and iron furniture. Most accepted socket bayonets, locking around the base of the front sight, though sword-bayonet bars appeared on some of the rifles purchased by individual states.

The Sharps rifles of the 1850s and 1860s attained their greatest fame in the hands of the United States Sharp Shooters, formed in the autumn of 1861 under the supervision of Hiram Berdan. A thousand 1859-type 'Sharpshooter's Rifles' with a special two-trigger mechanism – to give a finer release – and 30-inch barrels were ordered on 27 January 1862, followed by a duplicate order on 6 February. Numbered in batches within the 54,500–57,800 block, they bore the cursive 'JT-in-cartouche' of government arms inspector John Taylor.

Sharps made about 5,000 rifles and at least 65,000 carbines in accordance with a revised 1863-type 'New Model', identified by 'MODEL 1863' on the barrel though the changes were comparatively minor – for example, the dismantling screw in the breech-block was improved and a robust back-sight bed replaced the standing block and flimsy 'spring bed' of the earlier patterns. The breech mechanism proved to be strong enough to handle some of the most powerful sporting cartridges available in North America until the collapse of the rickety Sharps empire in 1881 brought work to an untimely halt.

The Civil War witnessed another innovation: it was the first in which optical sights were used in combat in quantity, even though experiments had been made many years earlier. Optically sighted rifles used by Federal forces usually prove to have had the Malcolm sights described in Chapter One, but Chapman-type instruments were made in Vermont and elsewhere. These are customarily attributed to 'L. M. Amidon', but one surviving sight of this type is clearly marked 'L. AMADON' over 'B.FALLS.' and 'VT'.

Census and other records reveal that Leander A. Amadon (also listed as 'Amiden' or 'Amidon') had been born in New Hampshire in 1814. However, he was trading as a watchmaker in Bellows Falls, Vermont, in the 1850 and 1860 Federal censuses. Still listed as a 'watchmaker' in 1870, with real estate valued at $6,000 and personal effects of $3,000, Amadon died in Rockingham, Vermont, on 12 December 1878.[13] Amadon sights are well-made, but optical quality is far from that of the Malcolm designs, as they rely on simple lenses.

The picked marksmen to whom optically sighted rifles were issued played an important role in some campaigns, scouting for information, trying to eliminate high-ranking officers and field artillerymen, killing 'runners', to disrupt communications, and employing the good performance of their weapons to threaten camps and observation posts from across a river or similar protective boundary.

Perhaps the most noteworthy – and undoubtedly most influential – were the regiments raised by Hiram Berdan in 1861, each mustering a thousand men at peak strength. Born in Phelps in New York State in 1824, Berdan had been trained as a 'machinist' (according to the 1850 US Federal census) and went on to become a mechanical engineer. He had patented several innovative and highly lucrative ideas – a gold-amalgamating machine that separated gold from gold-bearing ore and a reaper – and was renowned as one of the best marksmen in the USA in pre-war days.

By virtue of his commercial success (not family background as sometimes claimed), Berdan had access to influential politicians.[14] But he was not a popular figure. His business methods could be ruthless, and Major Alexander Dyer of the US Army, later to become a respected Chief of Ordnance, characterised Berdan as unscrupulous. There was even a suggestion that the 'Berdan' cartridge, now world-renowned, was really the work of Stephen Benét,

Hiram Berdan (1824–93) wearing the uniform of a colonel in the Federal Army, at a time when he was commanding the 1st US Sharpshooters. *US National Archives, Washington, DC*

commanding Springfield Armory, and members of the Armory staff. Berdan, it was alleged, had been shown the idea on a visit to Springfield in 1869 . . . and then, realising that Benét (a government employee) could not apply to protect the case-design, had simply patented it in his own name.

When the Civil War began with the attack on Fort Sumter, Hiram Berdan was able to persuade the War Department to authorise a corps of marksmen. With effect from 2 August 1861, therefore, he was duly appointed 'Colonel of US Volunteers' to command the 1st Regiment of United States Sharp Shooters ('1st USSS'), ten companies being raised in New Hampshire, New York, Michigan, Wisconsin and Vermont between August 1861 and March 1862. Eight companies of a second regiment commanded by Colonel Henry Post were raised in Maine, Michigan, Minnesota, New Hampshire, Pennsylvania and Vermont in the last three months of 1861.[15]

As soon as his ideas had been accepted, Hiram Berdan circulated an advertisement to anyone who could put 'ten bullets in succession within five inches from the centre at a distance of six hundred feet from a rest or three hundred feet off hand.' This was a standard of shooting practically unknown in military service at the time, but there was no shortage of men willing to try to meet the requirements. In practice, the 'string of 50' was relaxed so that the average of ten shots was five inches or less. A miss meant instant disqualification.

Initially, men attending the Camp of Instruction brought their own rifles with a promise – never honoured – that anyone who was passed for service would be paid $60 for his rifle.

The guns were customarily heavy-barrelled sporting and target guns. One observer recorded that they could weigh between 17 and 50 lb, but that 25–30 lb

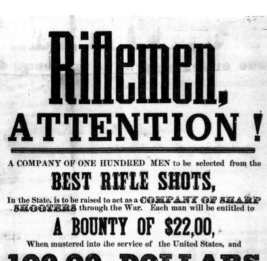

Riflemen,
ATTENTION !

A COMPANY OF ONE HUNDRED MEN to be selected from the

BEST RIFLE SHOTS,

In the State, is to be raised to act as a **COMPANY OF SHARP SHOOTERS** through the War.   Each man will be entitled to

A BOUNTY OF $22,00,

When mustered into the service of the United States, and

100,00 DOLLARS

at the close of the War, in addition to his regular pay.

No man will be accepted or mustered into service who is not an active and able bodied man, and who cannot when firing at a rest at a distance of two hundred yards, put ten consecutive shots into a target the average distance not to exceed five inches from the centre of the bull's eye to the centre of the ball ; and all candidates will have to pass such an examination as to satisfy the recruiting officer of their fitness for enlistment in this corps.

Recruits having Rifles to which they are accustomed are requested to bring them to the place of rendezvous.

Recruits will be received by   **JAMES D. FESSENDEN,**
Adams Block, No. 23, Market Square, PORTLAND, Maine.

Sept. 16, 1861.   Bridgton Reporter Press.—S. H. Noyes, Printer.

The flyer posted by Hiram Berdan's representatives to encourage men to apply to join the Sharpshooters. Note that applicants were allowed to use their own rifles when shooting to qualify, and that, at 200 yards, they were required to put ten shots on a target at an average of no more than five inches from the bull's eye. *Author's collection*

could be considered as average. Most of them fired comparatively small-calibre projectiles, and some even had full-length telescope sights; they were generally very finely made, capable at their best of shooting that would not disgrace a twentieth-century sniper's rifle. Guns made by Edwin Wesson (1811–49) and his successors were greatly prized, but there were many other gunmakers capable of supplying good-quality weapons.

Unfortunately, accepting the rifles of individual marksmen for military service complicated the supply of ammunition; calibre varied too greatly for any attempt at standardisation to succeed. Berdan was well aware of this handicap, and had determined to order Sharps' breech-loaders – a move that infuriated not only General Winfield 'Old Fuss and Feathers' Scott, the ageing Chief of the Army, but also Chief of Ordnance Brigadier-General James 'Old Fogey' Ripley, whose hostility to breech-loading was well known.

Berdan neatly sidestepped this challenge by demonstrating the skills of his men to Abraham Lincoln, and an order for 1,000 rifles (soon to be followed by another) was rushed through. Unfortunately, Sharps was unable to supply the rifles from stock. Berdan had asked for changes which included a set-trigger, and time was needed to develop tooling. It has been claimed that Berdan then attempted to acquire Spencer repeating rifles, but that the manufacturer was too hard-pressed supplying carbine orders to deliver rifles altered to Berdan's requirements. The answer was found in an issue in January 1862 of 2,000 .56-calibre Colt revolver rifles – which proved to be disastrous, as the sharp-shooters threatened to mutiny: they had been promised Sharps rifles, and the Colts were unpopular.

The revolver rifles had proved to be accurate – in 1861, pickets of the 12th Kentucky Infantry Regiment had killed four Confederate cavalrymen with just eight shots at a distance of 300 yards – but were prone to chain firing, when the powder-flash, travelling radially across the front face of the cylinder, ignited some or all of the other four chambers. Men had learned to fire the Colts, which were long and heavy, by placing the supporting hand beneath the stock under the cylinder; if the gun chain-fired, such a firer was much less likely to lose his hand.

The Sharpshooters subsequently used Colts in combat, but the Sharps rifles were delivered in the summer of 1862 and the revolver rifles were withdrawn into store. Subsequent engagements, which included the bloody battles of Chancellorsville and Gettysburg, were undertaken with the Sharps. One of the greatest advantages of the breechloader was the ease with which it could be loaded when prone or behind cover, and the Sharpshooters were trained to make best use of whatever natural cover was available.

To Berdan's disappointment, as he preferred his Sharpshooters to be used as a brigade, many of his men saw service as pickets or as 'detached snipers', seeking their own targets or working under the direction of officers from other units to eliminate specific threats such as Confederate officers, sharpshooters or gun crews. In these roles, especially if extreme accuracy was desired, Sharps rifles were occasionally replaced by heavy telescope-sighted cap-locks, acquired from individual recruits, which were carried in the baggage train.

Berdan's success encouraged the formation of other marksmen's units, including 'Company K' of the 1st Regiment of Michigan Volunteer Sharp-shooters, attached to the Army of the Potomac, who had been drawn almost exclusively from the Odawa, Ojibwa and Potawotami tribes.

The 66th Illinois Veteran Volunteer Infantry Regiment (the 'Western Sharp-shooters'), originally 'Birge's Western Sharpshooters' and then finally known as 'Western Sharpshooters, 14th Missouri Regiment of Volunteers', had been raised in St Louis by Major-General John Fremont. Most of its recruits came from states such as Ohio, Michigan, Illinois and Missouri where hunting and tracking were ways of life. The qualification standard is said to have been ten shots placed inside a three-inch circle at a range of 200 yards, which would not have disgraced a Second World War sniper rifle.

The Western sharpshooters carried half-stock target and deer rifles made by the renowned gunsmith Horace Dimick of St Louis. The 'Dimick Rifles' usually had Lawrence patent sights, and fired a special 'Swiss Chasseur' Minié-type expanding ball specially selected by Dimick to give optimal performance. Towards the end of 1863, the Dimick rifles gave way – gradually – to new lever-action Henry magazine rifles, many of which had been purchased by individual soldiers. Similar rifles were also acquired for men of the 64th Illinois Volunteer Infantry Regiment by Governor Richard Yates of Illinois (the unit was, as result, often known as 'Yates's Sharpshooters').

Spencer and Henry magazine rifles increased firepower considerably and had the even greater advantage that a second and subsequent shots could taken almost instantly. However, neither chambered ammunition that performed well at long range – the charges were comparatively weak, and the blunt-nose bullets were ballistically inferior – and so the best of the muzzle-loading cap-locks and the single-shot cap-lock Sharps were often preferred for long-range shots.

Captured Sharps were often favoured by Confederate sharpshooters, even though they lacked the telescope sights that equipped not only some of the Federal cap-lock target rifles but also some of the Confederate Whitworths equipped with Davidson sights – as the testimony of Lieutenant Isaac Newton Shannon, CSA, makes clear.

Born in Center Point, Tennessee, on 9 June 1833, Shannon enlisted as a private in E Company of the 1st Consolidated Regiment of Tennessee Volunteers on 23 May 1861.[16] He then joined the Whitworth Rifle Sharpshooters of Cheatham's Brigade – subsequently known as 'Maney's Brigade' – shortly after the Battle of Chickamauga (19/20 September 1863) and was eventually paroled out of

Men of 1st Massachusetts Sharpshooters pose for the camera some time in 1861. Note that they all carry heavy barrelled cap-lock target-style rifles with optical sights. *Courtesy of James D. Julia, auctioneers (www.jamesdjulia.com)*

service at Greensboro, North Carolina, on 1 May 1865. Shannon's wife Bettie Lee and their children settled in Center Point in 1875, but he died on 3 August 1883 in Dickson. His reputation was cemented posthumously by an article, 'Sharpshooters in Hood's Army' which was published in *Confederate Veteran* in March 1907.[17]

Shannon's recollections testify to the anti-Union sentiments that were widespread in Britain, for reasons which ranged from memories of the War of Independence to widespread concern in the 'Cotton Towns' that a Union victory would deprive the industry of vital raw material. Consequently, not only did many blockade-runners come from British shipyards, but large quantities of English-made arms and ammunition found their way to the Confederacy. They included large numbers of Enfield rifle-muskets, from the London Armoury Company and others.

The introduction of the P/53 'Enfield' rifle-musket into British service during the Crimean War had taken place against a background of uncertainty. The new .577-calibre gun was clearly an improvement on the .702 P/51, which though itself a great improvement over the Brunswick Rifle and the smooth-bore P/42 musket, was too heavy and had a vicious recoil. However, during the trials of what was to become the P/53, the Lancaster oval-bore rifles often outshot the conventionally rifled Enfield; and there were many who saw the slow-twist three-groove rifling of the P/53, which made a turn in 78 inches, as a compromise to facilitate loading. Experience in the Crimea then showed that the Enfield had a tendency to foul quicker than the authorities wished.

In addition to Charles Lancaster, whose oval-bore system was approved for the P/55 Sappers & Miners rifle, many other experimenters had tried to provide better rifling. Among them was a Royal Artilleryman, Sergeant-Major Robert Moore, who had proposed pentagonal and hexagonal bores in the expectation that the absence of conventional grooves would eliminate fouling. A Moore-type gun was made in Enfield as early as 1843, but subsequent testing came to nothing. Yet Isambard Brunel – probably unaware of Moore's work – commissioned an octagonal-bore rifle from Westley Richards in 1852, and it became clear, even as the war in the Crimea raged, that beneficial changes could be made in the basic British rifle-musket.

In 1854, therefore, the Board of Ordnance approached engineer Joseph Whitworth (1803–87), who had already established a reputation as a leading proponent of precision measurement and the values of standardisation. The expectation was that Whitworth would advise on the installation of production machinery in the Royal Small Arms Factory, and on the development of barrel-rifling machinery in particular.

Whitworth immediately drew attention to the lack of a satisfactory consensus among gunmakers, and the absence of scientific methods of assessing the value of the many types of rifling that were being promoted. Well aware of his own

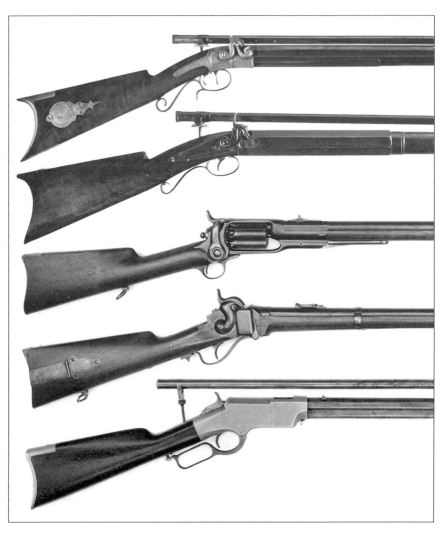

Guns used by sharpshooters and marksmen during the Civil War.
Top to bottom: two heavy-barrel target rifles fitted with optical sights; the despised Colt
Revolver Rifle; an 1859-type Sharps, lacking the double triggers characteristic of the
US Sharpshooters' rifles; and an 1860-type Henry repeater with a Malcolm sight.
Courtesy of iCommerce and James D. Julia, auctioneers

ignorance of the subject, Whitworth immediately sought advice from Charles
Lancaster, Westley Richards and many other leading English gunmakers, then
persuaded the Board of Ordnance to fund construction of a 500-yard covered
testing range (which had been completed by March 1855).

By 1857, Whitworth had finally produced a 'service calibre' (.577) rifle-
musket with a hexagonal bore, and, in addition, a hexagonal-body projectile

This well-known painting of a Civil War sniper by Winslow Homer shows one of the heavy-barrel caplocks that took a terrible toll of Confederate soldiers. An engraving based on the painting was duly published in *Harper's Magazine*. *Author's collection*

matched to the rifling. Tests showed that, in ideal conditions, the experimental gun could outshoot the P/53 by a very considerable margin. *The Times* reported results that included engravings of 500-yard targets, 6 ft 6 in. square, across which the Enfield had scattered shots wildly while the Whitworth placed all but one within a twelve-inch circle.

Though service-bore Whitworths were acquired in small numbers, it became clear that a smaller bore promised better results. Lancaster had experimented with .451 and .500 in 1848, and a .500-calibre P/53 derivative had been tested in 1856. All these results had been communicated to Whitworth, who, in 1859, finally produced a hexagonally bored barrel measuring .450 across the sides and .490 from corner to corner. The rapid twist, a turn in merely twenty inches, was, he claimed, particularly noteworthy. Ironically, when his rifle was tested against a .450 'small bore' P/53 with rapid-twist rifling, Whitworth protested that it infringed his patent rights – even though he had drawn heavily on the work of Moore, Brunel, Lancaster, Westley Richards and others.[18]

Substantial numbers of .450 P/62 and P/63 Whitworth rifles were made to allow exhaustive field trials to be undertaken across the British Empire, but the last reports, from India, were not submitted until 1865. By this time, it was clear to all that the days of the muzzle-loader were numbered; the Dreyse needle rifles and the vast numbers of rifles and carbines used during the Civil

War showed the great advantages of self-contained ammunition and loading from the breech.

In addition, the Whitworth bore had fouled so rapidly in hot, dry conditions that only a few shots could be fired between cleanings. In one trial of 1,000 shots 'uncleaned', a Lancaster oval-bore rifle, which tended to shoot better when foul, actually gave smaller shot-groups than the much-vaunted Whitworth. Consequently, the Whitworth rifle passed into history, overlooked in favour of the Snider but with its reputation for exceptional accuracy largely unsullied.

In the USA, however, the story had been different. The reputation of the Whitworth rifle had rested not on progress in the British Army – which was negligible in 1860 – but on the rise of the Volunteer movement, allaying public concerns over a feared French invasion. The rise of the Volunteers, many of whom were drawn from the aristocracy and the upper middle class, was accompanied by a huge growth in target shooting.

Immediate results included the formation in 1860 of the National Rifle Association ('NRA') and establishment of publicly accessible shooting ranges. The inaugural NRA competition, held at Wimbledon, was opened by Queen Victoria firing a Whitworth rifle clamped in a gigantic machine-rest. When she pulled the lanyard attached to the trigger, the first shot of a new era struck approximately 1¼ inches from the centre of the bull's-eye at a distance of 400

The Whitworth rifle, held in a machine rest, with which Queen Victoria fired the first shot of the inaugural National Rifle Association competition in 1860.

yards, a performance that many of today's sniper rifles would be hard-pressed to match.

When the Civil War began in 1861, therefore, many British gunmakers were offering 'Military Match' and similar small-bore rifles. Some of these were to take a terrible toll on Union soldiers. Confederate purchasing agents such as Caleb Huse began scouring Britain and Europe for the weapons that would make good the dreadful imbalance in production – virtually no large North American gunmaking businesses worked south of the Mason–Dixon line – and, it has been claimed, Whitworth (or his allies) arranged for large quantities of rifles to be supplied.

There is no evidence to show that as many as a thousand Whitworths were acquired. Probably only about 250 were actually purchased by agents acting on behalf of the Confederacy, and the interception of blockade runners by the Federal Navy probably cost the CSA a hundred of these guns.[19] However, substantial numbers of additional Whitworth, Kerr and other rifle-muskets were purchased privately, particularly by individual officers. Some of these were accompanied by Davidson sights, though Davidson himself, a deeply committed Christian and apparently also an opponent of slavery, knew nothing of Whitworth's transactions until a decade after the war had ended.

Evidence provided by surviving guns suggests that officially purchased Whitworths were all 'Second Quality Military Match Rifles', with two barrel bands and a nose-cap placed too close to the muzzle to allow a socket bayonet to be attached. The guns had a heavyweight hammer to enhance durability, a tangent-leaf Whitworth back sight, pivoting at the front of the bed, and a plain blade-type front sight which could be adjusted laterally in a dovetail. They lacked the safety bolt customarily found on the lock plate of 'First Quality' guns and their stocks were usually plain-grained.

The Confederate purchase seems to date from the autumn of 1861, according to serial numbers in the 'B' and 'C' number-groups. The guns are generally marked either 'WHITWORTH RIFLE C<u>o</u> MANCHESTER' (rare) or 'MANCHESTER ORDNANCE & RIFLE C<u>o</u>' (common) on the lockplate, and have '2<u>ND</u> QUALITY' on the rear strap of the trigger guard.[20]

The Confederate Army usually organised its marksmen into independent units, which were then sent in groups wherever they were needed. These men concentrated on traditional roles, including fire-suppression and the elimination of field-artillerymen, but also deliberately targeted officers when the opportunity arose.

The first Confederate sharpshooter units were created on 3 May 1862, when, inspired by Colonel Robert Rodes, 'General Orders Number 34' authorised each brigade to create a sharpshooter's battalion 'composed of the men selected from the brigade'. However, training the marksmen, however good their hunting skills may have been, took time; the first results were unpromising. This was at least partly due to the effect of manning the new battalions with super-numeraries and reservists, selected more for ready availability than marksmanship credentials, and by the reluctance of some of the more conservatively inclined commanders to hazard positions by employing 'pesky sharpshooters'.[21]

Standards of marksmanship began to improve once the Confederate sharpshooters had seen action, which was essentially a survival of the fittest and most able, and once an appreciation grew of the value of precision marksmanship promoted by the Whitworth and similar rifle-muskets. At the Battle of Cross Keys, the 1st North Carolina Sharpshooters advanced far enough to threaten the Federal position, before returning safely to their own lines.

Blackford's Battalion of Rodes's Brigade engaged Federal pickets prior to the main Confederate attack at the Battle of Seven Pines, and the idea that the marksmen made efficient skirmishers soon gained credence. At the Battle of Gettysburg, sharpshooters of McGowan's Brigade, who had been training under the watchful eye of Captain William Haskell since the beginning of 1863, proved their value by capturing Bliss Farm House. Marksmen attached to Davis's and Archer's Brigades, after repulsing Federal cavalrymen, managed to displace infantry from Herr's Ridge before the main Confederate attack took place. Gettysburg proved to be the first major battle in which these sharpshooters performed to expectations.[22]

When men such as Major William Dunlop ensured that only the best shots were recruited as sharpshooters, results improved until it could be claimed (somewhat optimistically unless a telescope-sighted Whitworth rifle was used) that each individual could hit a man-sized target with the first shot at distances as great as 800 yards.

The highest-ranking Union officer to be killed during the Civil War, fifty-year-old Major-General John Sedgwick, fell to a Confederate sniper during the Battle of Spotsylvania Court House on 9 May 1864. His chief of staff, Captain (later Brevet Major-General) Martin McMahon, recalled how:

> I gave the necessary order to move the troops to the right, and as they rose to execute the movement the enemy opened a sprinkling fire, partly from sharp-shooters. As the bullets whistled by, some of the men dodged. The general said laughingly, 'What! what! men, dodging this way for single bullets! What will you do when they open fire along the whole line? I am ashamed of you. They couldn't hit an elephant at this distance.' A few seconds after, a man who had been separated from his regiment passed directly in front of the general, and at the same moment a sharp-shooter's bullet passed with a long shrill whistle very close, and the soldier, who was then just in front of the general, dodged to the ground. The general touched him gently with his foot, and said, 'Why, my man, I am ashamed of you, dodging that way,' and

repeated the remark, 'They couldn't hit an elephant at this distance.' The man rose and saluted and said good-naturedly, 'General, I dodged a shell once, and if I hadn't, it would have taken my head off. I believe in dodging.' The general laughed and replied, 'All right, my man; go to your place.'

For a third time the same shrill whistle, closing with a dull, heavy stroke, interrupted our talk; when, as I was about to resume, the general's face turned slowly to me, the blood spurting from his left cheek under the eye in a steady stream. He fell in my direction; I was so close to him that my effort to support him failed, and I fell with him.[23]

Marksmen of the 12th and 15th South Carolina regiments subsequently claimed credit for the fatal shot, which the distinctive whistling approach showed to have been a hexagonal Whitworth bullet, but individual cases have never been proven. Men of the 6th Vermont Regiment, who mounted an immediate attack to avenge the well-respected 'Uncle John', reported on return to Union Lines that they had killed a sharpshooter during the advance. The range remains unknown, but some contemporaneous commentators suggested that the marksman had fired from about 800 yards.

Brigadier-General William Haines Lytle was another victim, shot from his horse while leading a counter-attack during the Battle of Chickamauga. Material said to have been presented to the United Daughters of the Confederacy[24] avers that the sniper was Hillary Garrison Waldrep of the 16th Alabama Regiment of Infantry, but there is almost no independent confirmation. The long-range shot was said to have been ordered by General Braxton Bragg, and so Waldrep had to raise his sight by 200 yards above its customary setting.

The small-bore Whitworths, and the more conventionally rifled .451 Kerr rifle-muskets supplied by the London Armoury Company, took a terrible toll on the Union forces. But there were many other types of cap-lock 'sniper rifles' in Confederate service.

Another notable Southern sharpshooter was John W. 'Old Jack' Hinson (1807–74), a 'plantation owner' or farmer of Stewart County, Tennessee, who took up arms after his sons George and John had been captured while on a hunting trip by Federal troops. Accused of being 'bushwhackers', the Hinsons were duly executed and their heads spiked outside Hinson's farmhouse.

Hinson had been uncommitted to either cause, living as he did in territory which was generally ambivalent to slavery. The death of his sons understandably changed his outlook, and he became a roving assassin owing allegiance to none but himself (though subsequently claimed as a 'Confederate hero').

His gravestone records CAPT JOHN HINSON [of] HINSON'S TENN SCOUTS CSA, but this may be an after-the-fact supposition; alternatively, it could simply suggest links with the guerrilla band led 'Between-the-Rivers' by Robert

Hinson, until he was killed in action on 18 September 1863. John Hinson's gun survives: a specially made .50-calibre cap-lock Long Rifle with a 41-inch barrel. Doubt has been cast on the validity of claims that he killed as many as a hundred men at distances up to a highly questionable thousand yards (see remarks about ballistics in Chapter One), but John Hinson was never caught, despite the price put on his head and the presence of Union troops in his area of operations. He died in 1874.

The description of Hinson's methods could have come from virtually any twentieth-century training manual: act independently; prepare well by prolonged observation; choose the shooting-site with due regard to terrain and light; make the best use of natural camouflage; strike swiftly at valuable targets; do not remain in the vicinity longer than necessary. Hinson's story has been told in two books by Lt.-Col. Tom McKenney, *Jack Hinson's One Man War* (2009) and *Battlefield Sniper: Over 100 Civil War Kills* (2016).[25]

The march of technology

When the Civil War ended, the commercial market was flooded with surplus military weapons of all types. Many Sharps rifles and carbines were among those sold out of service at the end of hostilities, though the US Navy still had 2,351 of them in January 1866 and the Army inventory approached 50,000. But there would be no return to the muzzle-loader once the advantages of the breech-loaders were appreciated by target shooters and buffalo hunters alike. On 2 November 1867, the US authorities signed a contract with Sharps to convert cap-lock guns to take the standard .50–70 centrefire cartridge (.50-calibre, 70 grains of powder). Some had already been converted to accept a .50–67–87 rimfire cartridge, similar to the .56–50 Spencer pattern, but fitting a striker and an extractor had presented difficulties.

Eventually, Richard Lawrence perfected an S-shape striker which could be fitted within the existing breech-block design and work began; 31,100 carbines were altered in February–October 1868, followed by 1,086 rifles in July–October 1869. The changes to the action were made by Sharps, but the barrels were bored out, lined and re-rifled in Springfield Armory. These conversions served until displaced by the 1873 model Springfields in the mid-1870s; the US Government was still selling cap-locks at auction as late as 1890.

The small-calibre single-shot breech-loader had established itself as the standard infantry weapon in virtually all the leading armies of the world by 1880. Performance had improved and certainty of ignition, which had not always been trustworthy in the earliest cartridges, also increased with every technological advance.

The near-universal introduction to military service of breech-loading rifles firing metal-case ammunition brought an immediate reduction in specialism.

This was partly because the new rifles offered good performance, even at long range, and sharpshooters and marksmen were no longer perceived to be useful. Since the days of the Baker and the Brunswick, the British had never issued specially developed rifles to picked marksmen. But even the Prussians and the Austrians, who had persisted with *Jägerbüchsen* (with special sights and sometimes also set-triggers) long after other armies had settled on the universal issue of Minié-type rifle-muskets, abandoned the concept in the 1870s.

Hard on the heels of the US Civil War came the war of 1866, the 'Seven Weeks War' between Prussia and Austria, the latter backed by the Bavarians and some other southern German states. Most military observers confidently predicted an Austrian victory, but reckoned without the Dreyse breech-loading *Zündnadelgewehre* ('needle-guns') of the Prussian troops.

Though the Austrian rifle-musket, the M54 Lorenz, was capable of firing more accurately to a greater range, the Dreyse could be fired much faster – particularly when the firer lay prone. The major battles were Prussian victories, and even the Austrian successes came at the expense of heavy casualties.

The consensus was that the Austrian field artillery had proved to be more effective, even though some of the Prussian guns were breech-loaders, but that musketry had greatly favoured the needle-guns. Some observers claimed that the Prussians had sometimes fired six shots for every Austrian one, and had occasionally run short of ammunition (the greatest fear of the opponents of breech-loading). The M.41 and M.62 Dreyse infantry rifles were long and clumsy, but the Prussian skirmishers, the *Jäger-Bataillone*, had newly delivered *Zündnadelbüchsen* M.65 with short heavy barrels, better sights and set-triggers to enhance accuracy.

The unexpected Prussian victory upset European equilibrium by eclipsing the once-powerful Austrian Habsburg empire. In 1870, the Prussians, bolstered by the Bavarians and others who had changed sides since 1866, picked a fight with France. If the Seven Weeks War had been the first in which one army largely equipped with breech-loading rifles was pitched against rivals armed with muzzle-loading cap-lock rifle-muskets, the Franco-Prussian War of 1870–1 was the first in which both sides took the field armed largely with breech-loaders.

The Prussians retained the Dreyse that had performed so well in 1866; the French had the Fusil d'Infanterie Mle. 1866, the bolt-action Chassepot which had been under development for almost a decade prior to adoption. Fighting ended in the virtual rout of the French Army, which had been badly led. The multi-barrel de Reffye *Mitrailleuse*, the 'secret weapon' of which so much had been expected, could not hope to be an effective artillery piece: its true role, customarily ignored, was that of infantry support.

The Prussian field artillery had been remodelled since the Seven Weeks War, in which it had been seen to perform poorly, and comprehensively out-performed the French guns it opposed. But the French infantry rifles were much more than a match for the Dreyse. Though handicapped by the inclusion of an india-rubber obturating washer to seal the breech – which became brittle and unreliable in the heat of combustion – the Chassepot fired a bullet which was ballistically far superior to its rival.

The Dreyse breech rarely failed in service, even though minor gas leaks were commonplace once guns became worn, but the combustible cartridge contained a strange ovoid bullet set in a papier-mâché sabot. The result was comparatively low muzzle velocity and a high looping trajectory ending at surprisingly short range. By contrast the smaller-diameter French bullet, with a better ballistic coefficient, flew faster, flatter and farther. Tests undertaken in Britain later in the 1870s suggested that the Chassepot was not greatly inferior to the leading metallic-cartridge weapons of its time, except that it could not sustain fire as effectively.

The major battles between the French and the Prussians and their allies were closely fought, and French infantrymen often took a dreadful toll of their opponents. Unfortunately, the French high command was paralysed with fear: the army in Metz, which numbered more than 140,000 men who could have changed the course of the war, surrendered almost without firing a shot.

With the demise of the French Army, in German eyes at least, the war was over. But they had not reckoned with the Armée de la Loire (mostly raised from the Gardes Mobiles) and the *francs-tireurs*, irregular sharpshooters, often civilians, who constantly harried German units. As many as 300 'armed bands' roved central France, according to one estimate, firing a shot or two at their enemies and moving on. Their ranks included experienced hunters, with the ability to observe, shoot, and then merge into the landscape. The Germans executed captives, burned villages and applied the harshest punishments: this was all to no avail, as the *francs-tireurs* continued their struggle to the very end of the war.

The guns used by the marksmen ranged from sporting guns and the service-issue Mle 66 Chassepot to those that had been acquired to equip the Gardes Mobiles. Most of these had been supplied from Britain and the USA, including Sharps, Spencer and Remington rolling-block rifles. Any of these could strike effectively at long range.

There is no particular evidence to suggest that *francs-tireurs* subscribed to the idea of sniping in its modern sense, deliberately eliminating officers, messengers and field-gun crews. Most of their attacks were simply aimed at causing disruption. However, a few important deaths can probably be attributed to a 'sniper's bullet'.

Jäger battalions were maintained after the formation of the 'dual monarchy' of Austria–Hungary in 1867 and then of the Deutsches Reich early in 1871. However, the Germans were armed with a shortened but otherwise near-identical version of the M1871 Mauser infantry rifle and the Austro-Hungarians had the Werndl *Infanterie- und Jägergewehre.*

The use of the optical sight during the US Civil War had passed if not unnoticed in Europe, at least unheralded. Attempts were occasionally made to test sights, but rarely with much enthusiasm. The British had fitted two Martini-Henrys with telescope sights clamped into brackets, on the side of the action body, which could be pivoted to adjust range. A photograph reproduced by Barry Temple and Ian Skennerton in *The British Military Martini* shows how little the promoters had learned about the use of a telescope on a rifle: the sight tubes are mounted much too high and have practically no eye relief.[26]

The official report was predictably unenthusiastic, and so fragile, cumbersome and clearly ineffectual bar-sights were preferred. The ability of an optical sight to magnify the image, greatly facilitating aim, seems to have been lost in poor lens performance. 'Colonel Davidson's Rifle Telescope', made by Edinburgh-based opticians D. & J. Fisher, was tried on Martini-type rifles in July and September 1887 with equally disillusioning results.[27] The testers concluded that the sight permitted a better aim to be taken, but that it was 'liable to damage' and 'to injure the eye of the firer'. This merely repeated the opinion of the Superintendent of the Royal Small Arms Factory, commenting on 11 January 1881 on the trials undertaken in 1880 by the Special Ordnance Committee: 'I have no remarks to offer with regard to the telescopic sights; it is not a practical idea, as I know of no means by which a blow in the eye can be avoided.'

The British thereafter lost interest in optical rifle sights until the First World War, and there was always a substantial part of the game-hunting fraternity which regarded the use of such aids as 'bad sportsmanship'. The French and the German armies were also reluctant to move forward, which is somewhat surprising in view of progress that was being made – particularly in Germany – in the manufacture of better optical glass, in the introduction of refined lens-grinding techniques, and in the development of range-finders and binoculars.

There were comparatively few large-scale conflicts between 1871 and 1899, with the exception of the brief (if conclusive) Spanish–American War of 1898. But the Second Anglo-Boer or South African War, hard on the heels of the Spanish–American War, was a study in greed, a struggle for control of the diamonds of Kimberley and the goldfields of the Witwatersrand.

The southernmost tip of Africa had been settled by the Dutch East India Company in the seventeenth century, but in 1806 the British had defeated the colonists at Blaauwberg and annexed the colony in 1814. The abolition of slavery in the British-controlled areas in 1834, and the annexation of Natal in 1848, had persuaded many Afrikaners to make the 'Great Trek' in search of new lands.

The ZAR (Zuid-Afrikaansche Republiek, South African Republic, also widely known as 'Transvaal') was proclaimed in 1852, followed in 1854 by the OVS (Oranje Vrijstaat in Dutch, 'Orange Free State'). These the British recognised until an attempt to re-take the Transvaal was made late in the 1870s.

The First Anglo-Boer or South African War of 1880–1 in general – and the catastrophic Battle of Majuba Hill in particular – showed the ineffectiveness of British tactics, which would not have seemed unfamiliar at Waterloo or in the Crimea. Such methods were inappropriate if a skilful enemy was determined to avoid set-piece battles, and in the upshot the independence of the ZAR and OVS was reaffirmed.

Boer riflemen occupy a shallow trench outside Mafeking in 1900 (if the original caption is to be believed). Most of them are armed with Mausers, though a Martini-Henry or two can also be seen. *Author's collection*

Or so it seemed until the discovery of gold in 1886 brought an influx of *uitlanders* (white non-Afrikaners, usually British or American) to work the mines. Paul Kruger and the government of the ZAR were well aware that the ultimate effect of this influx could be loss of sovereignty, and so refused to give *uitlanders* any citizens' rights. The ill-fated Jameson Raid into the ZAR from Rhodesia in 1895 then had a disastrous effect on political stability.

In September 1899, the British government demanded voting rights for the *uitlanders* in Transvaal. This was predictably rebuffed and, on 9 October, Kruger ordered all British troops stationed on the borders with the ZAR and the OVS to be withdrawn within forty-eight hours – conveniently overlooking that it was his men who manned the borderline. The British refused to comply and so the two states declared war on 11 October 1899, the Boers immediately striking deep into Cape Colony and Natal to encircle Ladysmith, Kimberley and Mafeking.

The British had the better of the first battles, but their casualties were worryingly high. Then came the 'Black Week', 10–17 December 1899, with three catastrophic defeats in succession: Stormberg, Magersfontein and Colenso. Still worse was the Battle of Spion Kop (24 January 1900), when the British took a hill-top position shrouded in fog only to find, as the weather cleared, that they were overlooked by nearby hills and would be shot to pieces by cleverly handled artillery and accurate rifle fire.[28]

Spion Kop effectively ended the career of General Sir Redvers Buller, the commander-in-chief of British forces in South Africa. He was replaced by Field Marshal Lord Roberts, 'Roberts of Kandahar', who immediately appointed Major-General Lord Kitchener as Chief of Staff and Frederick Burnham, an American given captain's rank, as Chief of Scouts.[29]

Heavy losses incurred during the Battles of Tugela Heights, 14–27 February 1900, though the British had been victorious, persuaded even Buller that lessons could be learned from the tactics employed by the Boers. It was important, he said, 'In the firing line – to advance in small rushes, covered by rifle fire from behind; to use the tactical support of artillery; and, above all, to use the ground, making rock and earth work for [us] as it did the enemy.'

This would be as familiar to the snipers of the world wars as to Afrikaners. The problem was that the British were still hidebound by geometry: the tradition of lines and squares that had served them well in countless campaigns, but not in Africa, where there had been a long history of disasters such as Isandlhwana.[30]

The second phase of the war saw gradual British domination, largely owing to reinforcements and a growing appreciation of intelligence-gathering, scouting and accurately mapping the ground over which the campaigns were being fought. In June 1900, the British captured Pretoria, capital of the ZAR, and the war seemed to be running to a close. Roberts handed command to Kitchener in December, returning to Britain in triumph, but the Boers had other ideas and a guerrilla war lasted almost until peace was signed at the end of May 1902.

There had been atrocities on both sides, and some remarkable incidents such as at Tweefontein, also known as the 'Battle of Groenkop', where Boers led by Christian de Wet scaled a gully on an 'unclimbable' slope and then fired down on the British from a ridge. Not only were there many British casualties (68 dead and 77 wounded, compared with 11 dead and 30 wounded raiders), but the Boers had also seized munitions and many rifles.

A scorched-earth policy, a system of easily defended blockhouses, better methods of sweeping territory and – controversially – penning Boer civilians in detention camps eventually gave the British an upper hand. When the war finally ended, the ZAR and OVS were integrated with the British colonies. A promise of eventual federation became a reality when the Union of South Africa was formed in 1910.

The cost in men and money had been horrendous. Of about half a million British troops involved in South Africa, 8,882 had died in combat and 14,210 of disease; 934 were missing, 22,828 had been wounded. Of about 40,000 Boer combatants (excluding volunteers from many nations), about 6,000 were dead.[31] In addition, the 'concentration camps' had cost the lives of more than 26,000 men, women and children and done incalculable harm to British prestige. The war marked the end of worldwide British industrial domination, as the export trade declined perceptibly.[32]

There were also many lessons to be learned militarily. The Boers had handled their artillery with ever-increasing skill; and the benefits of highly mobile cavalry had been largely ignored by the British until the value of 'mounted infantry' in South African conditions finally became obvious. Far too many British lives were lost by the failure to appreciate the utility of trenches in a defensive battle, which had been made abundantly clear in the US Civil War more than thirty years earlier; and Boer marksmanship was generally acknowledged to be better.

Yet the Boer soldier, considered individually, was not trained as well as the British regulars. When war began, all the Boer needed to provide was a rifle, a horse and provisions. But he was often a very good observer, an excellent shot – shooting matches were commonly held in Afrikaner communities – and could live off the land with ease. Consequently, small numbers of Boers were able to defeat British contingents many times their size simply by catching their enemies unaware, or mounting a night attack from an unexpected direction.[33]

There is as yet no evidence to show that sniping was used deliberately to eliminate officers or gun crews, but several battles swung the way of the Boers simply because of their high standards of marksmanship.

There was little to choose between the basic infantry weapons. The British adopted the .303 Lee-Enfield only four years before the fighting began, and so large numbers of older Lee-Metfords, differing principally in the design of their rifling, were also pressed into service. The front-line rifles of the ZAR and the OVS were variants of the 1893-type Spanish Mauser, chambering the 7×57 mm cartridge. About 37,000 Mausers are said to have been ordered from Ludwig Loewe & Co. and then, after 1 January 1897, from Deutsche Waffen- und Munitionsfabriken.[34] Full-length rifles were most numerous, but some short rifles and a few carbines were included in the deliveries. Most are the so-called '1895' pattern with a cylindrical bolt-head instead of the original flat-bottom type and a cut in the left side of the receiver wall to allow the thumb to press cartridges from the charger into the magazine.

General Tobias Smuts (1861–1916) one of the Boer commanders at the Battle of Tugela. Subsequently reduced to the ranks after burning the village of Bremersdorp in defiance of orders, he nonetheless fought until the end of the war. Smuts carries a 6.5 mm Norwegian M/94 Krag-Jørgensen rifle. *Author's collection*

The bolt handles of some guns are straight, but others are turned slightly downward towards the stock. This may simply indicate that bolts have been mismatched, as there seems to be no real correlation – at least on the basis of limited samples – between serial numbers and bolt types.

The Lee action can be cycled faster, and the ten-round detachable magazines held a theoretical advantage over the internal five-round case of the Mauser. However, the Lee magazines were prone to jam if left loaded to capacity for any length of time, and the Mauser could be reloaded rapidly from chargers (when available).

The 7 mm round had a flatter trajectory and was more accurate at long range if conditions were favourable, but the heavier .303 bullet rode the wind and retained energy somewhat more efficiently. In addition, though rarely used, the British rifles had special auxiliary sights that allowed volleys to be fired at exceptionally long range.[35]

Boers often took to the field with sporting rifles, rarely fitted with the optical sights that most men regarded more as an expensive liability than an asset. There were never enough Mausers, and captured Lee-Enfields were used in large numbers. The Boers made much better use of them than the British, simply because practised riflemen would instantly detect the unforgivable mis-setting of sights that plagued the service-issue Lee until inspection and acceptance were tightened. Lee-Metford and Lee-Enfield front sights, formed integrally with the base, could not be adjusted laterally and so the Boer simply 'led' or 'lagged' the target appropriately.[36]

Other weapons pressed into ZAR and OVS service included the Westley Richards Martini and some M1894 .44–40 Winchesters. Single-shot 1871-type 11mm Mausers which had been purchased and the British Martini-Henry rifles that had been captured during the First Anglo-Boer War were brought out of store; and a large consignment of otherwise unwanted Portuguese single-shot 8mm Guedes rifles were bought from Österreichische Waffenfabriks-Gesellschaft through intermediaries in Britain and southern Africa. None of these guns were ideal sharpshooters' weapons, and so, therefore, most would have been confined to second-rank troops.

Photographs taken during the war sometimes show Norwegian-type 6.5mm Krag-Jørgensen rifles (M/94) and cavalry carbines (M/95), which are assumed to have been acquired from commercial sources – most probably the German-controlled Fabrique Nationale d'Armes de Guerre, which had been making them for the Norwegian government.

Marksmanship skills attributed to the Boers, usually with justification, had a considerable long-term effect on the British Army. In addition, yeomanry and irregular units such as Thorneycroft's Mounted Infantry drew attention to the value of mobility, and the Lovat Scouts highlighted scouting and observation.

The Lovats had been raised by Simon Joseph Fraser (1871–1933), 16th Lord Lovat[37] and 23rd Chief of Clan Fraser, who had been commissioned into the Queen's Own Cameron Highlanders shortly after leaving Oxford University in the summer of 1890. Fraser transferred to the Life Guards in 1894, but resigned his commission three years later to return to his estate. When the Boer War began, however, Simon Fraser could see that the gamekeepers, stalkers, beaters and farmers of his and surrounding estates in the north of Scotland could render good service to the British Army.

The Lovat Scouts were ready for service by January 1900, embarking for South Africa, where Captain Fraser served as second-in-command. Attached to the Black Watch, the Scouts rendered invaluable service observing, surveying, laying telegraph lines and operating a heliograph system. Their role did not include sniping in what was to become the accepted sense, but they took every chance to shoot at Boers. Northern Scotland at the end of the nineteenth century was still almost feudal, with power in the hands of a few landowners, and the men of the estates faced the Boers on an equal footing. Both groups had learned to live off the land, and each had a tradition of marksmanship.

The Lovat Scouts were disbanded in July 1901 once the fighting was deemed to be over. Ironically, their stalking experience would have suited the guerrilla warfare pursued thereafter by the Boers. Two companies of Imperial Yeomanry were formed from the Lovat men, but even these were disbanded at the end of the South African War.[38]

Among the most interesting of characters was Frederick Russell Burnham, who was appointed not only Chief of Scouts by Roberts, but also commanded the Lovat Scouts in the field. Born on 11 May 1861 in Tivoli, Minnesota, on a Lakota Sioux reserve where his father was a preacher, Burnham began life without the benefit of formal education. When his father died, Frederick, aged just twelve, elected to stay in California while his mother and younger brother returned to Iowa.

Scarcely out of his teens, Burnham became involved in the early stages of the so-called Pleasant Valley War (1882–92), which was essentially a murderous feud between two rival families and their hirelings in Arizona. He then served the US Army as a civilian during the Apache Wars, learning as much as he could from long-established trackers, often Native Americans. Adding constantly to his own experience, the peripatetic Burnham moved to Tombstone, was made a deputy sheriff of Pinal County, married Blanche Blick on 6 February 1884, moved to Pasadena to farm oranges, returned to prospecting, and then set sail for Durban in 1893 in search of adventure.

He settled in Matabeleland, taking part in the Matabele Wars of 1893 and 1896 in the service of the British South Africa Company. There he met Robert Baden-Powell,[39] with whom he discussed scouting and survival techniques,

Major Frederick Russell Burnham (1861–1947), friend of Baden-Powell, Chief of Scouts to Lord Roberts and field commander of the Lovat Scouts. Burnham was an influential teacher of military scouting techniques. This photo was taken shortly after he received the DSO from King Edward VII. After enduring personal loss – a daughter had died during the Second Matabele War, and his younger son had drowned in the Thames – Burnham and his elder son made a fortune in the 1920s from an oil strike.

and killed the spiritual leader of the Matabele, Mlimo, after a classic 'concealed crawl' into the cave that had been used as a sanctuary.

When Roberts began to assemble his staff shortly after arriving in South Africa, he asked General Frederick Carrington to recommend a suitable chief scout. Carrington commended Frederick Burnham, whom he had known since the Matabele Wars, but the American was prospecting in the Klondyke and had to be summoned back to Africa.

Arriving on 15 February 1900, a few days before the Battle of Paardeberg began, Burnham was given captain's rank. He was to spend much of his time reconnoitring behind Boer lines, gathering priceless information about the enemy's dispositions; he was captured twice, once deliberately to gain knowledge, and escaped twice. He was severely injured on 2 June 1900 during a sabotage mission, his horse falling on him as they evaded the Boers.

Once he had recovered sufficiently, Burnham was ordered back to Britain. Promoted to the rank of major, highly regarded not only for his gallantry but also his devotion to the British cause, he was decorated with the Distinguished Service Order by Edward VII. And though the work Burnham had done in southern Africa was soon forgotten, the scouting techniques pioneered with Baden-Powell, Selous and others were to make a huge contribution to the development of sniping in the British Army during the First World War.

A widespread perception arose that the British shot spectacularly poorly whereas the Boers were all first-class marksmen. Of course, there were excellent marksmen among the British, regular and volunteers alike, but their rifles were often sighted badly enough – few had ever been tested before issue – to prevent individual targets being engaged effectively. But the war did highlight the value of marksmen, and, by implication, snipers. The 'farmers' army' had held a vast advantage over the British: local knowledge.

The Boers used natural features to hide attacks, took cover very effectively, and tried to exploit the potential of individuals – not simply put faith in sheer weight of numbers. Mass attacks could work in certain conditions, but only when the terrain favoured movement. With a couple of notable exceptions, this was never the situation in South Africa. The casualty figures usually testified eloquently.

Improved optics

The development of the telescope sight had stagnated even before the South African War began. The Malcolm sights of the American Civil War, even though they performed surprisingly well, were usually adjusted to suit the individual's eyesight. This had made them expensive and also greatly restricted production. Shooters in North America continued to demonstrate the advantages of optical sights on the target range and in the field, where the ability to enlarge an image and, to some extent, create this image even if conditions were dark or misty, was beneficial.

Renewed enthusiasm arose at least partly from the emergence in Europe and the USA of entrepreneurs prepared to exploit designs on a truly commercial scale, which soon reduced the price of a telescope sight to levels that most shooters could afford. Focus then switched from Britain and France to Germany, where Ernst Karl Abbe, concerned by a lack of lenses suitable for high-magnification microscopes, was persuaded by the inventor of borosilicate glass, Friedrich Otto Schott, to join Carl Zeiss[40] to form Glastechnische Laboratorium Schott & Genossen in 1884. Commercial production began in 1886, and though Zeiss died in 1888, the foundation of Carl Zeiss Stiftung in 1889 not only ensured the Laboratorium's future but also marked the beginning of a long-lasting supremacy.

German optical-equipment manufacturers were responsible for several very important innovations. The first Abbe prism binocular, giving stereoscopic vision by separating the objective lenses as far as possible, was patented by Zeiss on 9 July 1893 and marketed commercially in 1894.[41] On 14 April 1905, Wetzlarer Optische Werke M. Hensoldt & Sohne[42] was granted German patent 180,644 to protect an 'image reversal prism system made of a two-piece roof prism', which allowed good optical performance to be provided in a slim body. Introduced commercially in 1906, these binoculars were sold under the brand name 'Dialyt', registered trademark No. 149,303, accepted on 30 September 1911.

Another innovator was Walter Gérard[43] of Berlin, who developed the first 'belled' or multi-diameter telescope-sight body. The subject of several patents, such as French No. 413,618 – 'Enveloppe pour lunettes de visée' – sought on 15 March 1910, this design allowed large-diameter objective (and sometimes also ocular) lenses to improve light-gathering capabilities. It was destined to be exceptionally successful.

By 1914, therefore, German manufacturers, considered as a group, made the best mass-produced optical instruments that could be obtained. In Britain, conversely, few opticians were prepared to follow. Indeed, when the value of the sniper was finally accepted in the British Army of the First World War, there was an undignified scramble to acquire enough sights; the values of standardisation and unified training were lost.

In the USA, the army experimented with a Krag-Jørgensen rifle fitted with a sight made by the Cataract Tool & Optical Company of Buffalo, New York. Cataract had been founded in 1895 to make lathes and machine tools, but had rapidly diversified into optical instruments. Among these were telescope sights patented by Henry De Zeng Jr,[44] to whom US Patent 681,202, 'Telescopic Mounting for Guns', was granted of 27 August 1901; 691,248 followed on 14 January 1902, though an application had been made as early as 28 May 1900.

Convened by orders dating from 17 February 1900, the Board of Officers – Major John Greer, Captains Frank Baker, John Thompson and Odus Horney of the Ordnance Department – met at Springfield Armory at the end of May and duly reported on 8 June. Three Cataract sights had been tested, 8×, 12× and 20×, in the presence of De Zeng, who 'explained the working of the telescope sight submitted by his company'. The report noted that the sights had been:

> Tested by actual firings up a range of 2,000 yards ... As a result of this test the Board is of the opinion that the use of this telescopic sight appears to be of especial value in hazy or foggy weather and at long range ... The Board is of the opinion that this sight is suitable for use in the US service, and recommends that a number of them be purchased for trial by troops in the field. If found to be satisfactory, a sufficient number should be purchased to supply such a number of the sharpshooters of each organisation as experience in the field shall indicate to be desirable.

The report was approved by Lt.-Col. Frank Phipps, the commander of Springfield Armory, and forwarded to the Ordnance Board for a decision. Six M1898 Krag-Jørgensen rifles, including Nos 278,148, 278,769 and 278,777, fitted with 8× Cataract sights, were acquired in February 1901. Four were sent to the Philippine Islands, one to Fort Monroe in Virginia, and one to Fort Leavenworth in Kansas.

The reports were inconclusive. Some firers could see advantages in the sights, which allowed comparatively small targets to be engaged at long range or in poor light. But others fretted that the sights were cumbersome, and that offsetting them to the left side of the breech, forcing the firer to lift his cheek away from the customary position on the butt, made shooting uncomfortable. Like most bolt-action rifles, the Krag-Jørgensen ejected upwards and so the bolt-way had to be kept clear of obstructions.

The reticle of one sight had been broken by recoil – Cataract's claim to fit 'non-breakable cross-hairs' notwithstanding – and windage and elevation adjustments had occasionally worked loose, but the overall impression was largely favourable. However, a decision to replace the Krag-Jørgensen had already been taken; trials of the first Mauser-Springfield rifles were under way, so the telescope-sight experiments were temporarily abandoned.

The success of the De Zeng/Cataract sights, not only in the US Army but also commercially, had inspired others to compete. When military trials began again, the optical-sight division of Cataract had been purchased by Stevens[45] and preference was being given to the Warner & Swasey sight, made by machine-tool manufacturers who had also diversified into optics.

The Warner & Swasey Company was founded by Worcester Reed Warner (1846–1929) and Ambrose Swasey (1846–1937), who had met in 1866, fellow-

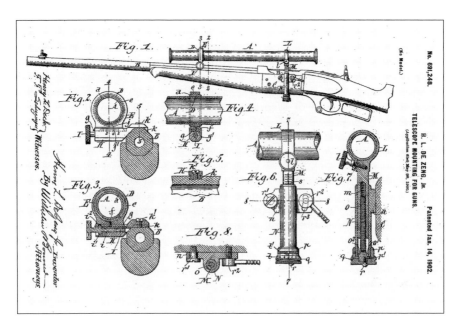

The De Zeng-designed Cataract optical sight, mounted on the then-new Savage M1899 lever action rifle. *US Patent Office, Washington, DC*

apprentices in the Exeter Machine Works of Exeter, New Hampshire, and then went together to Pratt & Whitney of Hartford, Connecticut. Leaving Pratt & Whitney in 1880 to form their own toolmaking partnership, Warner and Swasey decided to settle in Cleveland, Ohio, on the grounds that Chicago was too far west to satisfy their intended market. A workshop on Carnegie Avenue, Cleveland, began operations in August 1881. Products included turret lathes and telescopes, owing to Worcester Warner's passion for astronomy. Incorporated in Ohio in 1900, Warner & Swasey were to make much more profit from lathes than the business ever did from optics.

Patented by Ambrose Swasey,[46] the Warner & Swasey telescope sight was a radical departure from the traditional 'tube' type exemplified by the Malcolm and De Zeng designs. Derived from rangefinders developed by Swasey for military and naval use, the rifle-sight included a prism to allow the patented multi-part ocular lens to be offset from the line of the bore. In addition, the sight-body could be made much more compact than existing 'straight-line' patterns.

The design was not as innovative as sometimes claimed, as a prism-type rifle sight had been designed by Adolf König for Zeiss; made in small numbers for military trials and commercial sales, the sight was protected by US Patent 708,720, sought on 14 November 1901 and granted on 9 September 1902. The Zeiss *Prismatischen Zielfernrohr*, which often had the ocular lens aligned

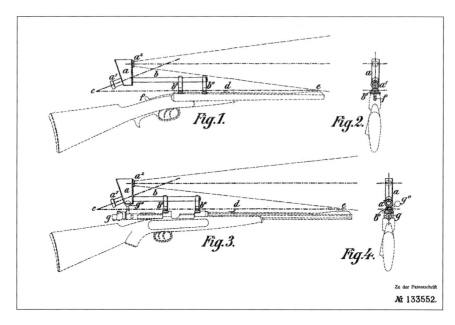

German Patent 133,532, granted to Carl Zeiss on 27 July 1901, protects the improved form of the prismatic sight. Note that the ocular lens is set at an angle; sights of this type were used in small numbers during the First World War, most notably by the Briton Leysters Greener (*see page 118*). *Deutsches Patent- und Markenamt*

with the bore axis, was much sturdier than Swasey's complicated solution to a comparatively simple problem.

Yet the Warner & Swasey sight was extensively tested by the US Army Infantry Board in 1905–7, with the new M1903 or 'Springfield'; indeed, the first was tested on the original rod-bayonet form of the rifle. The reports proved to be generally favourable, but the cost of the sight, $80, was substantially greater than that of the rifle. Swasey agreed to simplify the design as much as possible, and the 6× 'Telescopic Musket Sight, Model of 1908' was duly adopted for what was effectively a large-scale field trial.

Warner & Swasey delivered 1,000 sights in 1909. They were attached to 'Star Gaged' rifles (specially selected for the fit and finish of their parts) and issued to soldiers who had qualified as expert marksmen. A rail-like bracket held by three screws running into the left side of the receiver allowed the sight to be offset sufficiently from the centreline to allow the magazine to be reloaded from a charger. But the sight and its mounting arm were unnecessarily complicated.

An elevating wheel optimistically graduated to 3,000 yards and a 'windage wheel', graduated 38 points to the left and 46 points to the right, lay on the left side of the mount ahead of the box-like telescope body. The telescope was attached to the arm in such a way that it could pivot vertically and horizontally

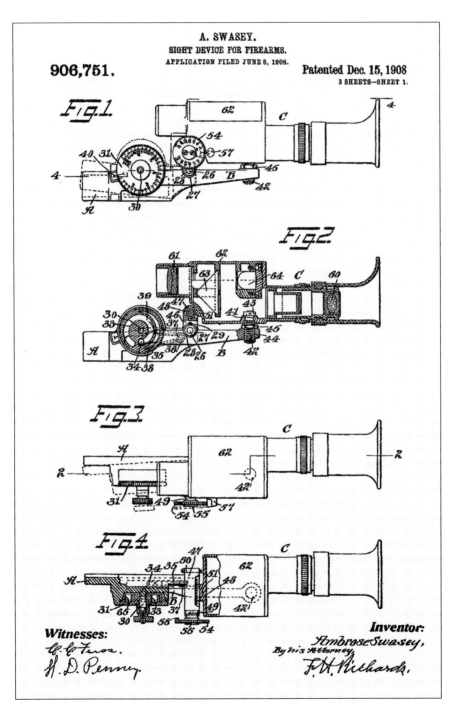

A. SWASEY.
SIGHT DEVICE FOR FIREARMS.
APPLICATION FILED JUNE 8, 1908.

906,751.

Patented Dec. 15, 1908.
3 SHEETS—SHEET 1.

The patent drawings show the finalised form of the Warner & Swasey sight, as only relatively minor modifications were made in service. *US Patent Office, Washington, DC* when adjustments were made. The steel mount and the telescope unit, which was made of bronze and brass, were painted black.[47]

The reports were initially very encouraging, as the sights were well received by men who had been used only to open sights. In 1910, therefore, the M1903 rifle and M1908 sight were issued on the basis of two to each infantry company or cavalry troop. The guns were also fitted with Maxim silencers, until these were reassessed as ineffectual.[48]

Once the Warner & Swasey sights had been in service for some time, however, problems began to appear. Though magnification was acceptable, and the field of view was surprisingly wide (7 per cent of range), the inclusion of an erector prism compromised optical performance. The relative brightness – the index of light-transfer[49] – was merely 11.1, and the prism absorbed light even before the image reached the ocular lens.

The optical system in general and the prism in particular were too delicate to withstand the battering of recoil, and the method of attaching the sight to the mount and the mount to the bracket was nowhere near rigid enough to 'keep zero'. Even though there were two notches on the under edge of the bracket to receive the spring-latch (or 'plunger') on the telescope mount, allowing the firer to place the sight in either of the positions, the eye relief of merely 1 inch was much too small. The rubber eyecup that lessened the blow that occurred on firing was not a long-term answer.

An improved Warner & Swasey sight was accepted by the US Army in 1913, though the changes were minimal: the magnification was reduced to 5.2×, which improved relative brightness to 14.8, the top-plate of the telescope sight was strengthened, a cruciform knob replaced the knurled finger-wheel of the elevation adjuster, and a clamp screw was added to the eyepiece-adjusting collar.

When the Canadians left for Europe in the summer of 1914 and the American Expeditionary Force left the USA in the spring of 1917, the M1913 Warner & Swasey sights were still standard sniping equipment. Combat showed that though they were improvements over open sights, and snipers, especially the Canadians, made good use of them, the basic design was flawed. Finding a comfortable shooting position was made difficult by the unwise combination of the standard M1903 stock and the offset telescope eyepiece (the Ross rifle, with a straight-line butt, fared a little better). Parts broke with monotonous regularity, and the insecurity of the mounting system required perpetual adjustment.

Attempts had been made prior to 1914 to promote the Winchester and Stevens sights, which were conventional 'in-line' designs, but none were successful.[50] Many types of mount were tried, but the appreciable 'kick' of the .30–06 cartridge was often enough to loosen windage and elevation settings and, in extreme cases, fracture the reticle ('cross hairs').

The eye-relief of many early telescope sights was comparatively short, owing to inefficient lens design, and the rear of the sight-tube was likely to strike the firer in the eye. The rubber eyecup of the original Warner & Swasey sights even had a reputation of glueing itself to the eye socket (by effectively creating a vacuum) until ventilation holes had been added.

Notes to Chapter Two

1. Crimean War casualty figures from Trevor Royle, *Crimea: The Great Crimean War, 1854–1856* (1999). Many alternative summaries have been offered.
2. Michael Springman, *Sharpshooter in the Crimea*, quoting information supplied by the renowned muzzle-loading marksman W. S. 'Bill' Curtis.
3. At the time of the Crimean War, many British Army officers bought their commissions, which explains why so many of them came from the aristocracy and the landed classes. In addition, Guards officers held 'double rank': one within their regiment, and another higher rank in the Army as a whole.
4. Goodlake recovered one 'M/54' Russian rifle-musket personally, which he called a 'needle rifle'. This has been questioned by some writers who have taken a term generally to applied to the Prussian *Zündnadelgewehr* ('Dreyse needle rifle') to suggest mistakenly that the Russians had something similar.
5. Royle, *Crimea*, quoting the memoir of Timothy Gowing, *A Soldier's Story, or a Voice from the Ranks* (1883)
6. The three original officers were Goodlake of the Coldstream Guards, Cameron of the Grenadiers – who had soon been wounded – and Baring of the Scots Fusilier Guards, who was removed from his position pending an enquiry into his conduct.
7. Major Theodore Laidley, 'Breech-Loading Musket', *The United States Service Magazine*, Vol. III (1865), p. 69, claimed that of 27,574 discarded muskets retrieved from the battlefield, 24,000 were loaded, 12,000 had been loaded twice, and 6,000 from three to ten times. But Laidley gave no supporting evidence.
8. According to the *Summary statement of purchases and fabrications from January 1, 1861, to June 30, 1866*, the Federal authorities purchased 35 Ballard rifles, 4,612 Colt revolver rifles, 900 Greene rifles, 1,731 Henry rifles, 583 Merrill rifles, 9,141 Sharps rifles, 12,471 Spencer rifles and 1,575 Hall rifles; many others had been purchased privately.
9. US Patent 9,308, 'Method of Priming Firearms', granted on 5 October 1852.
10. Two hundred were purchased in 1857 by the anti-slavery Massachusetts–Kansas Aid Committee. These were subsequently acquired by John Brown to arm anti-slavery factions; many ended their careers serving the Confederacy.
11. US Patent 14,554, 'Improvements in Breech-Loading Fire-Arms', granted on 1 April 1856 to Hezekiah Conant, 'of Hartford, Connecticut'.
12. US Patent 26,504, 'Improvements in Breech-Loading Fire-Arms', granted on 20 December 1859 to Richard S. Lawrence, 'of Hartford, Connecticut'.
13. His death records and will, which include a detailed inventory of his workshop, both give the spelling of his name as 'Amadon'. Amadon's identity has been confused with that of L[eland] M[organ] Amidon, born in Vermont in October 1836. However, Leland had moved with his family to Wisconsin by the 1850s.
14. Hiram Berdan was the son of John Berdan and Hannah Eldred. His father, of Dutch descent, was listed in the 1850 Federal census as a 'farmer' with an estate worth $5,000.
15. Widely regarded within the Federal army as a poor field commander, Berdan resigned

his commission on 2 January 1864 and left the army to concentrate on his business interests; the sharpshooters' regiments were mustered out of service from August 1864 (1st) and February 1865 (2nd) onward.

16. Shannon's unit seems to have been either renamed or almost immediately subsumed into the 9th Tennessee Regiment of Infantry commanded by Colonel Douglas.

17. 'Sharpshooters in Hood's Army', *Confederate Veteran*, 15, no. 3 (March 1907), pp. 124–5.

18. Sir Joseph Whitworth was a fascinating man, and his desire to impose standards of measurement underpinned the technological progress that was made in the middle of the nineteenth century. But he was also ruthless, and could be unscrupulous. Approached by David Davidson in the 1850s, Whitworth had been sufficiently impressed to acquire rights to the design, which had not been protected.

19. The total number of Whitworths serving the Confederacy is now impossible to confirm. In addition to those purchased by government agents such as Huse, others were undoubtedly acquired privately from commercial sources.

20. Whitworth was not made a baronet until 7 October 1869 and his Arms were not granted (or so it seems) until 1 November 1869. However, the trademark of a crest-coronet on the lock-plate had been used unofficially for several years.

21. The reluctance of some commanders to use snipers was usually based on a fear of retribution; even in the Civil War, a spur-of-the-moment attack or cannonade could well result.

22. Major-General John Fulton Reynolds, according to his sister, was killed by a bullet that struck him in the neck and had a downward trajectory. This has been interpreted as the work of a sharpshooter firing from a tree or a barn, but others blame a random shot from a volley fired by the 7th Tennessee Infantry at the end of its flight. Brigadier-General Stephen Hinsdale Weed, mortally wounded near the guns of Battery D, 5th United States Artillery, may also have been the victim of a Gettysburg sharpshooter.

23. From the obituary of Major-General Martin Thomas MacMahon published in the *New York Times* on 22 April 1906.

24. The United Daughters of the Confederacy was formed on 10 September 1894 in Nashville, Tennessee, by amalgamating several smaller groups. Among its remits are 'to record the part taken by Southern women in patient endurance of hardship and patriotic devotion during the struggle and in untiring efforts after the War during the reconstruction of the South'. The society still publishes *UDC Magazine*.

25. Some family members, reviewing these books through Ancestry and similar websites, have been critical of the accuracy of the genealogical content.

26. The poor positioning of the sight suggests that the optics were still of the original Chapman-type design, and not the vastly improved orthoptic pattern that had been promoted in the USA by William Malcolm since the 1850s.

27. B. A. Temple and I. D. Skennerton, *A Treatise on the British Military Martini: The .40 and .303 Martinis 1880–c. 1920*, p. 294.

28. When the survivors were withdrawn, about 350 men lay dead and about a thousand had been wounded; Boer losses are believed to have amounted to 300 dead and wounded.

29. Roberts also appointed men who had seen service in Afghanistan and India, where conditions were somewhat like southern Africa.

30. Fought on 22 January 1879, Isandlwana pitted 1,837 Britons and an unknown but comparatively small number of civilians against 12,000–15,000 Zulu warriors. British losses were 52 officers, 727 regular soldiers and 471 others killed; Zulu dead are usually reckoned as about a thousand. The battle virtually wiped-out the 1/24th Foot.

31. The population of the ZAR and the OVS totalled only a little over a quarter of a million

when the war began, so the loss of 6,000 men in battle (and 26,000 people in the concentration camps) was significant.

32. Britain's share of worldwide trade contracted greatly in the period 1900–14. Germany, in particular, had seized the opportunity not only to back the Boers but also extend trading links throughout those parts of the world that regarded the British as pariahs.

33. At the Battle of Colenso, a Boer force of about 8,000 men defeated 21,000 British troops caught trying to cross the Tugela river. Artillery and small-arms fire caused 145 British deaths and about 1,200 wounded, missing and captured; Boer loses were reportedly 8 dead and 32 wounded.

34. Fighting began while the last consignments were still in Germany, and these guns were eventually sold to Chile. Survivors still bear 'O.V.S.' markings.

35. The dial-sight on the left side of the fore-end was graduated to 2,900 yards. Trials undertaken in South Africa showed that suppressing fire at such a distance could be useful as long as volleys were used to overcome the inaccuracy of individual shots.

36. The front sight of the Lee-Enfield was offset 0.23 inches to the left to compensate for the tendency of the bullet to drift leftward. *List of Changes* Paragraph 10,393 of 22 October 1900 suggests that rifles that had been repaired in Britain could be fitted with removable front sights pinned into the sight-block. These could theoretically be altered to correct the point of aim, but there is no evidence that any reached South Africa.

37. It is sometimes claimed that Fraser was 14th Lord Lovat, arising from a misunderstanding of the attainder pronounced on the 11th Lord in 1698. The direct line died out after two additional generations (retrospectively counted as the twelfth and thirteenth), but inheritant Thomas Fraser, Baron Lovat from 1837, had the Act of Attainder reversed in 1857 to become 14th Lord Lovat. Simon Fraser – his grandson – was, therefore, legitimately '16th Lord'.

38. Only to be re-formed in 1903 as the 1st and 2nd Lovat Scouts. They were to put their experiences to good use during the First World War.

39. Major Robert Stephenson Smyth Baden-Powell, as he ranked when meeting Burnham for the first time, had already written two military manuals: *Reconnaissance and Scouting* (1884) and *Cavalry Instruction* (1885).

40. Born in 1816, Zeiss had opened a workshop in Jena on 17 November 1846 to make lenses for optical instruments. He made his first microscope in 1847, and introduced one of the first successful compound microscopes in 1857.

41. Abbe was not the first to design an erecting prism. Though his prism telescope had been exhibited in Vienna in 1873, the Italian optician Ignazio Porro had patented a prism in 1854; the first roof prism, exploited by Hensoldt, may have been the work of Frenchman Achille-Victor-Émile Daubesse in the 1880s.

42. Part-owned by the Carl Zeiss Stiftung at this time. Zeiss obtained a majority holding in 1928, but Hensoldt traded independently until 1964.

43. Walter Josef Édouard Gérard, of Huguenot descent, was born in Berlin on 7 September 1883. He was granted several patents protecting optical and electrical equipment, but it is still by no means certain if he made the sights that bear his name. It is suspected either that the Gérard business was succeeded by Oigee in 1912 or that the sight had always been made by Oigee.

44. Henry Lawrence De Zeng, Jr, was born on 7 January 1866 in Geneva, New York State, to a father of the same name. He claimed descent from a member of the aristocracy of Saxony who had settled in the USA at the end of the War of Independence and founded a factory to make coloured glass. De Zeng trained as an engineer with the Standard

Optical Company, and qualified in 1890 as a doctor specialising in optometry. He is now widely regarded among the pioneers of the ophthalmoscope and other optical and electro-optical medical instruments.

45. The Cataract Tool & Bicycle Company and the Cataract Tool & Optical Company were both incorporated in Buffalo in 1895, sharing premises and officers. In 1902, in a bid to acquire rights to the De Zeng-type telescope sights, J. Stevens Arms & Tool Company of Chicopee Falls, Massachusetts, purchased Cataract. Stevens retained the former Cataract manager F. L. Smith to develop the optical sights, initially continuing to market them under the 'Cataract' name (the brand lasted until 1940), but sold the machine-tool business almost immediately to Hardinge Brothers.

46. Swasey, of English ancestry, was granted several US patents, including 737,794 of 1 September 1903 for a 'Depression Range-Finder' and 812,464 of 6 February 1906 (jointly with Gottlieb Fecker) protecting the use of Porro prisms in telescopes. US Patent 964,709 of 19 July 1910, sometimes linked with the M1908 and M1913 sights, actually protected a much-modified design that was never fully exploited.

47. The paint was applied internally as well as externally, in an attempt to reduce glare; however, recoil tended to loosen paint particles which often stuck to the etched reticle. The comparatively large magnification made the problem worse.

48. Silencers made by the Maxim Silent Firearm Company of Hartford, Connecticut, had originally been tested as aids to recruit training. The Type 'J' was subsequently approved, but largely as a way of obscuring the direction of a sniper's shot by suppressing the sound of discharge. Trials undertaken in 1912 with Maxim and rival Moore designs then convinced the US Army that the advantages were illusory.

49. Relative brightness can be calculated simply by dividing the effective diameter of the objective lens (in millimetres) by the magnification, then squaring the result. Consequently, a conventional 4×20 sight (with a 20 mm objective) would have a relative brightness of 25 ($[20 \div 4]^2$). It is widely agreed that the diameter of the iris of the eye, which controls the passage of light, is 3 mm in daylight, 5 mm in 'dark dusk' and 7 mm at night. A shaft of light with a diameter greater than that of the iris is wasted; but, in dark-dusk conditions, the 4×20 ($\text{RB } 5^2 = 25$) would allow virtually all available light to pass. But an RB score of 11 is very poor, and so the M1908 Warner & Swasey sight would perform inefficiently even in daylight.

50. A German $4 \times$ Goerz sight tested in 1913 was judged to be far superior to the Warner & Swasey design, and even to the Winchester and Stevens patterns.

CHAPTER THREE

The First World War: The Sniper Comes of Age

When fighting began in Europe in the summer of 1914, most people believed that the campaigns would be short: 'All over by Christmas' was a common belief. Armies went to war confident in their prowess, and also often in their cause. But even the opening stages of conflict showed such optimism to be misplaced. The Germans had tinkered with the Schlieffen Plan, and their hope of seizing Paris, and by extension winning the war, came to grief at the Marne. Before that, though, a combination of experienced regular troops, the British Lee-Enfield rifle and rapid-fire training had on occasions disrupted the German advance.

It is interesting to speculate if the course of battle would have run differently had the Germans carried Lee-Enfields and the British the cumbersome German Mausers. For perhaps the last time, the design of infantry rifles could have changed history. Rapid-fire trials undertaken in 1915, with experienced men, suggested that the Gewehr 98 could fire 12–15 rounds a minute; the SMLE, however, could fire 24–27 rounds if charger-loaded (giving a direct comparison with the Mauser), and sometimes even more if several pre-loaded magazines were to hand.

The principal drawbacks of the Mauser, judged as a rapid-fire weapon, were the cumbersome bolt-stroke, which was longer than that of the Lee-Enfield owing to the position of the locking lugs being on the bolt head instead of behind the magazine well; a badly placed bolt handle, which forced the firer to raise his head after each shot; and a magazine that held only five rounds, compared with ten for the SMLE type. Judged purely as a marksman's weapon, however, the Mauser had an important edge: it fired a more powerful cartridge, with a bullet that was better ballistically than its British equivalent, and was accurate to a longer range.

By the end of 1914, the front line snaked from the Belgian North Sea coast across north-eastern France to the border with Switzerland. Any hopes of a fluid battlefield had vanished; trenches were created and, particularly on the

German side, fortifications, strongpoints and deep-dug shelters became the norm. The value of rapid-fire training and the individual soldier's rifle decreased in such circumstances, just as artillery and the machine gun grew greatly in strength. Volunteers and ultimately conscripts had neither the training nor the attitude of the rapidly diminishing pool of regulars who had marched off to war in August 1914, but the biggest problem was leadership. The British had a particular problem: an officer corps drawn largely from the upper classes, a love of tradition that verged on obsession, and a contempt for technology that would be repaid by hundreds of thousands of needless casualties. Too many senior commanders were cavalrymen, blind to the ineffectiveness of horsemen against well-sited machine guns or in the mud of a Picardy winter.

It was soon obvious that the most dangerous weapons were the field gun and the howitzer. The machine gun was at its best repelling assaults, which were all too often mounted across open country, suicidally, at a walking pace.

Many a German memoir testifies to the surprise with which British advances were observed – admiration for the bravery of men trudging ever onward through a hail of machine-gun bullets, incredulity that the commanders could order such sacrifice. But neither artillery nor machine guns were discriminating in a way that individual riflemen could be.

This first became clear to the Germans, almost as soon as the fighting had stagnated and 1914 had drawn to a close. The answer was to recruit marksmen from the ranks of the *Jäger*, the legions of foresters and hunters of proven tracking skills who lived to shoot. The ready availability of such men was due in no small part to topography. The woods and forests of Central Europe, still largely unspoiled when the First World War began, were full of game: wild boar, deer, wolves, even bears in some districts. In Britain, by contrast, though game shooting was widely practised, it concentrated more on birds than large mammals. Bears had been eradicated in Britain in the Iron Age; the last wolf was said to have been shot in Scotland in 1742; and wild boar had also disappeared from the landscape. The Briton seeking adventure travelled abroad, to hunt game across the Empire: bear and elk in Canada; tigers in India; lions, leopards and Cape buffalo in Africa.

With the rise of trench warfare, the Germans quickly became aware that they needed suitable snipers' rifles. The first to be acquired, almost as soon as the fighting had begun, were several hundred hunting rifles (*Jagdgewehre*) fitted with commercial Gérard, Goerz and Zeiss telescope sights, although a few standard Gewehre 98 were fitted with 4× Goerz 'Certar Kurz' sights as an experiment. Hans-Dieter Götz recorded that sniper rifles were first issued to the Bavarian Army in December 1914.

The 'civilian' rifles all chambered the obsolescent Patrone 88 rather than the improved S-Munition and so, to prevent accidents, a stamped tin plate was added to the left side of their butts; the plate bore a silhouetted Patrone 88, with its distinctive round-nosed bullet, and the warning 'NUR FÜR PATRONE 88 KEINE S MUNITION VERWENDEN' ('only for Patrone 88, unsuitable for S Munition'). These guns were not well designed for trench usage, since many were fitted with delicate commercial set-triggers which reacted badly to mud.

The value of sniping was soon apparent, but suitable rifles were in short supply. But then Viktor Amadeus, Duke of Ratibor,[1] sponsored an attempt to collect as many sporting guns as possible. The commonly accepted '20,000' may have been the intended total; there is no evidence that as many guns saw military service, and alternatives as low as 5,000 have been proffered. In addition, the acquisitions ranged from conventional bolt-action Haenels, Mannlichers and Mausers, in a variety of calibres, to double barrelled break-open shotguns, rifles and even *Drillinge* and *Vierlinge* (three- and four-barrelled guns with rifled and smooth-bore barrels).

Service issues seem to have been confined to 8 mm-calibre magazine rifles, even though these were restricted to the original Patrone 88 as they were deemed to be insufficiently strong to fire the more powerful 'S' cartridge. However, in the earliest days of the war, it is possible that rifles chambered for ammunition which was readily available from commercial stocks – 6.5×54, 7×57, 7.65×53, 9.3×72 – also saw limited use.

Said to have been photographed in March 1916, this sniper of Reserve-Infanterie-Regiment nr. 91 is equipped with a Scharfschützen-Gewehr 98 and a typical 4× optical sight in separate mounts. It is suspected that the picture is posed: despite the presence of an observer using a trench periscope, the embrasure through which the rifle is being fired is too large to offer sufficient protection against counter-sniping.

Bayerisches Hauptstaatsarchiv, Munich

The long-term solution was clearly to modify the Gew. 98, and so, towards the end of 1914, the Gewehr-Prüfungs-Kommission ordered 15,000 rifles – known as Zielfernrohr-Gewehre 98 or, later, Scharfschützen-Gewehre 98.

The Gew. 98 was the culmination not only of extensive trials but also a long-running feud between Mauser and the GPK, which dated back to the adoption of the *Reichsgewehr* (the 'Commission Rifle') in 1888. The Gew. 88 was little more than a copy of French Lebel rifling and a turn-bolt Mannlicher with a clip-loaded magazine, and had proved to have serious shortcomings.[2]

The Gew. 88 was clearly a better weapon than the contemporaneous Mauser C/88, which was clumsy and ineffectual. However, little more than a year later, Mauser developed a rifle that was accepted in Belgium; then came various improvements culminating in the 1893-patent rifle adopted in Spain, with the magazine within the stock. The Germans realised that the Gew. 88 had to be replaced, even though huge quantities were being made (many served throughout the First World War), and experiments with Mauser-type rifles began in the early 1890s. The initial outcome was the Gew. 88/97, designed largely by the GPK, but trials showed that even this gun was flawed. On 5 April 1898, though, the Gewehr 98 was formally adopted so Mauser had had the last laugh.

The Gew. 98 was typical of its day, long-barrelled and clumsy, but powerful and passably accurate against individual targets at 600 metres. The essence of its bolt-action, patented in Germany on 30 October 1895 (DRP 90,305), was generally comparable to the 1893 pattern but a third locking lug, under the body ahead of the bolt handle, supplemented two on the bolt head. This gave the action enough strength to handle the 8×57 mm round in safety, though the depth of the cut in the left receiver wall (to allow the firer to press cartridges from the charger into the magazine) was a potential weakness.

The Scharfschützen-Gewehr 98 was little more than a specially selected and finished Gewehr 98, capable of great accuracy. Its bolt handle was bent down, and a recess was cut in the stock and sometimes also milled out of the right side of the body; alternatively, the underside of the bolt handle was ground away, not only to allow the marksman a proper grip but also to let the bolt-handle stem pass back under the optical sight.

It had been estimated that 18,421 guns were needed for the armies of Prussia, Saxony and Württemberg. Bavaria ordered 750 guns for the state army, reckoned to be half the requirement, at about the same time. Sniper rifles seem to have been issued – in Bavaria at least – on the scale of one to each infantry and *Jäger* company, rising to three per company by August 1916. The Prussians seem to have favoured wider scales of issue, using their snipers as independent operators detached from their units to give freedom of movement.

Initially, none of the German armies regarded snipers as specialists who needed to use nothing but an individual rifle/sight pairing. The British sniper enthusiast H. V. Hesketh-Prichard remarked that:

It would appear that the telescopic-sighted rifles of the German army were served out in the ratio of six per company [in 1915], and that these rifles were issued not to private soldiers who shot with them, but to N.C.O.'s who were responsible for their accuracy, and from whom the actual privates who used the rifles obtained them, handing them back at intervals for inspection.[3]

Most guns were fitted with 4 × telescope sights in two ring mounts. It seems that the authorities initially specified Goerz and Zeiss 4 × sights, but demands of war forced them to use virtually any sight that could be obtained. Among the principal manufacturers were Otto Bock of Berlin; Ernst Busch of Rathenow ('Ebra'); Walter Gérard of Charlottenburg; C. P. Goerz of Berlin-Schönberg; M. Hensoldt of Wetzlar; Kahles of Vienna; Landlicht Zielfernrohrbau of Berlin; Optische-Industrie, Berlin ('O.I.G.' or 'Oigee', 'Luxor'); Voigtländer & Sohn of Brunswick ('Heliar', 'Skopar'); and Carl Zeiss of Jena. Magnification ranged from 2¾ × to, uncommonly, 6 ×; most surviving sights are rated 4 ×.

It is by no means certain if the sights and mounts, which were usually acquired separately, were 'set up' by gunsmiths or in military workshops. Mounts will be found with marks such as 'S. & H.', most probably Suhl gunmakers Schmidt & Habermann, but these could simply identify the parts-makers. More research is needed before conclusions can be drawn.

The mounts also vary. The most popular are claws which simply slot into the bases, to be locked in place by buttons, latches or wheels. Access to the magazine was essential for reloading, so the cranked back mount is attached to the left side of the body. A few sights were offset to the left, but these were attached to standard Gewehre 98 and are not true S.-Gew. 98.

Telescopes were set high enough to allow the firer to reload and cycle the bolt, and the mounts were usually cut away so that the iron sights could still be used. This was something of a compromise; the ability to use the standard back sight could be very useful, but raising the optical sight reduced the ease with which a sniper could conceal himself.

The S.-Gew. 98 was otherwise identical to the standard infantry rifle, but is now rarely seen since their owners often destroyed the guns before capture. The sights had knurled range drums, originally graduated for 200/400/600 metres in Bavaria, and from 100 to 1,000 metres in 100-metre increments in Prussia, Saxony and Württemberg. Few of the early sights had windage adjustments, other than by winding a key or screw in the rear mount, in effect to warp the sight-tube laterally – an unsatisfactory method of moving the point of impact, as it could play havoc with focus and, in extreme cases, cause permanent damage.

Snipers were also issued with the first supplies of armour-piercing bullets, introduced to penetrate the comparatively poor-grade iron used by the British and the French as loophole shields. An official document published late in 1915 stated:

> With the manufacturing of the K [S.m.K.] bullet being difficult and expensive, this cartridge must only be used for precision shooting when great penetration is sought. The S.m.K. cartridge is only being distributed to marksmen supplied with the Gewehr 98 and telescope sight ... There are two kinds of telescope sighted rifles in the German army: (a) the standard Gewehr 98 to which a telescope has been fitted. The first order [placed in late 1914] was for 15,000 rifles. (b) Hunting rifles with telescope. All of these ... have been requisitioned. These rifles are less sturdy than the Gewehr 98 and can only fire the Patrone 88. The weapons ... are very accurate up to 300 metres. They must only be issued to qualified marksmen who can guarantee results when firing from trench to trench, and especially at dusk or during clear nights when ordinary weapons are not satisfactory ... The marksman will use his telescope sight to watch the enemy front, recording his observations in a notebook, as well as his cartridge consumption and probable results of his shots. Marksmen are exempt from additional duties.[4]

A few optical sights were attached to the Karabiner 98 AZ, the short rifle that had been approved on 16 January 1908 for cavalry and artillerymen, for whom the infantry rifle was unhandy. This was probably due to a growing shortage of S.-Gew. 98, production of which may have been confined to 1915. However, the Kar. 98 AZ did not make a good sniper rifle. It was not as accurate as the Gew. 98 (and particularly the specially selected S.-Gew. 98) and the muzzle blast was so severe that the bayonet-hilt, which projected past the muzzle when attached, had to be given a sheet-steel flash guard to prevent the grips charring. Though photographs have shown optically sighted Kar. 98 AZ in use prior to 1918, some actually date from the era of the Freikorps and the early days of the Reichswehr.

German snipers were soon taking a toll of unsuspecting British soldiers. The ever-rising casualties due to head wounds alarmed the authorities. They were partly due to shrapnel and shell fragments, particularly from air-bursts, but medical staff also noticed a rise in the number of bullet wounds. The French issued a thin-steel *calotte protège-tête en tôle d'acier* or steel skull-cap, designed by a committee led by Sous-Intendant-Général August-Louis Adrian. The *calotte* was supposed to be worn beneath the *képi*, doubtless to maintain appearances, but proved to be efficient enough for a better solution to be sought. This proved to be the Adrian helmet or *Casque Adrian*, adopted in May 1915; by the summer of 1915, more than a million had been issued.

Though an inauspicious design, too thin, comparatively easily penetrated and made of too many parts to be truly rigid, the new Adrian helmet was greeted enthusiastically.[5] British officers purchased commercially made helmets for their own use, sometimes attaching a regimental badge instead of the French arm-of-service identifier, and nearly 500 'Adrians' were delivered into store in Hooge at the end of July 1915. These are said to have been issued to 'bombing parties' in the Ypres Salient.

The lack of protection conferred by the standard British fabric cap could no longer be ignored, and the introduction of a protective helmet (the Brodie

A group of Scharfschützengewehre 98 from the First World War. The optical sights (*top to bottom*) – 3× by Otto Bock, 4× by Walter Gérard, 3× Oigee 'Luxor', and a 4× Landlicht Model 'O' – are typical of their era, in mounts high enough to expose snipers to countermeasures. *Courtesy of James D. Julia, auctioneers (www.jamesdjulia.com)*

pattern) in 1916 was not only a great step forward but also reduced fatal head wounds by 80 per cent. The Germans had had similar experiences, as the leather or felt *Pickelhaube* was no answer to a bullet or shell fragment, and their steel helmet was introduced in 1916.

Eventually, sufficient reports reached the British high command of German sniper activity to change the view of sniping – which, like the U-boat (and even the machine gun at one time) had been considered as unsporting: not the thing a British gentlemen would do.

But many men within the army were ever-increasingly concerned by the spiralling casualty rate in the trenches. They included officers who had served overseas – particularly in India – or who had had broad experience of game or target shooting. They could see at first hand the effects the German snipers were having, but were rarely of high enough rank to influence decisions at command level. In an army where cavalry charges were deemed essential even as thousands of infantrymen were losing their lives to the Maxim machine gun on the Somme (one such charge was actually tried), it would take something or someone special to have an impact.

Many well-placed officers, drawn from the ranks of the aristocracy and from landowning stock, took advantage of rules which allowed them to acquire their own weapons by taking sporting guns to the front in the hope that the 'thrill of the chase' would include 'bagging a Boche before breakfast'. The military career of Julian Grenfell, now remembered largely as a war poet who saw a nobility in combat which few other wordsmiths shared, illustrates the improvisational approach taken in the earliest days of trench warfare.

Commissioned into the Royal Dragoons in 1910, shortly after leaving Oxford University, Grenfell had soon left Britain for service in India. A sportsman, athlete and accomplished shot, who adored 'the chase', he went with his regiment to South Africa in 1911. When the First World War began, the Royal Dragoons embarked immediately for England, arriving in September 1914, then left for northern France on 5 October.

Lieutenant Grenfell was elated. 'I cannot tell you how wonderful our men were', he wrote, 'going straight for the first time into fierce fire. They surpassed my utmost expectations. I have never been so fit or nearly so happy in my life before. I adore the fighting.' A few weeks later, however, reality had dawned: 'it is beastly', he admitted.[6] Mindful of the effect that sniping was having on morale, Grenfell, like many of his contemporaries, simply took matters into his own hands:

> We had been worried by their snipers all along, and I have always been
> asking for leave to go out and have a try myself . . . They told me to take
> a section with me, and I said I would sooner cut my throat and have
> done with it. So they let me go alone. Off I crawled through sodden clay

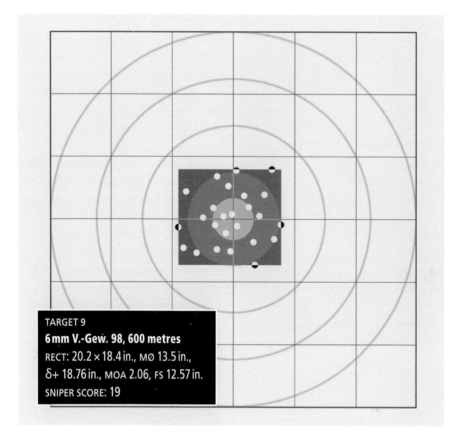

TARGET 9
6 mm V.-Gew. 98, 600 metres
RECT: 20.2 × 18.4 in., MØ 13.5 in.,
δ+ 18.76 in., MOA 2.06, FS 12.57 in.
SNIPER SCORE: 19

Accuracy of the Gewehr 98

Excepting the very occasional gun in which quirks of tolerance-limits gave exceptionally good results, it likely that the standard rifles generally did not out-perform the M1903 Springfield, which gave 2.3–2.7 MOA at distances up to 500 yards. Trials undertaken in 1896–9 with experimental versions of what became the Gew. 98 gave mean diameters of 13.5 in. at 600 metres (656 yards) with 6 mm-calibre gun No. 74, and 29.23 in. at 1,200 m (1,312 yd) with 7.65 mm gun No. 9. These equate to 2.06 MOA and 2.23 MOA respectively.

and trenches, going about a yard a minute, and listening and looking as I thought it was not possible to look and listen. I went out to the right of our lines, where the 10th were, and where the Germans were nearest. I took about thirty minutes to do thirty yards; then I saw the Hun trench, and I waited there a long time, but could see or hear nothing . . . Then I heard some Germans talking, and saw one put his head up over some bushes, about ten yards behind the trench. I could not get a shot at him; I was too low down, and of course I could not get up. So I crawled on again very slowly to the parapet of their trench . . .

Julian Grenfell, son of William Grenfell (1855–1945, First Baron Desborough 'of Taplow, in Buckinghamshire', from 1905) and his wife Ethel Priscilla Fane (1867–1952), was born in Westminster, London, on 30 March 1888. Julian and his younger brother Gerald William 'Billy' Grenfell had had the privileged upbringing of the élite: Eton College, Balliol College, Oxford, and then in Julian's case, 'passed in to the Army First of all the University candidates'. He is best known for the controversial war-poem 'Into Battle', published soon after his death in 1915. *Author's collection*

I peered through their loop-hole and saw nobody in the trench. Then the German behind put his head up again. He was laughing and talking. I saw his teeth glistening against my foresight, and I pulled the trigger very slowly. He just grunted and crumpled up.[7]

On 1 January 1915, already twice mentioned in despatches, Julian Grenfell was awarded the DSO for his gallantry. His calculated recklessness was not universally admired, but his superiors, suitably impressed, 'have made a ridiculous fuss about me stalking ... It was only up to someone to do it, instead of leaving it all to the Germans, and losing two officers a day through snipers.'

On 13 May, while engaged in observing the enemy positions, Grenfell was struck in the head by a shell fragment that lodged in his brain. He died on 26 May 1915. Little more than a month later, less than a mile from where his elder brother had been struck down, Billy Grenfell was killed by a German machine-gun bullet while leading an ill-fated attack across open ground.

Julian Grenfell's methods owed much to his experience of hunting and stalking. They were very effective but could not hope to overcome the sniper menace by themselves. He took great risks in approaching the German trenches so closely, though parallels can be seen in the way in which the Native American snipers of the Canadian Expeditionary Force often operated in the days before the idea of pairing a sniper with an observer, acting as a team instead of individually, gained widespread acceptance.

It is not known with any certainty what type of rifle Grenfell used – a .256 Mannlicher has been suggested, on no real evidence – but it is clear from his letters that it had open sights.

Among the guns purchased privately by British officers were many bolt-action magazine rifles. However, these created problems. They rarely chambered the service .303 cartridge, but, instead, one of a surprisingly wide choice of sporting rounds. This was partly due to prejudice: military ammunition was seen not only as poor quality (with some justification in wartime), but also not powerful enough for use on dangerous game. A typical Mark IV .303 round fired a 215-grain bullet at about 2,150 ft/sec, giving a 'striking energy' at the muzzle of about 2,200 ft-lb; by comparison, the .275 Belted Magnum cartridge, developed by Holland & Holland and introduced commercially at the beginning of 1912, fired a 160-grain soft-pointed bullet at 2,700 ft/sec, giving about 2,600 ft-lb at the muzzle.

Unfortunately, the British gun trade of 1914 was fragmented. A perception that only hand-made guns were acceptable, competition between the leading gunmakers to secure aristocratic patronage, and the problems associated with the far from simple transition from black powder to nitrocellulose-base propellants – a unique combination of evils – had restricted output.

The spectacular success of the .280 Ross round, designed by F. W. Jones and made by Eley from 1906, reinforced the views of those who believed that the best cartridges fired comparatively small-diameter and often also lightweight bullets at prodigious velocity. The .280 Ross, for example, fired a 150-grain bullet at 2,800 ft/sec (muzzle energy about 2,610 ft-lb). Yet the runaway success of the Ross rifle and the .280 at Bisley in 1908 camouflaged many potential problems.

Though the cartridge had been conceived for military use at a time when the pedestrian performance of the Lee-Enfield was coming under increasing scrutiny – the Germans had introduced the S.-Patrone, with a pointed bullet, and the British could not be seen to fall behind – success on the target range and Sir Charles Ross's position in society seduced the sporting fraternity. There was no doubt that the .280 bullet was extremely effective on thin-skinned game, but, as many commentators observed at the time, men were losing their lives needlessly when the .280 failed to down an elephant or a rhino.

The .303 Ross rifle subsequently proved to be a very useful tool in the hands of the snipers of the Canadian Expeditionary Force, but there is no evidence that many privately owned .280 Ross sporting rifles saw service, even though their flat-shooting capabilities and peerless accuracy could have been very useful in the trenches.

An exception to this rule was provided by Leysters Llewelyn Greener, a Briton who has often been included in lists of Canadian snipers simply because he achieved his fifty-four kills with a Ross rifle. Greener was educated at Rugby

Leysters Greener was born in Aston, Warwickshire, in April 1893. He was the second son of Charles Edward Greener – of the well-known Birmingham gunmaking dynasty – and Harriett Hutton Lort, who had been born in the Herefordshire village that provided Leysters' unusual forename. William Wellington Greener, the renowned authority on guns and shooting, author of *The Modern Breech-Loader* and *The Gun and Its Development*, was his grandfather. *Photograph from* Memorials of Rugbeians who fell in the Great War *(Vol. VI, 1921)*

School, where, according to *The Times*, he captained the '[rugby] football fifteen' and the 'shooting eight'. He proved to be an excellent shot; in the 1910 Ashburton Shield competition, shot annually at Bisley by the public schools, Greener 'did not drop a shot'.

The London Gazette reported on 27 June 1913 that 'L. L. Greener (cadet corporal Rugby School OTC)' had been commissioned into the 5th and 6th Battalions of the Royal Warwickshire Regiment ranking as second lieutenant. When the First World War began, Greener was sent to the Western Front, apparently in March 1915 with 1/6th Battalion of his regiment (confirmation is lacking).

On 14 June 1915, the Germans exploded a large mine near trenches occupied by the Warwicks near Ploegsteert; for keeping the Germans at bay the next day, Greener was awarded the Military Cross. Shortly afterwards, however, a detonator exploded in his hand during grenade practice and a splinter hit him in the right eye. Invalided back to England in September 1915, he subsequently lost the sight in his eye and was not deemed fit for service until a year had passed.

Greener eventually returned to duty as sniping officer, ranking as 'temporary captain' (confirmed on 30 June 1917). On 5 December 1917, in the Cambrai sector, the Warwicks were desperately repulsing German efforts to take back trenches they had lost to the British some weeks earlier; Greener was killed during the hand-to-hand fighting, aged just twenty-four. His Ross rifle, fitted with a Zeiss prismatic optical sight, was retrieved from France shortly after his death.

Sniping in 1915–16 was not the exclusive pastime of individual officers, as there were many from the rank-and-file who could shoot very well. They included well-known target shots, farmers who were used to shooting rabbits and crows, and others who had lengthy experience of stalking in the Highlands of Scotland.

Though their careers were brief – this was, after all, the period of the Somme – and often passed unnoticed, the names of a few British snipers have survived owing to the efforts of Martin Pegler[8] and others. Among them were NCOs and other ranks of very different backgrounds, but similar outlook.

Arthur George Fulton had competed for Britain in the 1908 Olympics, winning a silver medal in the military-rifle team event; he repeated the feat in 1912, was sixth in the 300-metre three-position military-rifle competition, and finished ninth in the 600-metre free-rifle event. Serving during the First World War with the 16th Battalion of the London Regiment, Sergeant Fulton was an accomplished sniper with a score usually reckoned to be at least 130. He survived the trenches to return to the family business, and died in 1972.

Walter Henry Fox, born in Birmingham on April Fool's Day 1881 to George Fox and Fanny Stone, is another of the small but select band of British snipers to be credited with at least a hundred kills, but comparatively little is known about him and details of his service career are elusive. Fox joined the 2nd Battalion of the Royal Welsh Fusiliers in 1914, rising rapidly to company sergeant-major, was mentioned in dispatches in 1915, and commissioned as second lieutenant 'for services in the field' on 9 August 1916. Army lists reveal that he was still serving with the Fusiliers at the beginning of 1918, ranking as lieutenant.[9]

Private John Tippins of 2nd Battalion, The Essex Regiment, had been born on 10 March 1887 in Winsford, Somerset, to schoolteachers Luke Tippins and Rose Ann Ellingham. His father had served in the British Army, acquired first-class gunmaking expertise, and was to write *Modern Rifle Shooting in Peace, War and Sport* (1906) and *The Rifleman's Companion* (c. 1910). Father and son were both serving in the Essex Regiment, albeit in different Territorial battalions, when the First World War began. However, Luke Tippins was declared unfit for service, and only John Tippins crossed the Channel to France on 19 September 1914 with the 2nd Battalion.

A marksman who had competed internationally, the younger Tippins had qualified for the final stages of the King's Prize in each year from 1908 to 1913 excepting 1912; he had won the Wimbledon Cup in 1909 and the Service Rifle Aggregate in 1911. But, on 26 November 1914, near Armentières, he was killed while returning to his position with water for his Vickers Gun. Captain Gerald Binsteed – killed in action in April 1915 – remarked that Tippins had gained himself a reputation as 'a daring sniper and a splendid shot, which had spread far beyond his own regiment'. Tippins is reckoned to have achieved thirty-eight kills.

One of the few British snipers to enjoy celebrity status was Alexander 'Sandy' Macdonald, born in Golspie in 1884 to John and Margaret Macdonald of The Kennels, Ashintully, Kirkmichael, Perthshire. When the 1911 census was taken, Sandy Macdonald was living in Dunrobin Kennels, Golspie, serving as a gamekeeper on the Sutherland estate. He enlisted in Golspie when the First World War began, joining 1/5th (Sutherland & Caithness Battalion TF) of the Seaforth Highlanders.

An excellent stalker and a practised shot, Macdonald took a terrible toll of the German regiments opposing the Seaforths: so many that he became known as 'Sniper Sandy', and a popular song was adapted in his honour.[10] His tally is said to have reached ninety-seven, but Sergeant Macdonald was killed on 13 November 1916 during the advance on Beaumont Hamel (probably by a machine-gun bullet) and now lies in Mailly Wood Cemetery.

John Herbert Fairall, the son of George and Emma Elizabeth Fairall, was born in 1874. He had enlisted in a Volunteer battalion of Queen Victoria's Rifles on 24 October 1895, and was a successful target shot. It has even been claimed that he had won the King's Prize at Bisley, but a search of records suggests that he probably only made the final stage (a good enough indicator of his capabilities).

By the summer of 1915, Company Sergeant-Major Fairall had organised a sniper section within 1/9th Battalion, The London Regiment. He and his snipers soon proved their worth, but on 24 August 1915 Fairall was hit by a German's sniper's bullet as he looked out through a loophole. Shot through the head, he died instantly.

More is known about the career of John Kenneth Forbes, a graduate of the University of Aberdeen, owing to the publication in 1916 of *Student and Sniper Sergeant* ('A Memoir of J. K. Forbes, M.A., 4th Battalion, Gordon Highlanders, who died for his country, 25 September 1915').

Written by William Taylor and Peter Diack, who had been Forbes's university friends, the book is one of few such memoirs to survive. John Forbes had been born on 12 April 1883 in Old Machar, Aberdeen, the son of Alexander Forbes and Jessie Keith. 'Pupil teacher' in the 1901 Scottish census, he graduated from the University of Aberdeen with a divinity degree and taught in Rathven School, Banffshire. John Forbes landed in France on 19 February 1915 to make a name as a sniper-instructor. Martin Pegler commented that:

> Although the lives of these sniping pioneers were to be brief, the priceless lessons they learned were passed on to others, in Forbes's case with official blessing, for he was soon to form the regiment's first sniping company. He picked sixteen men, all of whom he knew well, and whose characters he understood. He had, with typical thoroughness and extraordinary prescience, already prepared lectures on sniping.[11]

Like Fairall's snipers of the London Regiment, the men of 1/4th Gordon Highlanders trained by Forbes were very successful on a local basis. Like Fairall, Forbes was to die in action; he was killed by a shell fragment on 25 September 1915, during what came to be known as the Second Battle of Bellewaarde. Like so many others, at least until *Student and Sniper Sergeant* was published (and then only for a short time), his reputation was lost to public gaze.

Indeed, only a single British sniper was to win the Victoria Cross during the First World War. Born on 5 May 1875 in Coseley, Worcestershire, to James Barratt and Sarah Ann Bayliss, Private Thomas Barratt of the 7th Battalion, South Staffordshire Regiment, was a renowned sniper-hunter. *The London Gazette*, No. 30,272 of 4 September 1917, recorded the details of his heroism.

For the British marksmen, acquiring suitable rifles could be very difficult. In Britain only the Birmingham Small Arms & Metal Co. of Small Heath ('Birmingham Small Arms Co.' from 1902 until 1919) and the London Small Arms Co. of Bow made bolt-action rifles on an industrial scale. However, the Lee action, despite its undoubted merits judged militarily, was rarely favoured by sportsmen. This was partly due to the prejudice against an 'American Design' regularly aired in *Arms & Explosives* and even periodicals such as *Engineering* and *The Engineer*, and partly to technical deficiencies perceived in it.

BSA offered Lee rifles commercially from 1892, using actions taken from regular military production runs. A change to Enfield rifling was made about 1896, but the guns remained outwardly identical with their Metford-barrelled predecessors. They were also much the same as the army rifles of the day, excepting for commercial proof marks and 'LEE SPEED PATENTS' on the action.

The BSA 'No. 1 Pattern High Velocity Sporting Rifle', one of several variants on the theme, was made from about 1902 until the First World War began.[12] It could be chambered for the 7×57 Mauser, .303 British, 8×51R Lebel or .32–40 Winchester cartridges. A typical gun was 43½ in. overall, had a 24-in. barrel, and weighed about 7 lb 8 oz empty.

Built on the action of the Lee-Enfield Mk I rifle, with a plain Lee-Metford-type cocking piece, the No. 1 had its bolt handle turned down and forward in the fashion of the military carbines. The shape of the pistol grip was shared with British service rifles, but the half-length fore-end had a tip of rosewood, horn or ebonite. A full-length rib on top of the barrel carried a multi-leaf Cape pattern 1,000-yard back sight; the front sight was usually a bead within protecting wings.

A few British gunmakers made sporting rifles on Lee actions purchased from BSA, which were usually distinguished by a small (if well hidden) Piled Arms trademark. They included John Rigby, Westley Richards, W. W. Greener and others. However, the Lee action was crucially flawed: it was simply not strong enough to withstand a battering from any cartridge generating greater power than the service .303.

BSA had also made a special 'Heavy No. 1' variant of the No. 1 Sporter for the .375 Flanged Nitro Express cartridge; compared with the 215-grain bullet associated with the .303 sporting cartridges, the .375 fired a 270-grain bullet at comparable velocity.

LEFT: Mauser-action sporting rifles of the type that were used on the Western Front by British, German and possibly even French officers. *Left to right:* (1) A Mauser-made 'Typ B' Pirschbüchse in 7×57mm, with a 60 cm-barrelled action dated 1912. (2) A very early .275 HV gun, dated '1900' on the action, completed by John Rigby & Co. as order-book number 1,783. The original telescope sight is numbered to the gun. The rifle was sold to William Albert Christy, a clerk in the Civil Service who went to Natal in April 1902 and is believed to have returned in 1914 or 1915 to serve on the Western Front. (3) A powerful .425 WR Express Mauser (with a modern stock) completed by Westley Richards & Co. for Walter Locke & Co., 'Calcutta, Lahore and Delhi', shortly before the war began. Note the 1909-patent extension magazine. Guns of this power proved to be very useful against German trench-shields. (4) A Mauser-action .303 'Mark VII Pointed Bullet' sporter by John Rigby & Co., order-book number 4,442. Fitted with an Aldis telescope sight in Rigby mounts, the gun was sold by Rigby on 10 April 1915 to South African-born Old Etonian Captain John Harold Cuthbert, DSO, of the Scots Guards. Cuthbert landed in France on 15 April, but was killed at Loos on 27 September 1915 while leading an advance. *All images courtesy of James D. Julia, auctioneers (www.jamesdjulia.com)*

This raised muzzle energy by about 25 per cent, but tested the Lee action to its limits. Westley Richards tried to mate the Lee action with the .375/303 Accelerated Express cartridge – which fired a 215 grain bullet at 2,500 ft/sec with 42 grains of Axite – but these guns also seem to have encountered the problems associated with high chamber pressures and few seem to have survived.

The reputation of Mauser and Mannlicher actions was much higher than that of the Lee. These European guns were seen as sturdy, reliable, and capable of adaptation to fire virtually any cartridge which would fit within their magazines. The first steps had been taken in the late 1890s, when a few 1896-pattern Swedish Mauser actions were stocked by George Gibbs, and then the inflated reputation gained by the Mauser in South Africa raised public awareness until a superstition that it was unlucky to light three cigarettes from a single match gained credence: the Boer, it was said, saw the first, aimed at the second and fired with deadly effect at the third. The truth was more prosaic. But the reputation stuck.

Barrelled Mauser actions were soon being imported from Oberndorf to be completed in Britain, creating some of the finest of all repeating sporting rifles. However, Mauser's commitment to military production, for the German Army in addition to the government-led export drive, restricted the availability.

Österreichische Waffenfabriks-Gesellschaft, by contrast, though part of the Mauser rifle-making cartel, could supply Mannlicher actions from stock. A few military straight-pull actions were finished as sporting rifles, but they are very rare; however, the first rifle designed by Sir Charles Ross was derived from the straight-pull Mannlichers he owned.

TARGET 10
Pirschbüsche C. 98/08, 400 metres
RECT: 9.2 × 9.9 in., MØ 6.02 in.,
δ+ 8.36 in., MOA 1.38, FS 5.25 in.
SNIPER SCORE: 98

Accuracy of the Mauser Sporting Rifles

These often performed much better than even the S.-Gew. 98: a test of a 7 mm Mauser-Repetierpirschbüsche C. 98/08, for example, gave a mean diameter of only 6.02 in. at 400 metres (437 yards), a MOA of 1.38. At this distance, 19 of 25 shots would have lain within the 8-in. 'head-hit' bullseye.

The basis for conversion was the turn-bolt introduced on rifles sent for trials in Romania in 1892, though the ultimate source was the German Reichsgewehr of 1888. Mannlicher had brought a patent-infringement lawsuit against the German government, and had been allowed as part of a comparatively amicable settlement to exploit the Reichsgewehr action commercially – to which, in part, he could lay justifiable claim. It provided the basis for the Romanian and Dutch infantry rifles and then, in conjunction with the Schönauer spool magazine, for the Greek equivalent.

The Mannlicher was a particular favourite among British sportsmen. It was generally agreed that the position of the bolt handle, turning down ahead of the receiver bridge, gave a smoother stroke than the Mauser type projecting behind the bridge. By separating the cartridges, the spool magazine fed more reliably than either the staggered-row Mauser type or the detachable box of the Lee, which relied on accurate shaping of the magazine lips. However, British purchasers often opted for the single-column box magazines characteristic of the Steyr-made rifles supplied to Romania and the Netherlands.

There is little to choose between the Mauser and the turn-bolt Mannlicher in terms of strength, though tests to destruction have since shown that the 1910 pattern Ross rifle was far stronger than either. British gunmakers invariably favoured the Mauser for the largest and most powerful Express cartridges – though not without taking a few constructional risks. As early as 1913, Waffenfabrik Mauser was complaining in *Arms & Explosives* about changes made by enterprising gunmakers. The object of ire was a .404 'Magazine Sporting Rifle with Improved Mauser Action' advertised by a well-known London-based dealer, R. H. Mueller of Southampton Row.

Yet many Mauser-action sporting rifles had been made successfully prior to 1914 chambering .404 Rimless Nitro Express (.404 Jeffery) and .416 Rigby rounds; the latter, firing a 410-grain bullet at 2,370 ft/sec, was notably powerful for its day.

The addition of folding sights to the cocking piece was a retrograde step. Particularly popular on Mannlichers, the additional weight retarded what was already a notoriously slow striker fall; the guns based on the 1888-type German service rifle had a lock time of 8–9 milliseconds, which the burden of the sight increased to as much as 11 milliseconds.[13]

When the First World War began, Mauser and Mannlicher sporting rifles disappeared from the British market almost overnight, apart from those that were awaiting sale and a few actions which were still to be barrelled and stocked. And there was, of course, something of a stigma attached to their Teutonic origins. The Canadian Ross, an exceptionally controversial design which had had considerable success on the target range and as a big-game rifle, was also lost to the commercial market as the production line struggled to supply the needs of the Canadian Army in general and the Canadian Over-Seas Expeditionary Force in particular.

The infantry rifles of 1914 were still largely those of 1900, except in Britain where the SMLE had superseded the long Lee-Enfield. War in South Africa, where mounted infantry had been put to good use, had shown the desirability of a universal-issue weapon combining the handiness of a cavalry carbine with the accuracy of an infantry weapon. The 'No. 1 Improved Pattern Rifle' appeared experimentally in 1901 and, after protracted testing, became the SMLE Mk I, adopted in December 1902. Unforeseen teething troubles, to be expected with any new design, were so serious that the original introduction was cancelled and an improved version was substituted in September 1903 and again in September 1906. A particular alteration concerned the addition of a cut-off, to hold the magazine in reserve, which had been a feature of Navy-issue rifles since 1903 but was only extended to land service in October 1906.

The SMLE differed considerably from the long Lee-Enfield. Though the peculiarly British form of butt was retained, the barrel and fore-end were shortened, a full-length guard was added to protect the barrel, and a heavy nose cap not only protected the front sight but also carried the bayonet lug on a rearward spur. An adjustable tangent-leaf back sight replaced the previous leaf-and-slider pattern.

Charger guides were added, but proved to be one of the worst features of the Mk I. One of the guides was supported on the bolt head and the other on the receiver behind the magazine well: a weak and potentially troublesome design which should never have been considered, especially as Mauser rifles had had two rigid receiver-bridge guides since the Spanish Model of 1893.

Production soon began in earnest, greatly helped by the approval of the Converted Mk II short rifle in January 1903. This was a transformation of Mk I and Mk I* Lee-Enfields and a few Mk II and Mk II* Lee-Metfords that had escaped rebarrelling. Many other conversions were attempted, and the nomenclature became increasingly complicated.

The Mark III, approved in January 1907, had bridge-mounted charger guides and a lighter nose cap. The introduction of the high-velocity Mk VII cartridge, in 1910, required the sights to be revised. Older guns either had the graduations on the long-range dial plates altered or entirely new plates. The Mk III* was a wartime expedient, approved in January 1916. The long-range sights were discarded, the cut-off eventually became optional, and a new flat-sided cocking piece was approved. Specially selected sniping or '(T)' versions of the Mk III and Mk III* rifles will be encountered with optical sights.

The Rifle, Short, Magazine, Lee-Enfield, 0.303 inch, Mark III, was made only by the Royal Small Arms Factory at Enfield Lock, Middlesex; by the Birmingham Small Arms Co. and then BSA Guns of Small Heath in Birmingham; by the London Small Arms Co. of Bow; by the Standard Small Arms Co. of Birmingham; by National Rifle Factory No. 1 in Birmingham; by the Australian government factory in Lithgow, New South Wales; and by the Indian government small-arms factory in Ishapore.[14]

The Canadian Over-Seas Expeditionary Force ('CEF') had the Ross. Perhaps the most vilified of all twentieth-century military rifles, though a first-class sporter, the Ross owed its brief glory to the refusal of the British to supply Canadian troops with Lee-Enfields during the Second South African War of 1899–1902. It stands as testimony to the ease with which national pride can overcome technological deficiencies in weapons design.

The design is now customarily credited to Sir Charles Ross, a talented if opinionated engineer who received his first British patent in 1893, but Ross

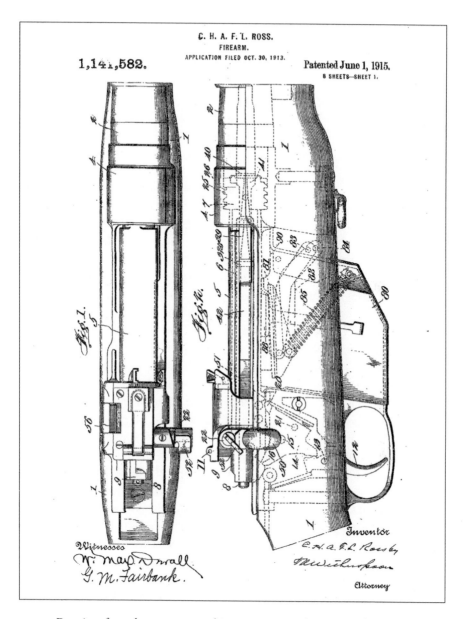

Drawings from the patent granted in 1915 to protect the Ross Mark III rifle.
US Patent Office, Washington, DC

had simply adapted the straight-pull 1890-pattern Mannlicher to suit his own purposes ... even though this Austrian design had performed badly in many military trials. Poor extraction was one of many regular complaints, but these had not stopped the Austro-Hungarians adopting the 1895 pattern 8 mm rifle as their standard infantry weapon.

There was no doubt that the Ross action could be operated slickly in perfect conditions; in addition, when properly locked, it proved to be unusually strong. However, quality control was poor and problems were soon apparent. Part of a bolt blew back into the face of a Royal North-West Mounted Police man during shooting practice in 1906, costing the firer an eye, and RNWMP Ross rifles were immediately recalled into store. Survivors were exchanged for improved Mk II rifles in 1909.

Made in Quebec by the Ross Rifle Company, the Rifle, Magazine, 0.303-inch Ross, Mark III, also known as 'Military Model 1910' or (misleadingly) as the 'Model 1912', was approved in the summer of 1911. A cumbersome and ungainly weapon which bore only a superficial resemblance to the gun that it had replaced, the Mk III was the principal weapon of the Canadians in 1914.

The 'triple-thread interrupted screw double-bearing cam bolt head' owed more to Ross's desire to create a suitable action for powerful sporting-rifle cartridges than military desirability. The bolt locked vertically instead of horizontally, in an attempt to improve the feed stroke, and a five-round in-line box magazine, which replaced the Harris platform type of the Mk II Ross, allowed the Mk III to be loaded from chargers. The magazine cut-off and bolt stop were combined in a single small lever on the left side of the receiver bridge.

The Mark III Ross had a protruding sheet steel magazine-housing for five .303 rounds, a shallow pistol grip, and – originally – a folding bridge-mounted aperture back sight graduated to 1,550 yards. The guns were 50.36 in. long, weighed 9.85 lb empty, and had 30.25 in. barrels with four-groove rifling. The long barrel was intended to enhance accuracy (but not combatworthiness), and a knife bayonet could be attached to the muzzle.

In September 1914, shortly after the First World War began, the British government had ordered 100,000 'Mk IIIB' Ross rifles from Canada. The subject of *List of Changes* paragraph 17,690 ('L.o.C. § 17690') of 21 October 1915, these had sights that differed from their Canadian prototypes. Deliveries were erratic and the contract was finally cancelled in March 1917, after 66,590 guns had been accepted, owing to the appropriation of the Ross Rifle Company's Quebec factory by the Canadian government. Most of the rifles supplied to the British were used for training purposes, but 45,000 were issued to the Royal Navy in the spring of 1917 to replace unwanted Japanese Arisakas. Survivors were brought out of store in 1940 to serve the Home Guard and the Merchant Navy.

Tests have shown the Ross to be one of the strongest of all military bolt-action systems (if not the strongest), but its weaknesses were mercilessly exposed during the First World War. The rifles worked reasonably well when clean – and firing good-quality ammunition – but the action was too easily jammed by mud, or by the heat generated during rapid fire. Many memoirs testify to difficulties encountered with recalcitrant bolts.

Some of the problems were traced to ammunition. The Canadian-made cartridges had been made to finer tolerances than British service issue (particularly of wartime manufacture) and so, as the chamber-diameter of Ross rifles was usually smaller than that of the SMLE, they performed measurably better when fired with Canadian ammunition. When supplies ran short and British-made ammunition had to be substituted, rounds that were over-size by Canadian standards often stuck in the chamber sufficiently hard to prevent efficient extraction. This was such a liability in the trenches that Sir Douglas Haig, with the backing of Lloyd George, ordered the removal of the Ross from the front line.

Snipers, of course, were allowed to retain the guns with which they were not only familiar but also taking a heavy toll on the enemy. It is likely that the most experienced snipers, often with time on their hands, simply 'blind chambered' British-made cartridges to identify any that, by Ross rifle standards, were either oversize or sufficiently misshapen to feed poorly. Merely chambering a round tended to re-form the case to suit the gun.

By far the worst feature of the Mark III Ross lay in the design and construction of the bolt head, which could rotate under the extractor after the bolt sleeve had been removed from the receiver. This was not serious in itself, unless the bolt sleeve was replaced in the receiver without putting the bolt head back to its correct position. When an incorrectly assembled gun was fired, the bolt, which had failed to lock, slammed open. If the bolt-stop failed under the unexpectedly harsh impact, the bolt flew out of the receiver into the firer's face with potentially fatal consequences.[15]

On 11 November 1916 details of new large-diameter bolt stops – to be substituted for the original design, which had proved to be too weak – were announced in L.o.C. § 18046. The Canadian armourers in France had added a rivet or a screw in the bolt sleeve to prevent misalignment, but the change, beneficial though it was, came too late to prevent the universal withdrawal of the Ross in favour of the SMLE.

Some guns were scrapped, some were sent to Russia and others were given to the Royal Navy, but more than 1,400 were retained for sniping, a role in which a clean, well-maintained Ross rifle excelled. Five hundred of these sniper rifles were issued with slightly modified 1913-model Warner & Swasey sights graduated to 2,400 yards for the .303 cartridge (one group in 1915 and the other in 1917), and 907 had 'A5' sights made by the Winchester Repeating Arms Company. The optical sights were offset to the left to clear the charger guides.

Ironically, the Ross, a dreadful failure in the trenches, was probably the most accurate of pre-1918 sniper rifles: one modern trial placed twenty consecutive .303 shots into a 4-inch diameter circle at 200 yards, the sniper Herbert McBride reported one in which forty shots fell inside a six-inch circle at 300 yards, and

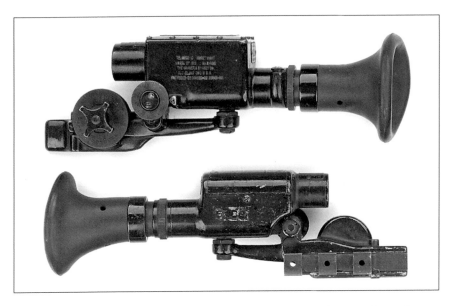

The Model 1913 Warner & Swasey sight was regulation issue in the US and Canadian armies. Though the inclusion of a prism gave a compact tube, the optics were poor and the adjustments were too complex to be effective in trench warfare.
Courtesy of James D. Julia, auctioneers (www.jamesdjulia.com)

a target reproduced in Ross literature showed a 4.2-inch group obtained with the .280 version at 500 yards.

The Canadian Expeditionary Force mustered many men from wilderness areas, to whom hunting, shooting and stalking were second nature, and even senior officers could see their potential – quite unlike the British, whose high-rankers often failed to attach any value to sniping. Among the high scorers in the Canadian Army, therefore, could be found Major Peter Anderson of the 3rd Battalion, CEF, who not only commanded the snipers of his unit but also accumulated a personal score said to have been in excess of 150 kills.

Peter Klaus Anderson had been born on 24 April 1868 in Denmark, to Niels Jorgen Anderson and Anna Clausen. The Anderson family emigrated to Canada while Peter was still a child. An accomplished hunter in the wilds of Alberta, he was also a successful businessman: owner of a large brickworks in Strathcona. When the fighting began in Europe, Anderson, who had married Mary Anne Allan in January 1895 and had been a militia officer with the 101st Edmonton Fusiliers, attested on 23 September 1914 and went to Europe with the 9th Battalion, CEF. In addition to his sniping duties, he had an interesting war which included being captured by the Germans during the Second Battle of Ypres and then escaping, to return to Britain by way of Denmark, Sweden and Norway.

His memoir, *I, That's Me: Escape from a German Prison Camp and Other Adventures*, relates an extraordinary story. He returned to the Western Front and was eventually raised to lieutenant-colonel's rank. He died in Vancouver in August 1945.

Canadian snipers rapidly made a name for themselves. Among them was William James Clifford (born on 19 January 1878 in Brampton, Ontario, to William and Ellen Mary Clifford), who, in 1911, became the first Canadian to win the King's Prize at Bisley. He enlisted in the 19th Battalion, CEF, attesting in Toronto on 12 November 1914, giving his 'trade or calling' as 'carpenter' and his wife Winifred, whom he had married in November 1910, as next-of-kin.

Arriving in England on 22 May 1915 and attached to 4th Infantry Brigade headquarters as an armourer-sergeant, William Clifford embarked for France in September 1915 to follow a successful career as a sniper. Credited with at least 150 kills, he was discharged in April 1916 to accept a commission in the Royal Flying Corps. On 25 April 1917, however, Lieutenant William Clifford of 48 Squadron was reported 'missing in action' over the Pas-de-Calais and was subsequently declared to have been killed in combat on 24 April. No trace of his body was ever found.

James Murdoch Christie, born in Scotland, came to Canada with his parents in the mid-1880s; his father, Joseph, settled in Manitoba, where he was listed in the 1901 and 1911 censuses as a farmer. The younger Christie had led a charmed life, as he had been attacked by a grizzly bear in October 1909, during a hunting expedition. Though Christie had shot the bear in the chest and head with his Ross rifle, which may have been a .280, the fatally wounded animal still managed to seize the hunter's head in its mouth, crushing his jaw and cheekbone, and seriously damaging an arm and a shoulder. As the bear died, James Christie escaped, blinded in one eye and with much of his scalp ripped away, yet he not only managed to stagger the eight miles to his cabin but also lived to tell the tale. However, his wounds were to trouble him greatly for the rest of his life.

When the war began, Christie seems to have enlisted in Winnipeg on 12 August 1914 and attested in Ottawa ten days later. He had lied about his age, probably realising that he would not otherwise be able to serve, claiming to have been born on 22 October 1874 when he had actually been born in Kinfauns, Perthshire, Scotland, on 22 April 1867.[16] Duly posted to Princess Patricia's Canadian Light Infantry ('PPCLI'), Christie arrived in France in the spring of 1915 to become part of a sniping unit created by Lieutenant W. G. Colquhoun on the instructions of Colonel Farquhar, commanding officer of the PPCLI. According to the war diary of the regiment, for 7 October 1915:

> A patrol of 8 snipers & 2 grenadiers went out late this afternoon under Sgt. Christie . . . and made their way, crawling through the grass to the German side of the marsh, with a view to intercepting a German patrol

believed to pass down the road from La Grenouillère to Curlu about 7 p.m. each evening. Our patrol got safely to the German side of the marsh and concealed themselves 20 yards from the road.

Just after dusk a strong German patrol came down the road (consisting of 30 men under an officer) marching in fours and with a flanking party in the marsh. Sgt. Christie seeing himself hopelessly outnumbered and in danger of being cut off between the two parties resorted to bluff and ordered the Germans to 'Hands Up'. The enemy not complying, our men opened rapid fire, the grenadiers at the same time throwing their bombs into the close mass of men in the road. The Germans threw themselves down and returned the fire of our men while the flanking party closed up. Our two right hand men faced around to meet them and one of our men killed a German who had come within a couple of yards of him. The enemy after throwing 2 or 3 bombs which did no damage began to crawl away, leaving several dead and wounded men groaning in the road.

Fearing a return of the enemy with reinforcements from La Grenouillère, Sgt. Christie took the opportunity of withdrawing, his whole party returning to our lines without a casualty.

For his gallant leadership, Christie was awarded the Distinguished Conduct Medal. He was wounded twice, on 22 April 1915, again on 16 July 1916, and ultimately – commissioned as 'temporary lieutenant' – took charge of the PPCLI sniping section. There he remained, receiving the Military Cross for gallantry on 18 January 1918, until deemed unfit for duty in France on 8 July. Christie was 'Struck-off the Strength as Medically Unfit' on 21 August 1918, but had already returned to Canada.

James Christie was just one of the many Canadians who used their shooting skills to good effect in the trenches. They included several Native Americans, exceptionally patient men, expert trackers, skilled shots, and often well used to the hardship and privation that was too often their lot. And yet, ironically, the Canadian government had deliberately excluded 'aboriginal peoples' and 'ethnic minorities' from military service almost as soon as the war had started; most of the restrictions had soon been lifted, but so-called 'Treaty Indians' were exempted from service.[17]

This did not stop Native Americans volunteering to join the army in large numbers. Major E. Penberthy, who had been commandant of the British Third Army Sniping School in northern France, wrote in 'British Snipers (i). An Account of the Training and Organisation of Snipers in the British Armies in France', published in the *English Review* in September 1920, that:

One of the finest snipers I ever met was a full-blooded Red Indian – John Ballantyne. He applied all the methods of the chase, so familiar

The principal sniper rifles of Allied forces during the First World War. *From left*, the Canadian Ross, the SMLE Mk III* (a later Australian heavy-barrel version with a cheek-piece), and the P/14. *Courtesy of the Canadian War Museum, Ottawa, and James D. Julia, auctioneers (www.jamesdjulia.com)*

to him in his beloved Canadian forests, to hunting the Boche sniper. He had been known to wait patiently for seven days in a wonderfully concealed and prepared sniper's post for a valuable target – a Hun officer, whom he finally killed.

This story of sniper 'Ballantyne' has been perpetuated by many writers who simply copied what Penberthy had written without realising that the name is misspelled. Census records identify him to be John Ballendine of Battleford, Saskatchewan, who came to be regarded as the 'champion Canadian sniper'.

Ballendine gave his date of birth on his attestation as 18 October 1884 (though some Canadian records give 9 August 1883). He apparently enlisted on 12 November 1914 and attested in Winnipeg on 23 December 1914, but records are still unclear. Serving with the 32nd Battalion, Ballendine landed in France on 3 May 1915 and served continuously until a serious knee injury admitted him to No. 4 General Hospital, Camiers, on 10 April 1916.[18] He was sent back to Britain for treatment, spending some months in Scotland, but a recommendation to discharge him from military service was made on 25 November 1916 (granted on 9 January 1917).

Returning to Battleford, John Ballendine recovered sufficiently by 21 May 1918 to re-enlist in the Engineers to serve with No. 10 Forestry & Railway Construction Depot. He was finally discharged from military service on 12 February 1919.

John's brother James Ballendine (1882–1938) had enlisted in 1911 to serve with the 22nd Saskatchewan Light Horse, subsequently attesting in Valcartier for service with the Canadian Over-Seas Expeditionary Force. Leaving for France with the first CEF contingent, with 6th Battalion (Fort Garry Horse), he transferred to the 8th Battalion (90th Winnipeg Rifles) where his company commander was Battleford resident Captain G. G. Smith.

James also proved to be an effective sniper, but was wounded by shrapnel during the Second Battle of Ypres on 13 April 1915. The fragment passed through his lung to lodge (and remain) next to his heart. After treatment, James Ballendine was discharged as unfit for service and sent home.

The *Saskatchewan Herald* reported on 9 March 1917 that Privates John and James Ballendine, 'who have returned from overseas crippled and unfit for further military duty have each been presented with a handsomely engrossed address in recognition of their services to the empire'. They were to find their return to civilian life unrewarding, however, as the plight of Native Americans had not improved. However, all eight of John's sons served in the Second World War.

Philip McDonald was an Akwesasne Mohawk, born in Lacie, Quebec on 1 February 1886, who attested in Valcartier, Quebec, on 21 September 1914. Serving as a sniper with the 8th Canadian Infantry Battalion (90th Winnipeg

Sniper Michael Mathew Ackabe[e] (1892–1938) of the Wabigoon Band enlisted in the Canadian Army on 26 January 1916. Photographed in Thunder Bay, Ontario, in 1918 or 1919, he carries a Ross Mk III rifle in half-stock form. Guns of this type were issued to many of the CEF marksmen, but their provenance is still unknown. *Courtesy of the Kenora Great War Project*

Rifles), he was killed by a shell which hit the battalion aid station near the Messines Road on 3 January 1916. By then McDonald had been credited with seventy kills.

Henry Louis Norwest (also known as 'North West' or 'North West Wind'), part French-Canadian, part Cree, was born in Fort Saskatchewan, Alberta, on 1 May 1884. Listed in the 1911 Canadian census in Edmonton, married and working as a 'harness maker', Norwest enlisted as 'Henry Louie' in Wetaskiwin on 2 January 1915, to serve with the 3rd Canadian Mounted Rifles. Discharged for drunkenness after a few weeks, he spent five months in the Royal Canadian Mounted Police before re-enlisting in Calgary on 8 September 1915.

The 50th Infantry Battalion left Canada bound for Britain at the end of the year, but Norwest does not seem to have reached France until the summer of 1916. Time spent in the Alberta woodlands had perfected his marksmanship techniques, and Norwest soon proved himself to be a very effective sniper. An entry made in the war diary of the 50th Canadian Battalion on 29 April 1918, recorded that:

Sniper H. Norwest M.M. today accounted for his hundredth German, and before night had added another to the score, a record enjoyed by few, if any, other snipers in the British Army. Norwest's methods are peculiarly his own. 'Wait until not a single Hun has a chance of seeing your rifle flash, then get your man', is his motto.

The Germans were only too well aware of his reputation, but Norwest, credited with 115 kills, survived unscathed until the summer of 1918 – when, having been stood down from front-line service, he volunteered to join an attack in the vicinity of Amiens to help neutralise German snipers and machine-gun nests.

On 18 August, however, according to the official record, 'While taking part in operations North-West of Hattencourt, he was hit in the head and killed by a bullet fired by an enemy sniper.' Oliver Payne, who was acting as spotter, reported that 'the instant Ducky [Norwest's nickname] fired, a German sniper fired. It missed me but got Ducky right through the head, coming out the other side.' He also claimed that Norwest's shot had killed the German near-simultaneously.

Lance-Corporal Henry Norwest was posthumously awarded a bar to the Military Medal bestowed in 1917 for his courage at Vimy Ridge.

Snipers of the Canadian Over-Seas Expeditionary Force pose for the camera in northern France in 1915. Henry Norwest sits on the right. All four of the rifles visible in the picture seem to be half-stocked versions of the Ross Mk III.

Johnson Paudash of Kawartha, whom the Canadian census of 1911 described as 'Chippawa' but is now generally considered to be Ojibwa (Anishinaabe), was another successful sniper, credited with eighty-eight kills while attached to the 21st Battalion of the CEF. Born in Peterborough, Ontario, on 29 January 1875, Paudash served with the 40th Volunteer Militia in 1900–1 and was acquainted with Sir Sam Hughes, Minister of Militia in the Borden government.

When war began, recognising his skill as a marksman, Hughes persuaded Paudash – whose family had a record of service dating back to the campaigns of General James Wolfe – to join the CEF. Enlisting on 11 November 1914 at Kingston, Ontario, having given his 'trade or calling' as 'farmer', Paudash underwent basic training in Canada before embarking for Britain in 1915. On 22 September, he was shot in the right leg while fighting near Messines, but recovered to rejoin his unit near Ypres. There the sharp-eyed and exceptionally patient Paudash demonstrated his abilities as a scout as well as a sniper, and was regularly used as a forward observer. Herbert McBride said of him:

> By birth, inheritance and inclination, he was a hunter. He never would have an observer, preferring to work alone as he made his devious way along behind the lines, waiting for a chance to take a shot. He seldom (perhaps never) had any permanent 'nests' but moved about continually. Each evening he would turn in his report, and I for one believed him, which is more than I could say for any other lone sniper.[19]

Paudash was awarded the Military Medal for discovering that the Germans were massing for an attack. Though the assault was made less than half an hour after Paudash had returned to make his report, sufficient time had been gained to allow the attackers to be repulsed with heavy casualties.

It has also been claimed that Johnson Paudash was recommended for the Distinguished Conduct Medal for 'saving the life of an officer during the Battle of the Somme', but the award was never approved; his Native American background most probably told against him.[20]

Francis 'Peggy' Pegahmagabow – part Ojibwa (Anishinaabe) – enlisted in the 23rd Regiment (Northern Pioneers) as soon as the First World War began, then signed an attestation in Valcartier, Quebec, on 15 September 1914. Enlistment papers show that he was not married, had no next of kin, and was a fireman (he had been a marine fireman in the Marine and Fisheries Department on the Great Lakes).

Landing in France in February 1915 as part of the 1st Canadian Infantry Battalion, Pegahmagabow was engaged in the Second Battle of Ypres (April 1915) when the Germans released chlorine gas in the first significant chemical attack and the strength of the Canadian battalion was reduced by more than 40 per cent in just three days of fighting. He was 'lightly gassed', though even this was to affect his health for the remainder of his life. Yet Pegahmagabow

Native American, or 'First Nation' Canadian sniper Francis 'Peggy' Pegahmagabow – part Ojibwa (Anishinaabe) – was born on 9 March 1891 in Shawanaga, Parry Sound, Ontario. He enlisted in the 23rd Regiment (Northern Pioneers) in September 1914 and served with distinction, becoming the most decorated of all the Native Americans in Canadian service after receiving the Military Medal three times. His 'kills' score, even though it cannot be independently verified, makes him by far the highest-scoring of all First World War snipers. *Canadian Military Museum, Ottawa*

became an extraordinarily effective sniper, crawling unseen into no man's land at night, where he waited until Germans showed themselves under the glare of flares or in the cold light of dawn.

His score mounted rapidly, in addition to prisoners captured regularly on his forays. A leg wound sustained during the Battle of the Somme was not enough to keep him out of action for long. An award of the Military Medal followed in 1916, for delivering messages in the face of enemy fire; a bar was granted for bravery shown during the Battle of Passchendaele in November 1917, and a second bar, most unusually, after the Battle of the Scarpe (1918).

When fighting ended on 11 November 1918, Francis Pegahmagabow was said to have 378 'confirmed kills' plus many others unconfirmed. If true, and it is difficult to find independent verification, he was by far the most successful sniper of the British and Empire Forces.[21]

Samson or 'Sampson' Comego of 21st Battalion was another of the Native American CEF snipers. Said to have been born in Alderville, Ontario, between 1861 and 1872, depending on the source of information,[22] Comego claimed to be a chief of the Rice Lake Band on the Alnwick reserve. He gave his 'trade or calling' as 'musician' on attestation, but was 'unmarried' and gave next-of-kin as his brother Peter. His records are contradictory (one gives his height as '6 ft 2 in', his attestation gives '5 ft 7½ in'), but there is no doubt that he was a big man weighing as much as 300 lb.

After enlisting on 6 November 1914, claiming prior service in the 40th Regiment, Comego arrived in northern France. On 10 November 1915, 'in the Line at La Clytte', with his tally at twenty-eight, according to Herbert McBride, Comego was killed by a German rifle grenade.[23]

Joseph Patrick Riel was classed as Métis, people of mixed race with Native American and early European settler ancestors. Riel was actually only one-eighth Métis, the rest of his ancestry being largely French and Irish. According to an attestation made in Valcartier, Quebec, Riel, who was illiterate, had been born in Chelsea, Quebec, on 17 March 1876. He gave no trade or calling and his next-of-kin was his daughter Catherine; his wife had died some years earlier.

Patrick, or 'Paddy' Riel as he was known, was duly sent to Europe in 1915 with the 8th Canadian Infantry Battalion. His colleagues in the 90th Winnipeg Rifles joked that their unit had fought against 'Patrick Riel's Uncle Louis' in the Red River Rebellion of 1868 and the Riel Rebellion of 1885, when the Métis had risen against the Canadian government.

Many writers have accepted the claimed relationship between Louis Riel and Joseph Patrick Riel at face value. The surname 'Riel' was common amongst Métis people, and it is possible that there was a family connection in an earlier generation; however, this has yet to be found, and there is little doubt that the story arose simply from 'barrack-room humour'.[24]

Patrick Riel proved to be an effective scout and a very good marksman. Credited with thirty kills, attached to 18th Infantry Battalion, 'While on duty in the front line trenches at PLOEGSTREET, on the afternoon of January 14th 1916, he was hit in the neck by a splinter from an enemy high-velocity shell, and instantly killed.' Riel was buried in the Rosenberg Château Military Cemetery.

Herbert Wesley McBride, another of the renowned Canadian snipers, had a career which fiction would have been hard pressed to rival. His obituary published in *The American Rifleman* in May 1933, reported that:

> He had a varied and colorful career. Gold mining in Alaska, logging, railroad construction work in the Yukon and northern British Columbia, exploring, hunting and fishing widely, wars – all these formed interesting chapters of activity in his life. Biology, geology,

ethnology, anthropology, botany and entomology, in addition to his long devotion to shooting, all claimed his attention . . .

Captain McBride had handled about every type of small arms extant during his lifetime. His fondness for guns was acquired in his tender years, and he had shot his first deer and wild turkey before he was 10 years old. As a member of the Indiana National Guard Team he shot at the National Matches from 1905 to 1911 inclusive, and he won the Indiana State Championship in 1905, 1906 and 1907. Also, he attended six National Matches after 1919. He was the organizer of the Indiana State Rifle Association and many rifle clubs, and served as N.R.A. State Secretary for Indiana for a number of years. Captain McBride always maintained a close association with some military organization from March, 1888, up until the time of his death.

In 1914 he was in command of a company of the Indiana National Guard, but resigned and went to Canada where he served as captain and military instructor in the 38th and 21st Battalions until the latter was sent overseas in May, 1915. He then resigned his commission and accompanied the battalion as a private machine gunner. He served overseas until early in 1917, when he was invalided home.

Reaching this country late in April, he was assigned to duty as an instructor. He served throughout the remainder of the war in that capacity. When the Small Arms Firing School was organized at Camp Perry in May, 1918, he was one of the first instructors. Captain McBride, before wounds cut short his war service, was decorated with the British Military Medal for capturing 12 machine guns at the Battle of St. Eloi in Flanders in 1916, the Medaille Militaire for invading the German lines and capturing a German flag, and the Croix de Guerre. He was wounded seven times.

In addition to his service in the World War, Captain McBride also saw service in the South African Boer War . . .

However, this only told part of the truth. The outline is broadly correct, but McBride's chequered career has been sanitised. Author of the well-respected *The Emma Gees* (1918) and *A Rifleman Went To War* (1933), Herbert McBride contracted tuberculosis in his youth, before travelling to Colorado and New Mexico, and then to the Klondike in the gold rush of 1898. Returning to Indiana, where the Federal census of 1900 describes him as a 'shipping clerk', he enlisted in the National Guard, probably in 1902, and had attained the rank of captain by 1907.

The 1910 Federal census still places McBride in Indianapolis, employed as 'travelling agent, powder mill'; in 1912, he returned to British Columbia to hunt game. On 1 February 1915, McBride – still a citizen of the USA – was com-

The Ross Mk III rifle, seen here with an M1913 Warner & Swasey optical sight, was rapidly replaced in the Canadian Army by the SMLE. However, more than a thousand Ross rifles were retained as snipers' weapons, a role in which they excelled.

missioned into the Canadian 43rd Infantry Regiment (Duke of Cornwall's Own Rifles) ranking as a lieutenant. Temporarily ranking as captain, he was almost immediately attached to the 21st Battalion, CEF, as a musketry instructor.

McBride's records reveal a self-destructive streak. With the 38th Battalion in Ottawa, he became drunk in public and was absent on parade. On 25 March 1915, his commission was rescinded for 'conduct unbecoming an officer'. Yet Herbert McBride had already volunteered to serve overseas. His attestation, dated 3 March 1915, describes him as a lawyer, and his military experience as twenty-one years in the US National Guard and two months' service in the Canadian 43rd Infantry Regiment. Reaching Britain on 15 May 1915, Private McBride, as he had become, embarked immediately on a period of insubordination and drunkenness before the 21st Battalion was landed in France in September.

Once the fighting started, however, McBride, firing a Ross Mk III fitted with an M1913 Warner & Swasey sight, began to demonstrate the aptitude for marksmanship for which he was to be renowned. On 31 May 1916, he was granted a temporary commission and attached to the 39th Battalion; on 3 June, he was awarded the Military Medal for 'conspicuous service and devotion to duty in connection with scouting and patrol work'.

Yet by August he was in trouble again. At first absolved of blame, he was then re-arrested and severely reprimanded by a court martial. Arrested once more

Herbert W. McBride was born on 15 October 1873 in Waterloo, Indiana, to lawyer Robert Wesley McBride and Ida Chamberlain. Herbert came from military stock: his father had served in the Union cavalry during the Civil War, thereafter rising to colonel's rank in the Indiana National Guard. On 25 December 1928 McBride petitioned for membership of the Indiana Society of the Sons of the American Revolution, claiming descent from Revolutionary War veteran Ebenezer Smith of York County, Pennsylvania; his application was approved early in 1929. *Image from* A Rifleman went to War.

in October, he was admitted to hospital with 'alcohol debility' and returned to England for treatment. Eventually, the Army lost patience and the *London Gazette* of 19 March 1917 noted that 'Lt. H. W. McBride is dismissed the Service by sentence of a General Court-Martial'.

In *A Rifleman Went To War*, reprinted many times, McBride recounts killing a substantial number of German soldiers not only as an independent sniper – a short-lived assignment – but also as a machine-gun team commander who used his rifle as a 'sniper's weapon' whenever possible. However, though he kept a diary, there seems to be no independent confirmation from military sources, and claims made by others on McBride's behalf should probably be treated with caution.

An application for a US passport, made from London on 2 April 1917, notes that Herbert McBride had left the USA for Canada on 14 January 1915. He sailed from Liverpool, arriving in New York on 17 April 1917, and then joined the US Army to serve as a marksmanship and sniping instructor until resigning in October 1918, once again in disgrace over perceived insubordination.

A Rifleman Went to War is not merely a sniping manual, but also a valuable memoir of McBride's service with the Canadian Army. Though rarely drawing attention to his inner demons, the book gives one of the clearest descriptions of war on the Western Front. Herbert McBride can be compared in some ways to Hesketh-Prichard, with the same love of adventure, nature and even history, but his attitude to killing was closer to the 'bag a Boche before breakfast' that drove Julian Grenfell.

The Germans (whom he rather oddly persistently calls 'Dutch') had started the war and so had to suffer the consequences of their actions. McBride was determined to take revenge for the loss of friends, and, in particular, for the stretcher bearers who had been shot down by a sniper while going to the aid of a casualty. And the memoirs, scathing of officers with neither tactical ability nor common sense, reveal the traits that were to wreck McBride's military career. Yet he was generally right, which is why the chapter on sniping (and much of the material devoted to concealment and observation) has been so highly regarded by post-1920 sniper instructors.[25]

The story of the Australian Imperial Force and the New Zealanders who together formed the Australia and New Zealand Army Corps ('Anzac'), is one not only of great comradeship but tragedy and unparalleled loss.[26] Thrown into action in Gallipoli in 1915, they faced an immediate challenge posed by Turkish snipers. Many of the Turks were experienced hunters, and also often veterans of the bloody Balkan Wars that had finished only a year before the First World War began.

Usually equipped with the 7.65 mm M1893 Mauser (though M1890 and M1903 rifles were also used), the Turks often had the advantage of high ground. But they were also exceptionally skilled in the use of cover, and often fired from undetected pits sited perhaps only tens of yards from Anzac lines. Observers remarked:

> The Turkish snipers, well supplied with ammunition and food, and frequently camouflaged with Hessian painting to assimilate with the scrub, watched their opportunity to pick off our men, and it was some considerable time before reports ceased to be sent into various headquarters that snipers were still inside our lines.[27]

Casualties mounted alarmingly, particularly among officers and 'runners'. The Australians were often excellent shots – many had made their living as kangaroo hunters to whom a shot-holed skin was anathema – but, at least initially, lacked the discipline of the Turks. They would happily blaze away at a Turkish attack from the parapet of their trenches or take unnecessary risks moving from trench to trench.

Casualties included the commanding officer of the 1st Division, Scots-born Major-General Sir William Throsby Bridges, who was shot by a sniper while inspecting a defensive position on 15 May 1915. The bullet severed the femoral artery in his right leg and he died of his injury three days later.

Spotting periscopes were created on the beach-head from mirrors cannibalised from the warships and troopers gathered in Suvla Bay (to be destroyed by Turkish snipers, at least initially, almost as soon as they had been made); periscope-sighted 'trench rifles' were made in some numbers; and a 'blind eye' turned by commanders to unofficial counter-sniping allowed anyone – cooks, clerks, even a Presbyterian minister, the Reverend Andrew Gillison (an experienced target shooter) – to take on the Turkish marksmen.

William Edward 'Billy' Sing
(1886–1943), the second-
highest scoring sniper of
the First World War after
Francis Pegahmagabow
of the Canadian
Army. *Australian War
Memorial*

Sometimes these initiatives were successful, but no real progress was made until the New Zealanders organised a 'sniper squad' of picked men, including Jesse Wallingford, who had shot for Great Britain in the 1908 Olympic Games. They began to take such a toll of Turks that the Australian commander ordered his men to follow the example of their Antipodean cousins. Into this cauldron came William Edward 'Billy' Sing (1886–1943), son of Shanghai-born John Sing and Mary Ann Pugh, born in Staffordshire.

At an early age, Sing and his two sisters encountered anti-Chinese sentiment and their upbringing was harsh. Something of a loner, Billy drifted from job to job, to become a successful kangaroo hunter and a prize-winning target shot for the Proserpine Rifle Club in Queensland.

Billy Sing enlisted on 24 October 1914 in the 5th Australian Light Horse Regiment. After basic training in Egypt, his unit was sent to Gallipoli in the spring of 1915. With Turks taking a terrible toll of Anzac troops, experienced marksmen were sought in the hope of a suitable riposte. Sing's rifle-shooting abilities had already been recognised and so he began his career as a sniper from Chatham's Post, firing a standard SMLE which is sometimes said, apparently without much evidence, to have been fitted with a Galilean sight (these are discussed later in the chapter).[28]

In no way handicapped by the roundnose Mk VI .303 ball ammunition issued to the Anzacs, which was inferior ballistically to the 7.65 mm Turkish Mauser round, Sing took steady, calculated and cold-blooded revenge; helped by his spotters, Ion 'Jack' Idriess and Tom Sheehan, his score mounted rapidly.

Despite the hindrance of a freak injury to his shoulder, by a bullet which ricochetted off Sheehan's spotting telescope, Billy Sing was credited with 119 kills by Brigadier Granville Ryrie, commander of the 2nd Australian Light Horse Brigade, as early as September 1915. On 23 October, General Sir William Birdwood, commander of the Anzac troops, noted Sing's kills to be 201 'including unconfirmed', and Major Stephen Midgeley, who had drawn Birdwood's attention to his sniper's performance, put the total as 'about 300' when the Allied forces withdrew from Gallipoli.

They were said to have included 'Abdul the Terrible', a renowned Turkish sniper killed by Sing in a counter-sniper duel, but the details rely greatly on an article written by Idriess in the 1950s and are now considered to be largely apocryphal. But there can be little doubt that Sing's reputation would have been well known in Turkish lines, and that the most effective snipers would have been charged with eliminating the Australian.

Billy Sing was mentioned in dispatches in February 1916 and awarded the Distinguished Conduct Medal on 10 March. He arrived on the Western Front early in 1917, but was wounded in the leg in March and returned to Britain to recuperate. Discharged from Harefield Hospital on 22 May 1917, Sing travelled north. After a whirlwind courtship, he married Gladys Elizabeth Addison Stewart in Edinburgh on 29 June 1917 but left for the Western Front on 3 August.

There is no evidence that he returned to sniping with the dramatic effect he had had in Gallipoli, as his health had declined owing to shrapnel wounds and the cumulative effects of gas poisoning. However, Billy Sing is known to have led a counter-sniping sweep in Polygon Wood in September 1917 – he was credited unofficially with many additional kills – and continued to serve effectively until hospitalised in November.

Sing was repatriated to Australia on 31 July 1918 as 'Permanently unfit for Service'. The remainder of his life passed largely in anonymity. His marriage had failed,[29] and he was unable to find anything other than short-term employment. His health continued to decline, until he died in Brisbane on 19 May 1943, unmourned and unnoticed. John Hamilton's *Gallipoli Sniper: The Life of Billy Sing* (published in 2008) is by far the best source of information about his service career.

Perhaps surprisingly, few other Anzac snipers have been identified, though there must have been many whose marksmanship contributed to success on the battlefield. Among them was a New Zealander, Richard Charles Travis (1884–1918), born 'Dickson Cornelius Savage' to an Irish father who had once served in the New Zealand Armed Constabulary, and an Australian–Irish mother. Travis had argued with his father, moved away from home, and, after a scandalous liaison with a woman (probably married), changed his identity.

Small and wiry, like Billy Sing, and an excellent horse-trainer, Travis enlisted in the 7th (Southland) Mounted Rifles in September 1914. After proving his skills at Gallipoli – where he was eventually allowed to stay after initially stowing away with combat troops – Richard Travis reached the Western Front in April 1916. There his scouting, sniping and counter-sniping skills were put to good use; in 1917, he was promoted to sergeant and placed in charge of a sniping and reconnaissance section composed of men of his own choosing.

Holder of the Belgian Croix de Guerre and the British Military Medal, Travis took part in the July 1918 'Hundred Day Offensive' in Rossignol Wood, near the village of Hébuterne. He destroyed two machine-gun emplacements single-handedly, and then killed the five Germans who had intercepted his retreat, but was himself killed by a shell fragment on 25 July. A posthumous Victoria Cross was gazetted less than three months after his death. It was perhaps ironic that Travis, a white New Zealander, received the highest of all British gallantry medals while Billy Sing, of mixed race, was denied even the Military Medal for which several senior officers had commended him.

Private Alfred Hugh Dillon of the Wellington Infantry Regiment was another New Zealander with a reputation for marksmanship. Arriving in France on 7 December 1916, Dillon was awarded the Military Medal for gallantry while acting as 'Battalion Observer' on Bon Avis Ridge from 29 September to 3 October 1918. His score as a sniper remains unknown, but his rifle, a Canadian Mk III Ross with an M1913 Warner & Swasey sight, survives in the collection of the New Zealand National Army Museum.[30]

At the beginning of 1916, a noteworthy attempt was made in South Africa to assist the battle against German snipers on the Western Front. The *Sydney Morning News* of 31 January 1916 was just one of many newspapers across the British Empire to report that:

> Sir Abe Bailey intends raising, equipping, and transporting 100 South African sharpshooters for service in Europe. (Sir Abe Bailey is one of the principal mine owners in South Africa, and was at one time M.L.A for Krugersdorp, Transvaal. He served in the Boer War, and holds the King's and Queen's medals with six clasps. He is also well known in sporting circles.)

Abraham 'Abe' Bailey (1864–1940) was one of the richest men in South Africa. Born in Cardock in Eastern Cape, to a father from Yorkshire and a mother from Scotland, Bailey made his fortune from gold. By 1884, at the age of just twenty, he was head of the 'Bailey Group' of mines and had become immersed in politics. Implicated in the Jameson Raid of 1895–6, though as a member of the 'Johannesburg [organising] Committee' and not a participant, he was given six month's imprisonment – a sentence immediately commuted to a fine, which to a man of his wealth was no more than a trifling irritation.

After enlisting on 3 January 1900 in the mounted division of the City of London Imperial Volunteers (formed in December 1899 and disbanded a year later), Abe Bailey greatly increased his fortune through canny investments in diamonds, gold and land. An accomplished sportsman who had played cricket for Transvaal, he was made KCMG in 1911 and served as Deputy Assistant Quartermaster-General of the 6th Mounted Brigade of South African forces in 1914–15 with the rank of major. On 12 February 1919, Bailey was made 'Baronet of Cardock in the Province of the Cape of Good Hope and the Union of South Africa' in recognition of his 'Services to the British Empire' during the First World War.

The South African Sharpshooters were chosen from the game hunters of southern Africa. Most of them came from Rhodesia to demonstrate their marksmanship skills: tests which included five rounds at 'Crossing Target No. 6' from a range of 200 yards 'over and around cover'; two series of five shots at 'Snap-Shooting Figure No. 3' placed 200 yards away; and five at a 'Slow 1st-Class Figure' at 800 yards. The qualification score was high enough to ensure that only candidates with the ability to shoot quickly and accurately were accepted.

Led by Temporary Second Lieutenant Neville Methven,[31] a respected big-game hunter, seventeen men set sail on 22 April 1916 from Cape Town, aboard RMS *Saxon*, bound for Britain. By the time they arrived at Bordon Camp on 10 May, one man had been transferred to the East African Rifles. Training with specially regulated SMLE rifles purchased by Bailey from Purdey, fitted with Aldis telescope sights, was completed quickly enough to allow the South Africans to leave for France – where, on 16 June, assigned to the 1st Infantry Division commanded by Major-General Peter Strickland, they were pitched into the Somme battles,

When the First World War ended on 11 November 1918, only twenty-three South African Sharpshooters had served on the Western Front, seven of whom were regarded as replacements. Two had been killed, two were missing (later accepted as dead), two had died of wounds, one had been captured by the Germans, another had been transferred, and eight had been classed as 'medically unfit' (one temporarily).

Methven claimed, in an interview conducted long after the event, that his unit had accounted for 3,000 German soldiers, with 'a hundred or more' for himself. Others have credited Methven, who was awarded the Military Cross for leadership, with 100, 103 or 108 kills. But it seems unlikely that the overall kill-total is accurate, as so few of the men served until the end of the war. If it is true, then some of them would have to have made have more than 200 kills – which could hardly have passed unremarked. It has even been suggested that '3,000' should have been '1,300', which would, perhaps, be easier to justify in the circumstances.

Creation of the South African Sharpshooters had been a praiseworthy attempt by an individual to support his country, but achieved little of note in the context of the scale of the First World War. Attempts were made in Britain and France to collect as many privately owned sporting guns as possible, but the results were disappointing; a much more successful appeal was made in Germany in the spring of 1915, allegedly gathering more than 20,000 rifles. At least some of these made it to the trenches.

Though the Germans ordered 20,000 Scharfschützen-Gewehre 98, impressments formed the bulk of sniping equipment until the summer of 1915. On the Allied side, big game rifles firing large bullets at considerable velocity landed a more telling punch than the Lee-Enfield service rifle ever could. Martin Pegler[32] notes that, in 1915, the War Office purchased more than sixty large-calibre sporting rifles, four of which chambered the legendary .600 Express cartridge that generated a muzzle energy of no less than 4,500 ft-lb. He also quotes Lieutenant Stuart Cloete, the celebrated South African essayist, novelist and biographer, writing in *A Victorian Son* (1972): 'when we hit a [loophole] plate we stove it right in, right into the German sniper's face. But [the rifle] had to be fired from a standing or kneeling position to take up the recoil. The first man who fired it in the prone position had his collarbone broken.'

About 200 optically sighted rifles made by BSA were issued to the British Army in May 1915. These are believed to have been full-length Lee-Enfields, possibly charger loading, and had telescope sights. A belief that the Long Lee-Enfield was more accurate than the SMLE persisted in many circles, despite trials undertaken in the early 1900s which had shown there to be very little difference. There seemed to be evidence that the long-barrelled rifle shot better at distances greater than 600 yards, but even this could be (and regularly was) contested. However, performance was undoubtedly affected by variations in ammunition.

At this time, it was by no means clear what type of optical sight should be standardised. When the first SMLE-type rifles were set up in 1916, sights had to be obtained from a variety of sources. This was largely due to the lack of a large-scale manufacturer in Britain, where development of optics had lagged behind even the USA The problem was much less obvious in Germany, where many well-established scientifically orientated companies such as Carl Zeiss were making not only fine-quality telescope sights but also binoculars and microscopes. The arbiter was the quality of the lenses. There was probably little to choose between best British and best German practice prior to 1914, but for each of the former there were a hundred of the latter. The problem was so acute that, *during* the war, the British tried to purchase optical sights and binoculars from Germany in exchange for rubber, with Switzerland as intermediary.

By 1915, the time was right for snipers and sniping to become a useful part of the military establishment, however much some higher-ranking officers despised such an 'un-British' concept. The question then became largely one of procedure. Marksmanship was encouraged within the Army largely on the basis of winning competitions with other regiments, which in practice meant that only a handful of the best shots actually had much practice. In addition, such practice was inevitably conventional target shooting with an eye on the ultimate goal: a medal at Bisley.

This was wholly understandable, and, indeed, laudable in peacetime. The efforts of the NRA and the many miniature-rifle clubs, inspired by the poor performance of the British Army in South Africa in 1899–1902, had helped to create the disciplined well-trained British Expeditionary Force that had performed so well at Mons. But it also showed that the skills of even the best target shot could be entirely inappropriate for the sniper who had to kill someone who had the same skills.

The biggest problems were to be lack of status and lack of direction. No one at the highest levels had enough foresight to lay down Army-wide rules for the training and employment of snipers, and so, predictably, the earliest results were far from noteworthy.

Attempting to organise snipers on a local basis inevitably led to inconsistency. Some units, especially those whose officers were experienced hunters, performed well; others regarded sniping and counter-sniping as an irrelevant nuisance, even when faced by the evidence of steadily mounting casualties. The ferocity of warfare had clearly caught many of the old guard unaware.

The British people had always loved successful mavericks such as Horatio Nelson, but men such as these often found promotion elusive in peacetime; only in war did their remarkable qualities come to the fore. Unsurprisingly, the British Army of 1914 had few of them.

Now that a century has passed, it is difficult to judge where credit for the introduction of effective sniper-training techniques is rightfully due. However, among the visionaries was Langford Newman Lloyd, a well-qualified doctor serving the Royal Army Medical Corps.

Born in London on 28 December 1873 to Colonel E. G. K. P. Lloyd and Lilian May Hooper (daughter of the Surgeon-General), Langford Lloyd had trained at Charing Cross Hospital, qualifying in 1898. Employment as assistant medical officer in an infirmary in Islington did not satisfy his ambition, however, and so he accepted a commission into the Army Medical Service on 27 July 1899. Sent immediately to South Africa as medical officer of 1st Battalion, Dublin Fusiliers, he was twice mentioned in dispatches and awarded the DSO, gazetted on 27 September 1901, 'in recognition of his services in South Africa'. The award was conferred in Bermuda in 1902, where he had been sent almost as soon as he had returned from the war.

Captain Langford Lloyd, as he had become in 1902, returned to Britain to become adjutant of the RAMC Volunteers, and then adjutant of the Territorial Force (1908–11). An active Freemason in London Irish Rifles and Arduis Fidelis Lodges,[33] he was promoted to the rank of major on 27 July 1911.

When the First World War began, Lloyd was dispatched to France on 18 August 1914 in command of the 2nd Cavalry Field Ambulance. Promoted to lieutenant-colonel on 1 March 1915 and made CMG in 1916, he served in France until 1 November 1917. A spell in Italy as a temporary colonel was followed by a return to France until April 1919. Then came a posting to India (1922–7), promotion to colonel on 3 January 1927, and 'retirement on pay' at the end of 1930. He died on 20 April 1956.

Hesketh-Prichard credits Langford Lloyd not only with the foundation of one of the earliest sniping schools, but also for his abilities as a teacher. It seems strange that a doctor should be credited in this way, but Lloyd was a keen shot. He had represented England many times at Bisley, serving as Captain of the Mother Country team of 1922. Consequently, he was only too aware of the deficiencies in marksmanship that were obvious throughout the British Army.

Another pioneer was Frederick Maurice Crum, author of, amongst other relevant works, *Memoirs of a Rifleman Scout*. Born in Cheshire on 12 October 1872 to wealthy Scotsman William Crum[34] and his wife Jean Campbell, and after finishing his education, 'Gentleman Cadet' Crum was accepted by the Royal Military College, passing out on 7 February 1893. He was duly posted to Rawalpindi in India with 1st Battalion, 60th Rifles (King's Royal Rifle Corps), ranking as lieutenant.

The battalion was moved to Cape Colony in 1896, and was in Dundee in Natal when the Second South Africa or Boer War was declared on 12 October 1899. Crum was severely wounded in the shoulder during the Battle of Talana Hill, and was taken prisoner. Eventually, he reached hospital in besieged Ladysmith. Released in June 1900, Crum was promoted to the rank of captain on 1 January 1901 but, still suffering from his wound, was invalided home at the beginning of 1902.

Then began a period of intermittent service, beginning with a return to De Aar in the summer of 1902 and then a move to Malta with 1/60th before returning to India. In the autumn of 1906 Frederick Crum was seconded to scout training at Rhaniket, a discipline in which he excelled. A short period in Britain to study for Staff College exams (which he failed) was followed by a return to India where, on 28 February 1908, he was seconded for service on the staff of the King's Royal Rifle Corps and posted to the Mounted Infantry School in Fategarh. There he became assistant commandant and then took

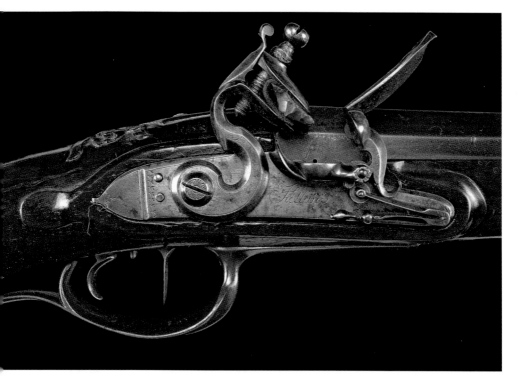

Made about 1750 by the renowned gunmaker Johann Andreas Kuchenreuter of Regensburg, this *Jäger* rifle is typical of mid-eighteenth-century sporting rifles. Military rifles were customarily plainer, but took similar form. *James D. Julia, auctioneers (www.jamesdjulia.com))*

The flintlock (*left*) of a typical Baker rifle of the Naploeonic Wars, the first of its type to be issued for general service in the British Army; the muzzle of the Baker Rifle (*right*), showing the shallow-pitched grooves that gave the rifleman a huge advantage over infantrymen armed with smooth-bore muskets. *Both courtesy Martin Pegler*

Advances in rifle design, 1750–1865. *Left to right*: the Kuchenreuter *Jäger* rifle; a comparatively short heavy-barrel rifle by gunmaker 'P. S.', made in Germany *c.* 1770 (but now fitted with an English-made lock signed 'Ketland'); a Revolutionary War-period flintlock longarm, possibly reverted from cap-lock, made *c.* 1770–5 in Lititz, Pennsylvania, most probably by German-born Andreas Albrecht (1718–1802); and a heavy-barrel cap-lock of the Civil War, with a full-length telescope sight. *James D. Julia, auctioneers (www.jamesdjulia.com)*

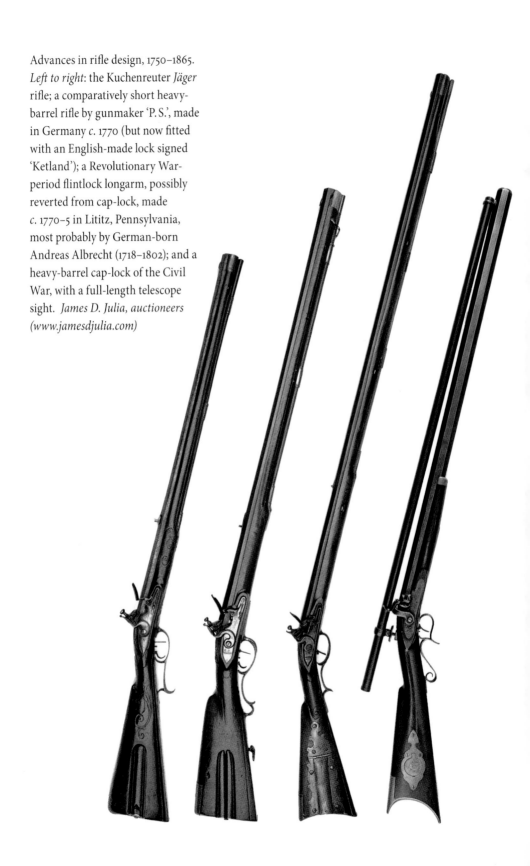

The death of British Admiral Horatio Nelson at the Battle of Trafalgar (21 October 1805), struck down on the quarterdeck of HMS *Victory* in his moment of triumph by a French sharpshooter. *Author's collection*

The battle between USS *Constitution* and HMS *Guerrière*, 19 August 1812, ended in victory for the American frigate. Topsmen armed with muskets and rifles took a terrible toll of the officers and men on the decks of each ship. *Author's collection.*

Above: 'Hancock at Gettysburg' (1–3 July 1863), lithograph *c.* 1887 by Louis Prang & Co. from an illustration by Swedish-American artist Thure de Thulstrup (1848–1930). *Library of Congress.*

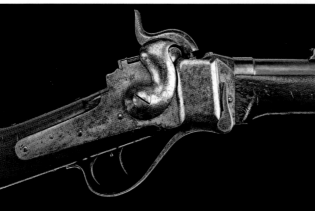

Above left: The breech of the Berdan-type .52-calibre Sharps New Model or M1859 rifle, no. 56,932, showing the double triggers that distinguished this particular sub-variety.

Left: The Sharps breech-block can be lowered to allow a combustible cartridge to be placed directly in the chamber. The action was very strong, but apt to leak gas.

Both, James D. Julia, auctioneers (www.jamesdjulia.com)

his cap-lock rifle, sold by C. B. Holden but also marked by James & Ferris of Utica, New York, was issued to the 1st Massachusetts Sharpshooters during the Civil War and bears the State seal on the patch box lid. Note the full-length Malcolm-type sight. *James D. Julia, auctioneers (www.jamesdjulia.com)*

The .44 RF M1860 Henry repeating rifle was a good short-range sniping weapon owing to the rapidity with which it could be fired. This gun, no. 3,283, one of 800 ordered on 30 December 1863 for the 1st District of Columbia Cavalry, also has a full-length Malcolm-type sight – unmarked, and possibly installed by Winchester when the gun was made. *James D. Julia, auctioneers (www.jamesdjulia.com)*

Four sniper rifles of the First World War. *Left to right*: a Steyr-made Austro-Hungarian M95 8 × 50R Mannlicher with a 3 × sight by C. P. Goerz of Berlin; a German Mauser-action 8 mm SG. 98 used prior to 1916 by Infanterie-Regiment Nr. 96, with a 4 × Gérard sight; a British Winchester-made .303 Pattern 1914 Mark I* with a BSA-marked sight, probably a replacement for an Aldis; and a Canadian .303 Mk IIIB Ross with the 5.2 × M1913 Warner & Swasey sight. *From left (three) James D. Julia, auctioneers (www.jamesdjulia.com), and, right, Canadian War Museum, Ottawa*

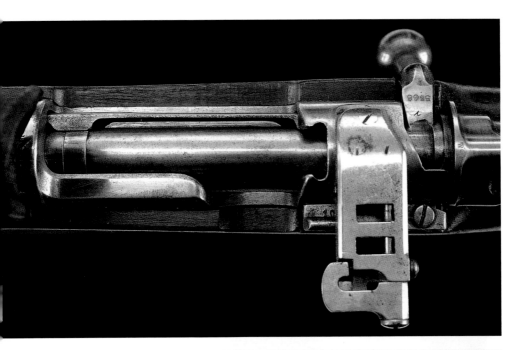

The breech of a German SG. 98, showing how far the telescope-sight mounts had to be offset to allow a charger to be used to reload the magazine. *James D. Julia, auctioneers (www.jamesdjulia.com)*

Red Army sniper Ziba Pasha qizi Ganieva (1923–2010), a one-time ballerina and drama student – later to have a minor role in a post-war Soviet film and then become a renowned philologist – poses with an Obr. 1891/30g. Mosin-Nagant rifle fitted with a PE optical sight of the type introduced in 1937. *Soviet official photograph; author's collection*

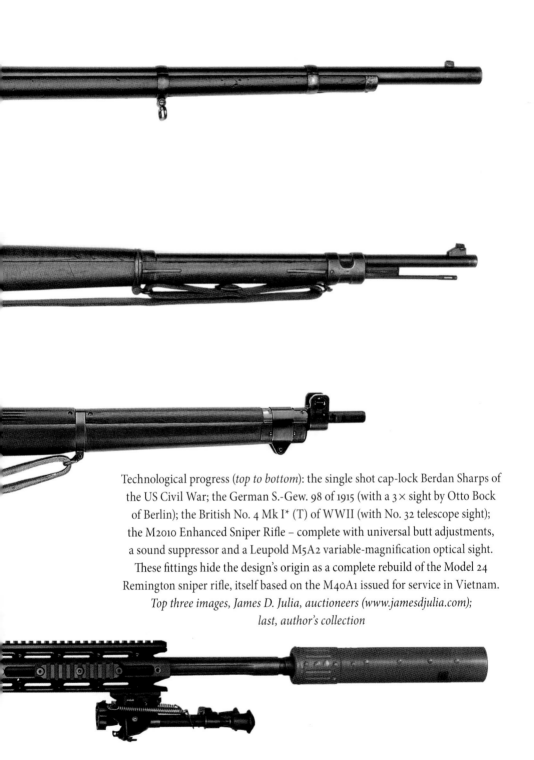

Technological progress (*top to bottom*): the single shot cap-lock Berdan Sharps of the US Civil War; the German S.-Gew. 98 of 1915 (with a 3 × sight by Otto Bock of Berlin); the British No. 4 Mk I* (T) of WWII (with No. 32 telescope sight); the M2010 Enhanced Sniper Rifle – complete with universal butt adjustments, a sound suppressor and a Leupold M5A2 variable-magnification optical sight. These fittings hide the design's origin as a complete rebuild of the Model 24 Remington sniper rifle, itself based on the M40A1 issued for service in Vietnam.
Top three images, James D. Julia, auctioneers (www.jamesdjulia.com); last, author's collection

A bolt-action Zf.-Kar. 98k no. 93,081 (*top*), made in 1944 by 'bcd'—Gustloff-Werke of Weimar—and the gun that was intended to replace it, the semi-automatic Zf.-Gew. 43. The Kar. 98 is fitted with a Dialytan 4× optical sight by 'bek' (Hensoldt Werk für Optik u. Mechanic of Herborn/Dillkreis) in the long side-rail mount, while the Gew. 43 has a rarely encountered Zf.-K. 43/1 4× sight made by Zeiss towards the end of the war with a machined cast-zinc body developed to accelerate production.

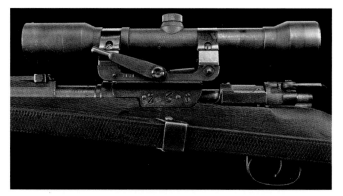

A close-up of the clamping mount used to attach the *Zielvier* to the Zf.-Kar. 98k. Note the proliferation of screws and pins holding the mounting-plate to the side of the gun body. The power of the German 7.9 mm cartridge was such that recoil often loosened the mount to a point where the gun 'lost zero'.

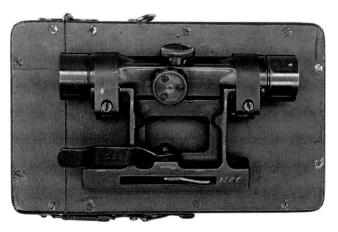

The 4× Zf. 4 sight, originally developed for the paratroop rifle, the FG. 42, was based on the Russian PU design. Optically acceptable, standardisation was compromised by the employment of too many manufacturers. This example, no. 66,037, made by 'ddx' (Voigtländer & Sohn AG of Brunswick), is shown with its protective case of wood-fibre.

A selection of US sniper rifles. *Top to bottom*: an M1903, barrel-date October 1908, with a Winchester A5 sight in patented ring mounts; an M1903, barrel-date April 1915, with a 5.2 × M1913 Warner & Swasey sight; a USMC M1903, barrel-date May 1939, with a Unertl 'u.s.m.c.—sniper' 8 × sight; and an M1903A4 with a variant of the M73B2 telescope sight made in France by Société Optique et Précision de Levallois (OPL).

Japanese sniper rifles: a 6.5 mm Type 97 (*top*) with the standard 2.5 × Type 97 telescope sight, and a 7.7mm Type 99, made in Nagoya in 1940, with a 4 × Type 99 sight. The Type 99 and the rarely seen Type 2 are often listed as '4 × 7'; however, markings show the '7' to be the field of view expressed in degrees and not the diameter of the ocular lens.

All items on this spread courtesy of James D. Julia, auctioneers (www.jamesdjulia.com)

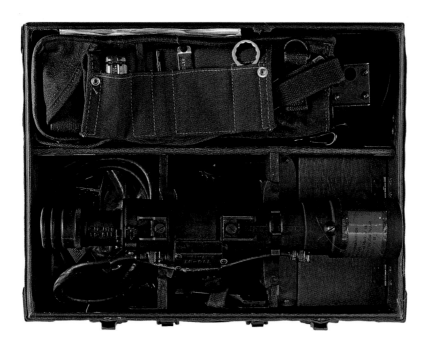

This 'Sniperscope Infrared Set No. 1, 20,000 volts' was issued with a .30 T-3 Carbine (an experimental variant of the M1). Cumbersome it may have been, but the Sniperscope allowed short-range shooting to be undertaken in very poor light. *James D. Julia, auctioneers (www.jamesdjulia.com)*

This dramatic photo is a reminder that the first-generation image-intensifying sights (this is a Rank-Pullin SS80 on an M16) were exceptionally bulky. But they were far better than nothing, and facilitated accurate shooting in circumstances where the human eye would be powerless. *Author's collection*

Above: The increasing sophistication of sniping equipment is reflected in an ever-widening choice of accessories and the care that needs to be taken in transit. The British L42A1, a 7.62 mm derivative of the .303 No. 4 Mk I (T), packs into a wooden case typical of those used during WWII. The numbers of the gun and sight are noted on the lid. The two pictures below show an Accuracy International AWR owned by the renowned sniper and sniper-trainer John L. Plaster (sold at auction with a golf ball he had struck squarely at 200 yards!), and a Heckler & Koch PSG-1. *James D. Julia, auctioneers (www.jamesdjulia.com)*

Opposite: John Plaster's AWR, with bolt open and bipod extended. The 1990s promotional leaflet for the Soviet/Russian Dragunov SVD shows the bulky NSPUM or 1PN58 passive infra-red and NSPU-3 (1PN51) cascade-type image-intensifying sights. Rifles of this type still pose a threat in the hands of insurgents. The US Army sniper pictured in Baghdad is firing a Remington M24 bolt-action rifle. *Main picture James D. Julia, auctioneers (www.jamesdjulia.com); leaflet, author's collection; photograph, US Army*

ОТКРЫТОЕ
АКЦИОНЕРНОЕ
ОБЩЕСТВО

ИЖМАШ

РОССИЯ
УДМУРТСКАЯ РЕСПУБЛИКА
426006
ИЖЕВСК
ПРОЕЗД ДЕРЯБИНА 3
телекс 755 113 URAN RU   255 113 УРАН
факс   (3412) 78-10-55   (3412) 78-70-10
телефон (3412) 78-17-42

7,62мм
СНАЙПЕРСКАЯ ВИНТОВКА
ДРАГУНОВА
CBD

The gas-operated M110 SS (Knight's SR-25) is effectively a re-engineered version of the old ArmaLite AR-10 combined with the best features of the AR-15/M16 series. It has proved to be accurate, but reliability in adverse conditions has been questioned. This rifle is one of only two to have been signed by C. Reed Knight, responsible for, amongst other things, a variety of sound-suppressing systems.
James D. Julia, auctioneers (www.jamesdjulia.com)

The introduction of ever-more powerful sniper rifles, in a quest for long-range accuracy, generally brings penalties of length and weight. This Accuracy International AX-50 is typical of the current generation of anti-matériel rifles (AMRs), suiting firers with frames robust enough to withstand the not inconsiderable recoil. *Accuracy International*

A US Army soldier tests the M110 SS on the firing range – 'practice makes perfect': a saying in which, for the sniper, there is a very real truth. *US Army*

command of the Mounted School in Poona, only to be given six months' leave on the grounds that he had been over-worked and was unwell.

In March 1910, Brevet Major Crum was 'restored to the establishment', and confirmed in the rank of major on 5 March. However, he was unable to take up his duties and left military service 'on retired pay' on 28 October 1911 with the intention of devoting his time to the Boy Scout movement in Scotland. This had arisen from a chance meeting with Baden-Powell.

When the First World War began, Major Crum immediately re-enlisted and, after organising enlistment of recruits in Lanarkshire, joined 8th King's Royal Rifles on 12 October 1914. His battalion was sent to France, which included a spell in the trenches at Hooge in July 1915. Major Crum, still suffering from ill-health, was sent to Acq in May 1916 to organise a brigade-level sniping school. This closed on 19 June, achieving its aims of training enough men to ensure that the skills filtered down to the rank-and-file, but General Skinner, suitably impressed by the work that had been done, requested that Crum and his staff move to the brigade headquarters in Arras to oversee the work of snipers and intelligence-gatherers.

Crum's health failed again on a march to the Somme, and he returned to Britain. On 10 October 1916, he was attached to the Senior Officers' School in Aldershot as an instructor, but his limitations soon restricted his efforts to reorganising the existing Sniping School.

On 17 May 1917, Crum took over co-ordination of scouting and sniper training throughout the British Army, the first time such a bold move had been made. In the spring of 1918, Frederick Crum returned to the depot of 5th King's Royal Rifles in Sheerness, where he laid the foundations of fourteen-day scouting training courses. Restricted to home service for the remainder of the war, he toured the country to lecture on sniping and scouting, speaking to groups ranging from 1,500 Guardsmen in Caterham to individual Scout troops.

On 1 November 1918, Major Crum retired from the army and returned to Scotland to promote the Boy Scouts. The renowned surgeon Sir Harold Stiles operated on his damaged arm in February 1919, largely curing the problems that had dogged him since the South African War, and his recovery was complete by the summer of that year. Frederick Maurice Crum of Kenmuir, regarded as a founder of the Boy Scouts in Scotland, eventually died in Rosneath, Dunbartonshire, on 8 October 1955.

Lloyd and Crum, serving officers, were bound from the outset by military conventions and their pioneering instruction efforts – undeniably effective – passed largely unnoticed. But there was always a maverick to take the spotlight: Hesketh Vernon 'Hex' Prichard, novelist, journalist, traveller, big-game hunter, and reputedly the 'best shot in England'.

Born on 17 November 1876 in Jhansi, in the Punjab, Prichard was the son of a British infantry officer – Lieutenant Hesketh Brodrick Prichard of the

Hesketh Vernon Hesketh-Prichard (1876–1923), now regarded as the most influential British sniper-trainer of the First World War. A member of the Royal Geographical Society and the Royal Zoological Society, he had named a river in Patagonia after his mother. Prichard played cricket for Hampshire on the recommendation of Arthur Conan Doyle (one of his mother's circle of influential friends), and took 339 first-class wickets at an average of only a little over 22 runs apiece. Often seen as an arch-Edwardian, almost as a pantomime villain, Prichard was a man of intelligence and very real sensibilities.

King's Own Scottish Borderers, acting as Quartermaster to the 24th Native Infantry Regiment – and Katharine 'Kate' O'Brien Ryall, daughter of a general. Unfortunately, his father had died six weeks before Hesketh Vernon Prichard was born, and his widowed mother returned almost immediately to Britain. The 1881 census shows that they were living with her parents.

Not without influence, Katharine's family helped to get Hesketh Vernon a scholarship to Fettes College, a public school in Edinburgh, where the boy soon showed an interest in sport in general and hunting and shooting in particular. It is said that he acquired a .410 cane-shotgun, which he disguised as an umbrella to ensure that he could shoot game without the school's knowledge.

Prichard left Fettes with a passable academic record and a reputation as a fine cricketer – 'one of the best bowlers the school had ever produced' in the words of one of his tutors. He seemed set for a career in the law, but he and his mother had begun to write adventure fiction which was serialised in some of the leading popular periodicals of the day.

The stories and novellas have not withstood the test of time. Yet they not only made Kate and Hesketh Prichard a good living, but also opened doors. A legal career long forgotten, Hesketh embarked on a new career as an adventurer-explorer, supported by the *Daily Express* on trips to places such as Haiti (where he reported on voodoo) and Patagonia, where a search for the giant ground sloth proved to be fruitless. However, his travel books justifiably gained many admirers and he was instrumental in the successful passage of a law that prevented the clubbing of seals of the British coast.

When the First World War began, Hesketh-Prichard as he was now known tried to join the Black Watch and the Grenadier Guards, but was rejected by both owing to his age – he was thirty-seven – and instead found a niche working for the Intelligence Office, chaperoning journalists to the trenches of the Western Front. Ever the opportunist, 'Hex' always took his personal telescope-sighted hunting rifle in case he could help the war effort by killing Germans.

In his classic book, *Sniping in France*, Hesketh-Prichard relates how on one occasion he and his party came across a British sniper, firing what was most likely a .303 Lee-Enfield fitted with an optical sight:

> The sniper was . . . shooting over the top of the parapet with a telescopic-sighted rifle. These rifles were coming out from England at that time in very small numbers, and were being issued to the troops.
>
> I had . . . possessed telescopic-sighted rifles, and had some under-standing of their manipulation as used in big-game shooting. In a general way I could not help thinking that they were unsportsmanlike, as they made shooting very easy . . .
>
> But to return to the sniper. Much interested, we asked him how he liked his rifle, and he announced he could put a shot through the loophole of the iron shields in the German trench 'every time'. As the German trenches were some six hundred yards away, it seemed to me that the sniper was optimistic, and we asked him if he would let us see him shoot. I had with me a Ross glass which I always carried in the trenches, and when the sniper shot I saw his bullet strike some six feet to the left of the plate at which he was aiming. He, however, was convinced from the sound that it had gone clean through the loophole. He had another shot, and again struck well to the left.[35]

Hesketh-Prichard realised that the sniper had no idea how to adjust his sights, or much grasp of the fundamentals of observation and target-selection. He was unimpressed by the telescope sight, which was carried in a tip-off mount and clearly no longer aligned with the bore.

Back at home, Hesketh-Prichard pondered the chances of defeating, or at least reducing the effectiveness of German snipers who continued to take a heavy toll of British troops. There were obvious problems: many British officers, with an unwise pride in the appearance of their trenches, insisted that sandbags were placed on the parapet with geometric precision, loopholes and embrasures placed at regular intervals, and polished cap badges cost British infantrymen their lives by catching the light of the sun or even the moon.

German trenches, by comparison, looked to be shambolic. Not only were the sandbags piled indiscriminately, but they were painted in many different colours (including stripes) and all kinds of spoil and the detritus of war were heaped on them. But there was method in this apparent carelessness; loopholes and embrasures, often formed with short lengths of pipe or even an artillery shell case with the base removed, could be buried in the parapet with little chance of detection by an enemy with very little conception of the value of long-term observation.

Hesketh-Prichard had even retrieved samples of the armoured shields the Germans issued, initially to machine-gun crews but then to protect snipers. These he elected to test to see what type of bullet could penetrate sufficiently easily to pose a threat to anyone sheltering behind them. His experiments were undertaken with a wide range of rifles, firing cartridges ranging from .256 Mannlicher (a popular sporting round) to large-calibre Express rounds used in big-game rifles. It was soon obvious that the .303 bullet did not have sufficient energy to penetrate the German shields, but that the more-powerful German 8 mm Mauser bullet would usually penetrate the shields used by the British.

It is interesting to speculate what would have happened if the British had succeeded in replacing the .303 Lee-Enfield with the high-velocity .276 Pattern 1913 rifle, as Hesketh-Prichard's experiments proved that his own preference for the .333 Jeffery cartridge was an ideal solution: powerful enough to defeat the German shields, but capable of being fired in a sporting rifle that was no heavier than the SMLE.

Hesketh-Prichard is known to have used several rifles, but apparently settled on .333 and then .350 Mausers, strong enough to handle the chamber pressures generated by what would now be considered as 'magnum' but were then generally labelled 'Express' cartridges. Convinced that he was finding effective answers to a serious problem, Hesketh-Prichard embarked on a difficult road. Like his contemporaries Langford Lloyd and Frederick Crum, Hesketh-Prichard had to impress upon senior officers that counter-sniping was practicable, and find sufficient backing to maintain his campaign in the face of official indifference.

When he was eventually given permission to return to France, with a variety of optically sighted rifles he had begged from friends 'for the duration [of the hostilities]', Hesketh-Prichard struggled to convince the upper echelons of command that sniping could make a valid contribution to the war in the trenches. Eventually, he was ordered to collaborate with Langford Lloyd whose 'telescopic-sight school' was attached to X Corps.

Hesketh-Prichard admitted that he had 'learned a great deal that I did not know about telescopic sights' and had contributed details of the tricks and ruses he had employed in his own area. The two men co-wrote a pamphlet explaining scouting, observation and sniping, but a change of command ensured that it was never published. A chance had probably been lost.

Hesketh-Prichard then went to Third Army, where, under a succession of sympathetic commanders, he was able to transform his role as roving instructor to something more permanent. British snipers were gradually gaining a stronger hand, and the Germans were increasingly 'beginning to

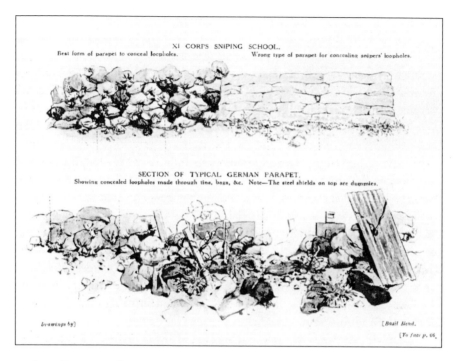

[To face p. 96.

These illustrations from Hesketh-Prichard's *Sniping in France*, the work of Basil Head (killed in action in 1918), show how the 'spit-and-polish' sandbagged trench-parapet beloved by many British officers (*top right*) is poor camouflage: the loophole invites a sniper's bullet. This is not true of the obscured loopholes in the 'pile at will' design (*top left*), or the randomly scattered impedimenta of the Germans (*bottom*).

cower' in their trenches. Fewer and fewer men were being lost to German riflemen, and British scores rose steadily.

Hesketh-Prichard proved to be a very good teacher, enthusiastic and keen to be as 'close to the Line' as possible. He approved of snipers and observers working as a team, and could see the value of prolonged and accurate observation in strategic planning. He had also become a passionate advocate of the telescope sight, but only if properly regulated. Few marksmen of 1916 understood optics; consequently, many sights were in lamentable condition by the time men attended sniper courses arranged by individual armies:

> When the object glass becomes dirty or fogged with wet, snipers often unscrew it. Unless they put it back in its exact original position, they of course alter the shooting of the rifle hopelessly. They also unscrew the capstan heads, which are for the lateral regulation of the sighting. I have seen telescopic sights which were 30 inches out at 100 yards, or about 25 feet at 1,000 yards.[36]

Eventually, in the summer of 1916, though still unofficially, Hesketh-Prichard was finally allowed to form the First Army School of Sniping. A range was created near the village of Linghem, in Pas de Calais, and an establishment of merely six officers and men began work. Hesketh-Prichard and his second-in-command, Lieutenant George Gray of the Scottish Rifles (winner of the King's Prize in 1908 and the Caledonian Challenge Shield of 1913), not only trained many successful snipers but were able to ensure that their weapons were effective.

The school grew, employing Canadians among its staff, though Gray was moved when his unit was transferred elsewhere – typical of an army in which sniping had no great standing. On 24 November 1916, however, the First Army Sniping School was granted 'provisional establishment' status. Hesketh-Prichard, Gray and others had suggested many innovations, including the use of dummy heads as a surprisingly effective 'sniper locator' and the 'Gray Board' method of inserting an undetectable steel-lined loophole in a sandbag wall.

There was no consensus until, on 21 July 1917, a three-day Conference of Experts in Scouting, Sniping and Observation was held in Boulogne, attended by representatives from the sniping schools attached to each of the five armies. Hesketh-Prichard records those present as himself from First Army, Lt.-Col. Sclater of Second Army, Major Pemberthy of Third Army, an unnamed major of Fourth Army and Major Michie of Fifth Army. The chairman was Major Frederick Crum, representing GHQ France.

Hesketh-Prichard refers to Crum only in other circumstances, and Crum, in his diary/memoir gives only a brief nod to Hesketh-Prichard's experiences, but the Boulogne conference was highly successful.

Frederick Crum had been invalided home to serve as a sniper instructor in Aldershot from 16 October 1916. This gave him something that was denied to Prichard and his contemporaries in the army sniping schools: the backing of the Chief of Staff, General Sir William Robertson, and an opportunity to acquire optical equipment by dealing directly with the Ministry of Munitions. At the beginning of May 1917, Crum had submitted a report in which he complained that there was no recognised training organisation and that rifles, periscopes, sniperscopes and dark glasses were being wasted simply by issuing them to untrained men.[37]

Major Crum was clearly right; so right, indeed, that he had been appointed 'Scouting and Sniping Expert G.H.Q. France'. The Boulogne conference allowed Crum, whose trench experience was comparatively meagre, to collect data from men who had been serving continuously on the Western Front for two years or more, and the publication of the first officially sanctioned handbook, *Scouting, Sniping and Observation* (SS 195) was among the beneficial results. What others thought of Crum's pre-eminence is not recorded, but it has to be said that Hesketh-Prichard's *Sniping in France* is a much more exciting read than SS 195.

The needs of the sniper were not so easily answered by an assortment of sporting rifles brought over from Britain by individual officers. Though the guns themselves were often very useful – especially if they had telescope sights – few of them chambered the regulation .303 cartridge and the resupply of ammunition caused perpetual problems.

Attempts were made early in 1915 to provide Long Lee-Enfield rifles with optical sights, but these were in perpetually short supply. Consequently, many inventors were keen to provide the answer to the soldiers' prayers – weapons and equipment which could be used without exposing the operator to danger.

These included trench periscopes of various types, including some which were adapted to firearms, 'rifle batteries' which harked back to the ribaldequin (and the Billinghurst Requa battery gun of the American Civil War), and special 'trench rifles' fitted with mirror sights and remote-firing systems.

Scottish-born architect William Youlten (1862–1929) was granted British Patent 5,971/03 to protect 'Improvements in and relating to Instruments for Sighting Ordnance, Rifles and other Small Arms from Cover'. Accepted on 14 March 1904, this showed a periscope sight attached to the left side of the SMLE fore-end and an offset head to place an angled mirror directly behind the centreline of the back sight.

Youlten also developed the 'Rifle Butt Hyposcope', protected by British patents granted from 1900–1 until No. 26,487/14, 'Improvements in and relating to the Sighting of Rifles and other Small Arms', sought on 28 December 1914, was accepted on 25 November 1915. Made in small numbers by Periscope Hyposcopes of Westminster, London, the Hyposcope was being sold in 1915 for £7. 1. od. It comprised a prismatic sight hood attached to a pivoting lever on the right side of the butt – shown as a collar on the patent drawings – which aligns with the sights while allowing the firer to observe from below and behind the gun. The Hyposcope predictably proved to be too flimsy to survive war in the trenches. It was necessary to find a way of supporting the gun while it was being fired if accuracy was to be guaranteed.

Another solution was offered in the form of the Galilean sight, nothing more than two widely spaced magnifying lenses. The idea was old, dating back to the very first telescopes of the seventeenth century, but was still being used for target shooting in 1914 and held promise for military service. The British Army purchased large numbers of these sights, largely because there were too few telescope sights to be had: many had come from Germany prior to the First World War, and supplies had been cut off.

William Youlten had designed a Galilean sight of his own, the subject of British Patent 11,227/15 sought on 3 August 1915, but it was not among the most successful of its type. The Contract Books reveal purchases of 7,000 sights of a design suggested by Army Captain Henry Lattey, 4,000 Caldwell-patent Neill sights, 575 Martin sights and 80 BSA aperture sights. There was also an unknown quantity of Gibbs sights. Many other Galilean sights were purchased privately, and it is has been suggested that as many as 75,000 saw service.

Protected by British Patent 788/15 of 1915, granted to John Elsden Martin (1868–1951), 'Gun Maker' of 20 Royal Exchange Square, Glasgow, the Martin sight was adopted for service on 1 May 1915. It was swiftly followed by a sight designed by Captain Lattey and that of Thomas Caldwell, an optician of Donegal Place, Belfast, protected by British Patent 1,850/15 (but known initially as the 'Neill sight', renamed 'Barnett' in 1916 and also known as 'Ulster'), which were both accepted on 28 September 1915.[38] However, the *Lists of Changes* reveal that issue of the Lattey and Neil sights began first.

'Trench rifles' allowed the firer to remain safely hidden, usually by fitting an auxiliary butt/trigger unit and by using a periscope so that the standard open sights could be seen. This is a Beech-type example used by Anzac soldiers in Gallipoli to counter Turkish snipers. *Australian War Memorial*

Service soon revealed the weaknesses of all the Galilean sights. The lenses were widely separated, but not protected by an intervening tube or body; dust, reflected light and the shimmering of barrel-heat could all degrade the image. Magnification was small – generally no more than $2.5\times$ – and the field of view was very narrow, little more than six feet at a hundred yards. The sights were easily damaged when in place on the rifle, especially in the hurly-burly of the trenches, and as easily lost when detached. Though some marksmen made good practice with them, and modern trials have highlighted great improvements on open sights, Galilean sights were no match for the true telescope.

Trench rifles, tried by all the leading combatants during the First World War, had a lengthy pedigree. One of the earliest patents was granted to the prolific but little-known William Youlten. British Patent 15,273/1900, accepted on 28 November 1901 (and its US improvement, No. 694,904 of 4 March 1902), illustrated periscope-fitted rifles which could be placed over a trench parapet without exposing the firer's head to snipers. Special butt-fittings and, in one case, a cantilevered butt allowed mirrors or periscopes to be used.

A later Youlten patent, US 813,932 of 27 February 1906, even shows a periscope attached to the breech of a Maxim Gun part-concealed behind a bank. And a US patent granted on 25 February 1908 to Hermann Geis of Regensburg, Germany, No. 880,378 ('Device for Aiming from Cover'), relied on offset mirrors attached to a rifle back-sight (the drawings show a Gew. 88) to let a soldier fire over an obstacle.

When the First World War began in earnest, and German snipers began to take a steady toll of Allied ranks, interest in periscopes and trench rifles grew steadily. Though few inventors were able to convince the conservative (and therefore highly sceptical) military authorities of the merits of their weaponry, this was no barrier to creation. Consequently, several types of trench rifle were tested on both sides of no man's land.

The one great success seems to have been at Gallipoli, where Anzac units, suffering terribly at the hands of Turkish snipers, made several hundred such guns. Who designed them is still contested, but General Sir William Birdwood, the corps commander, recorded that 'Our complete moral superiority over the Turk is partly due to a very clever invention of a man named Beech, who produced a periscope rifle'. Lance-Corporal William Beech of the New South Wales 2nd Battalion, said to have been born in Shropshire, England, was a builder's foreman. His design included a skeletal wooden frame with an elevated cradle for the rifle, and two mirrors forming a rudimentary periscope.[39]

The Turkish defenders are said to have captured one or two of these periscope rifles, which were then copied. However, there does not seem to be any evidence to show that they existed in large numbers.

The Germans made several periscope rifles of their own, and, unsurprisingly British and French inventors were also lured by the prospect of success. John

This Gewehr 98 'trench rifle' has the customary auxiliary butt/trigger unit
and a periscope. It also has a detachable twenty-round magazine.
Courtesy of James D. Julia, auctioneers (www.jamesdjulia.com)

Edgar Chandler, a 'Dealer' of Winchcomb Street, Cheltenham, Gloucestershire
(British Patent 2,582/15, accepted on 9 March 1916), and Guibert Gérard,
'formerly of Antwerp, Belgium, but now of 134, Fenchurch Street, London, E.C.,
a Director of the Comptoir Commercial Anversois' (British Patent 13,031/15),
both relied on mirrors, prisms, auxiliary butts and Bowden cables to allow the
firer to hide behind the parapet.

The Gérard pattern was the simpler, using a periscope sight instead of a train
of mirrors, but both suffered the same insoluble problem: while they allowed a
soldier to fire in comparative safety, they were tactically unhandy. The field of
view was poor, and reloading meant removing the gun from its position with
loss of target and leaving the firer vulnerable to an unseen attack.

Edgar Duerr of Bowdon, Chester, obtained British Patents 103,382 and
109,236, accepted on 25 January and 6 September 1917 (though an application
for 103,382 had been made in March 1916). These show a periscope fixed onto
the SMLE butt, ahead of the shoulder-plate, and a remote-firing rod attached
to the trigger; the improvement also allowed the gun to be reloaded remotely.
Much simpler than most trench-rifle designs, which verged on a 'plumber's
nightmare' of tubes, bars and rods, the Duerr conversion was probably too late
to be exploited; by 1917, the threat posed by German snipers had been largely
eliminated by more conventional means, and the Allies, bolstered by the
imminent arrival of the AEF, were committing more to attack than defence.

US Patent 1,300,688, 'Apparatus for Actuating the Bolts of Rifles or Other
Weapons', was granted on 15 April 1919 to Alfred Bellard of Paris, though an
application had been made in the summer of 1916. Drawings show a periscope
and remote-loading mechanism attached to the Lebel and Lee-Enfield rifles.
Bellard's system, used by the French Army in some numbers, can be identified

in photographs by the distinctive horseshoe or 'U'-shape frame attached to the rifle-butt.

A radically different approach was taken by Ebenezer Morris, a Briton living in Sydney, Australia, who simply strapped a mirror-block to the Lee-Enfield so that firer could take aim while holding the rifle laterally, with his right hand cupping the butt and his left hand operating the trigger. The subject of US Patent 1,264,133, granted on 23 April 1918 but sought in July 1917, the sight-block did not offer enough protection to be successful.

Several types of periscope rifle were tested in the USA, the earliest design apparently that protected by US Patent 1,174,282, granted on 7 March 1916 to Leroy Richard of Coalinga, California (but sought in June 1915). The specification states specifically that:

> The object of ... is to provide a smallarm adapted for use in the trenches, in that it is capable of being laid over the rim of the trench, while its user, by means of an altiscope with which the arm is fitted, is able to aim without exposing himself.

Richard's design is one of the first in which the stock is adapted so that a mirror system can be incorporated in the wrist: in many ways, it is the precursor of the Guiberson type tested by the US Army in 1918.

George Cordell of Kearney, Missouri (US Patent 1,260,385 of 19 March 1918), proposed another frame-supported rifle, and the 'Sitascope' tested by the Infantry Board in 1917 (US Patent 1,371,964) was the work, according to the similar British specification, of James Love Cameron, 'Architect', and Laurence Emerson Yaggi, 'Attorney', of East Cleveland, Ohio. Granted in March 1921, the patent had been sought in December 1915; consequently, the Sitascope preceded the complicated design, with a padded shoulder rest, that had been patented by Jenks, Cameron & Yaggi (US No. 1,371,899 of 1921, sought in August 1917).

Trench rifles usually proved to be ineffective, particularly those that relied on simple unprotected mirrors. Though Anzac forces in Gallipoli used substantial numbers of the Beech design for counter-sniping, tests have since suggested that they were difficult to shoot accurately. A few of the more sophisticated designs had magnifying-lens periscopes, but snipers found even these guns notoriously difficult to control.

In 'Making the Springfield Shoot Around the Corner', published in *Arms and the Man* on 7 April 1917, Edward C. Crossman, one of very few journalists actually to shoot any of the trench rifles, suggested that:

> From the first of the programme of digging in and in, and still farther in on the west front of the present European battle lines, there came into popularity various devices for letting the Tommy pot Hans without affording Hans or his side-kicker Fritz, a chance for reciprocation.

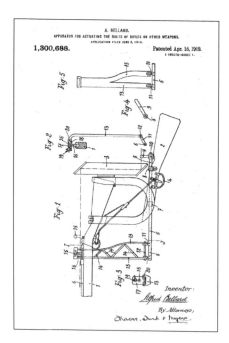

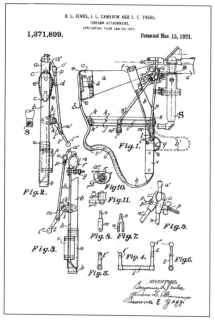

Those devices were known generally as periscopes, and were of fearful and wonderful construction. Tommy used to make them out of bits of boxes, and pocket mirrors and hitch them anywise to the Lee or to the Ross, while the plain ordinary periscopes for merely observing and not necessarily potting the gents in the other sets of trenches, grew and grew in popularity. Scores of different designs of rifle and plain trench periscopes are made commercially in England and brought and sent over to Tommy by his loving friends . . .

Queerly enough there is little kick to the new form of rifle, which has been submitted to the Ordnance Department and partly approved. The shape of the grip portion is not quite right, and several of us got nice raised and purple spots on the forehead from contact with this, but it is merely a matter of detail. Avoiding this, the rifle seemed to kick but little; less than the direct drive of the regular arm . . . The Guiberson rifle is of course not going to revolutionize the rules of warfare as one wild-eyed The arm in its present state is entirely practical, strong, and fool-proof, and the inventor claims to have received financial aid from the War Department to perfect it to its present stage.

None of the trench rifles tried in Britain, France, Germany and the USA had any lasting impact on sniping, which could only be done effectively by a marksman in direct contact with his rifle; and a brief obsession with luminous 'night sights', which proved to be useful only in overcast conditions and for a

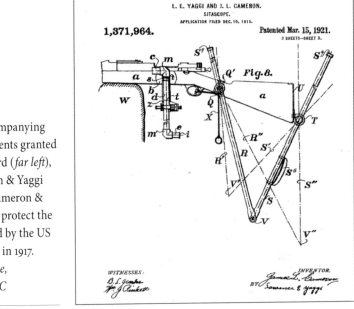

Drawings accompanying trench-rifle patents granted to Alfred Bellard (*far left*), Jenks, Cameron & Yaggi (*left*), and to Cameron & Yaggi (*right*) to protect the Sitascope tested by the US Infantry Board in 1917. *US Patent Office, Washington, DC*

short time after dusk, soon gave way to the use of optical sights. Anyone who had used a spotting telescope could see that the light-gathering abilities of a large-diameter ocular lens eclipsed the performance of the night sights.

The first telescope-fitted rifles reached the Western Front in 1915, but supplies were few and far between. This was partly due to the poverty of the British optical industry, which could not make items in large quantities, and partly due to a lack of direction. By 1 January 1919, 9,788 telescope-sighted rifles had been accepted and delivered into store. Substantial numbers of these – but by no means all of them – had been sent to the army. There were 4,830 rifles set-up by the Periscopic Prism Co. with sights of its own manufacture (latterly under government control); 3,196 rifles with sights made by Aldis Bros. of Birmingham, completed by a variety of contractors; and 908 Winchester sights, types A5 and B4, mounted by Whitehead Brothers.

The remaining 854 gun/sight combinations were provided from a variety of sources. These 'setters-up' included several leading gunmakers such as Evans, Holland & Holland, Purdey, Rigby, Jeffery and Gibbs. Many of the sights on these guns, however, often prove to have been of German or Austrian origin and had simply been in store when the fighting began. There were a few Norwegian sights, made by Fidjelands Siktekikkert of Oslo, and others made by opticians Watts & Co., but supply rarely matched demand once the benefits of sniping had been recognised.

The most popular sights were the Periscopic Prism and Aldis designs, though none was outstanding. They suffered the twin problems of low magnification,

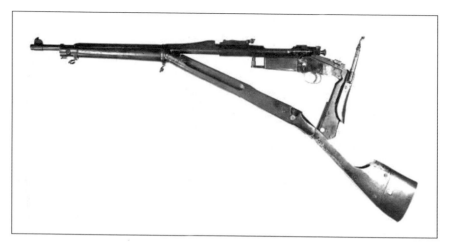

One of the Guiberson trench rifles tested by the US Army during the First World War. Less complicated than some rival designs, it consists largely of a specially articulated stock with an integral periscope sight. *From an original Springfield Armory photograph*

usually only 2.5×, and poor light-gathering capabilities as the diameter of the objective lens was customarily only 19–20 mm. The Winchester 5× A5 and 4× B4 were little more than target-shooting sights, with no range-adjusting drum on the tube (thimble-type lateral and vertical adjustments were incorporated in the rear mount), and usually had slender crosshair reticles instead of the picket-and-crosshair of the Periscopic Prism and Aldis types.

In addition, Winchester sights tended to slide forward when the gun was fired, owing to the effects of recoil and an ineffective anchoring system, and had to be returned to position before each shot. Even though the .303 cartridge was not the most powerful of military rounds, it still generated more than enough recoil to loosen a sight in its mount if there was even the smallest insecurity. Maintaining zero, so essential to efficient sniping, was a very real problem with Winchester-sighted rifles.

Excepting some of the 16-inch-long Winchesters, which were in centreline mounts (sometimes placing the ocular lens too close to the firer's eye), the telescopes were invariably mounted to the left of the bolt to allow charger-loading to be undertaken. This problem was shared with almost all bolt-action military rifles, even though the SMLE had a detachable magazine. Centreline mountings were preferred by the Germans, but had the effect of raising the sniper's head and, therefore, making his presence more conspicuous than a low side-mounted sight would do. However, offset mounts had a zeroing problem: the sight line and the bore met only at a single point, whereas the centreline sight was always in alignment with the bore axis. Opinions differed. Hesketh-Prichard was not convinced:

One of the greatest difficulties we had . . . was the fact that the telescopic sights were set, not on top, but at the left-hand side of the rifle. This caused all kinds of errors. The set-off, of course, affected the shooting of the rifle, and had to be allowed for, and the clumsy position of the sight was very apt to cause men to cant their rifles. Worse than all, perhaps, in trench warfare was the fact that with the Government pattern of telescopic sight . . . it was impossible to see through the loopholes of the steel plates which were issued, as these loopholes were naturally narrow . . .

Why the telescopic sights were set on the sides of the rifles was never . . . satisfactorily explained, but it was always said that it was done so that rapid fire should be possible . . . then surely whoever was responsible can have had no knowledge whatever of the use of telescopic sights.[40]

Herbert McBride took a very different view: that the offset mount of his Warner & Swasey sight was an advantage, as it not only allowed his Ross rifle to be reloaded from a charger but also gave ready access to the standard open sights. However, McBride, a machine gunner, usually sniped in the open and in locations where the restraints of narrow loopholes were uncommon.

The 3 × Periscopic Prism sights on the SMLE were usually carried on a rail attached to the left side of the body with screws and pins – customarily four of the former and two of the latter – and sometimes also solder. A typical sight

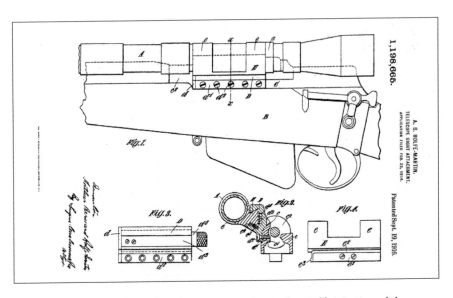

Drawings taken from the US patent granted to Arthur Rolfe-Martin and the Periscopic Prism Co. to protect the telescope-sight mounting – low, offset to the left of the bore. *US Patent Office, Washington, DC*

was about 12 in. long, with a 9.5° field of view and an eye-relief of about 3 in. The range-drum was graduated from 100 to 600 yards in 100-yard increments.

The mounting usually associated with the SMLE was protected by British Patent 3,027/15, 'Improvements in or relating to the Attachment of Telescopic Rifle Sights', granted on 24 February 1916 to 'The Periscopic Prism Company, Limited, Manufacturing Opticians, and Arthur Bernard Rolfe-Martin, Engineer' though sought exactly a year earlier.[41] US Patent 1,198,665 of 19 September 1916 was comparable. Offset to the left side of the rifle body, the mount consisted of a block dovetailed longitudinally, attached to the body by several screws (and also often solder), onto which the telescope could be slid until locked in place by the stop at the front end of the groove and a spring catch at the rear. The sight could be focused by moving a small lever protruding from the sight-body along its helical slot. The biggest weakness of the Periscopic Prism sights proved to be the stop-screws at the front of the mounting block, which often sheared under the shock of recoil.

The adoption of the SMLE had been greeted by the gun trade and many self-appointed 'experts' with howls of protest. The rifle was derided as an 'abomination' on the grounds that it would win no prizes on the target range; the value of combat-worthiness was evidently of no importance.

Trials showed the accuracy of the SMLE to be scarcely inferior to the long Lee-Enfield, but the bench-mark had become the best sporting guns, Goaded by their critics, but perhaps with a doubter or two within their own ranks, the War Office began trials with Mauser-type rifles. These included a hybrid 'Mauser-Springfield', and an inclined-bolt design submitted by George Norman of BSA which could trace its ancestry from the Godsal, Thornycroft and Gamwell rifles developed to provide a long barrel in an ultra-compact action. Most of the trials rifles chambered more powerful cartridges than even the Mk VI .303 (including the .276 Eley sporting round), and it was acknowledged that these were too robust for the Lee action.

Experimentation with .276 cartridges and adapted Mauser-type rifles continued until the P/13 was issued for field trials. These showed that the rifle would be acceptable once a few small changes had been made, but also that the cartridge would need much more development before adoption. Night-firing trials had shown that the flash was far too great, and the nitrocellulose propellant was unreliable.

The First World War began before the worst problems could be resolved. Unwilling to disrupt production of the established service cartridge, which was chambered by the Vickers machine guns in addition to the SMLEs, the British compromised, sensibly, by adapting the P/13 for the .303 round to create the P/14. The new rifle was introduced to service by *Lists of Changes* Paragraph 17,798 ('§ 17798') of 21 June 1916: 'Rifles, magazine, .303-inch, Pattern 1914 Mark Ie, Ir, Iw [and bayonet]'. (As weapons from different manufacturers were not truly interchangeable, appropriate suffixes were added to the nomenclature.)

Far too little additional production capacity existed in Britain in 1915, and so orders for the new rifles were placed in the USA The first deliveries from the Eddystone, Remington and Winchester factories were made in May 1916, though the guns were rejected by the inspectorate. A Mark I* was substituted in January 1917 (§ 18151), Variants distinguished with an 'F' suffix had fine-adjustment sights, and some Winchester-made guns, with 'T' suffixes, were issued for sniping with a 2.5 × Aldis telescope sight.[42]

Accelerating production of SMLEs persuaded the British to terminate the P/14 contracts in 1916, after 1,235,298 rifles had been delivered. However, the US authorities subsequently used the facilities to make the outwardly identical M1917 Enfield described later in this chapter. The M1917, lacked the long-range sights fitted to the P/14, but this could only be seen from the left side.

The P/14 could be identified by its distinctive shape, with a prominent swell in the underside of the stock to accommodate the magazine; the back sight was mounted within protective wings on top of the receiver; and the bolt handle was cranked rearward. A single longitudinal flute in the fore-end distinguished it from the experimental .276 P/13, which had four small diagonal finger grips.

Few men who had regularly fired both guns, excepting snipers, would readily exchange the wonderfully smooth bolt stroke of the Lee-Enfield for the theoretical strength of the P/14.[43] However, the performance of the .303 cartridge was still being widely questioned, particularly when the Germans began to protect machine guns and snipers with iron shields. Comparative trials were undertaken by the Machine Gun School with .333 Jeffery, .450 Express, .475 and .600 Express sporting-rifle cartridges, but results were uninspiring. The Jeffery cartridge, loaded with a bullet patterned on the then-experimental .303 Mk VII armour-piercing type, was considered to have some promise. However, a P/13 action rechambered for a .470 round did not perform as well as the .450 sporting cartridge had done and the experiment was abandoned.

A suspicion lingered that something better than the service .303 was needed. In February 1917, therefore, the Director-General of Munitions Design suggested to the Director-General of Artillery that development of a .276 armour-piercing cartridge could allow the 600–700 P/13 rifles that remained in store to be issued to snipers. Several types of bullet weighing between 128 and 151 grains were tried in the standard case, obtaining velocities of 2,845–3,000 ft/sec. However, even the perfected RD570 bullet, which had a tungsten-carbide core in a brass envelope, proved capable of penetrating only 9 mm of armour plate at 200 yards. This was deemed to be too little to justify continuing work, and so the .276 sniper cartridge was abandoned late in 1917.

The tactical situation had changed with the advent of tanks, generally to the detriment of sniping, and the introduction of a .303 armour-piercing round was deemed to be preferable.[44] However, though snipers were rarely able to obtain the perfected round introduced in 1918, boxes of the earlier and less successful 'Cartridges, S.A., Ball .303-Inch, VII.P' (dated 1917) have been seen marked specially selected for snipers.

Attempts had also been made to introduce 'Rifle Batteries' to service. A very old idea, harking back to the ribaldequin of medieval times, these were usually in the form of several SMLEs fixed in a frame so that they could be fired simply by pulling a lanyard. One such device, 'Battery Rifle, Mark I', was even introduced formally by L.o.C. § 17750 of 1916.

To some extent left behind by nineteenth-century industrial progress, the French had often led where ideas were concerned. The first successful internal-combustion engine had French origins, and the rise of the automobile and the aeroplane prior to 1914 was also evident more in France than in Britain. The influence of the Académie des Sciences permeated even the military mind, and so the French Army had adopted the surprisingly portable Hotchkiss machine guns even as the British Army was embroiled in the South African War.

Experiments with optical sights had begun in the 1890s, and a 2× design credited to the Arsenal de Puteaux ('A.Px.') was adopted for limited service in 1896. This sight had been improved in several respects to become the Mle 1907, and then, during the war, to become the 3×21 Mle 07/15. The sights had brass bodies, a focusing ring ahead of the ocular lens, and fine crosshairs. Eventually, the Lunette de Fusil Mle 1916 was approved. This was essentially a mass-produced version of the Mle 07/15 with a steel tube, a crosshair reticle, and an elevation drum adjustable to 800 metres.

The sights were issued to *Tirailleurs d'Élite*, marksmen, but the scale and distribution remains uncertain. One source[45] states that a sniper rifle was available to each section of an infantry battalion, to be issued when and where needed to those who held relevant qualifications. The French made good use of them, as the Germans involved at Verdun and elsewhere would testify. Yet the potentially fascinating story of French sniping has yet to be explored in satisfactory detail in English.

The sights were usually issued with selected Mle 86/93 Lebel rifles, which were sturdy and acceptably accurate. A few were tried with the Berthier rifles derived from the carbines, mousquetons and colonial-service rifles that had been introduced to service since 1890.[46] However, wartime Berthiers, Mle 07/15 and Mle 16, proved to be structurally weak: poor substitutes for the venerable Lebel. The massive body of the Mle 86/93 was much more rigid than the Berthier, and so had far greater resistance to torsional forces. Consequently, not only was the Mle 86/93 retained for many years as the standard sniper rifle, it was also the only French rifle deemed to be strong enough to fire grenades.

Inventiveness was not generally matched by productivity, and it is likely that the optical sights were not made in Puteaux but by optical specialists. There were many who were making binoculars and opera glasses, and could easily produce a few telescope sights. Homogeneity was lacking, but even the Germans acquired the output of many individual manufacturers during the war.

French sights, particularly the Mle 1907, sold commercially in small numbers (another sign of the involvement of individual opticians), and have even been encountered on Lee-Enfields. Whether these were acquired privately prior to 1914 or during the First World War, when the British authorities purchased almost any sight that could be found, has yet to be determined.

In April 1917, the USA belatedly entered the conflict. The standard rifle remained the .30-calibre M1903 Springfield. Derived from the experimental M1901, reduced in length to approximate with the British SMLE, the M1903 had been adopted to replace the M1892 Krag-Jørgensen in June 1903. The original rod bayonet, carried in a channel beneath the fore-end, predictably proved to be weak and was abandoned in April 1905. A 2,400-yard back sight was also approved at this time.

However, the advent in October 1906 of a pointed or 'spitzer' bullet in a new M1906 cartridge – better known as the '.30–06' – caused the sight graduations to be changed to 2,850 yards. A solid tubular back-sight mount replaced the skeletal pattern; a flute in the top surface of the barrel guard and a recoil bolt through the stock were added in 1910.

When the USA entered the First World War, the ordnance authorities realised that small arms were in perilously short supply: the inventory stood at merely 760,000 rifles, including a substantial quantity of obsolete Krag-Jørgensens. Work on the M1903 continued throughout the First World War, until the last rifle to be assembled in Rock Island Arsenal left the factory in June 1919.[47] A few thousand sniping-rifle derivatives, 'Marksman's Rifles', were issued with 6× M1908 (rarely seen in combat) or 5.2× M1913 Warner & Swasey telescope sights. However, the optical performance of these proved to be very poor. Herbert McBride, in *A Rifleman Goes to War*, is one of the few experienced commentators to have a good word for them.

Accelerating production in Springfield Armory and Rock Island Arsenal was only part of the answer. Though many Ross rifles were acquired for training and 280,000 Mosin-Nagants were purchased when the 1917 October Revolution prevented US manufacturers delivering any more guns to Russia, the ideal solution lay in the cancellation of contracts placed for the British .303 P/1914 rifle. This had left three new, large and efficient manufacturing plants lying idle, and so, in May 1917, the US Ordnance Department adapted the P/14 for the .30–06 Springfield cartridge.

The changes presented comparatively few problems, as the rifle had originally been designed to handle a 7 mm/.276 high-velocity rimless cartridge and had ample reserves of strength. By 12 July 1917, a prototype (externally all but indistinguishable from its British predecessor) had been successfully

A soldier of the American Expeditionary Force, with a British-style steel helmet, takes aim through a loophole. Note that he is firing an SMLE and not the M1903 Springfield. As the latter was in short supply when American units first saw action, British weapons were often used in sectors where the two armies fought alongside each other. AEF snipers, however, usually had the M1903 rifle/M1913 sight combination. *Author's collection*

tested and was immediately ordered into series production. By 9 November 1918, nearly 2.2 million M1917 rifles had been accepted, compared with just 313,000 M1903 Springfields.

US Army soldiers generally disliked the Enfield, which was clumsier than the Springfield and cocked, unfamiliarly, on the closing stroke of the bolt; however, the rifle was strong and very reliable, and its rifling unexpectedly resisted wear better than the Springfield pattern. The M1917 had been made in such large numbers that it nearly supplanted the M1903 as the official infantry rifle of the US Army at the end of the First World War. However, the authorities decided that it was better to keep production within government-owned facilities than rely on private contractors, and the Springfield prevailed.[48]

The involvement of the American Expeditionary Force in the First World War came at a time when the impetus of German incursions into northern France was largely spent. Among the Americans were truly excellent marksmen, and many who would have made outstanding snipers. However, the Germans had 'dug deep', forcing the Allies to attack if progress was to be made. Sniping was generally reckoned to be of value more in defence than attack, but there are many testimonies to the effectiveness of long-range shooting:

> The outpost was on the forward slope of the knoll where they had good fields of fire. The small group of Marines took the enemy formations under fire with their Springfield rifles. They picked off the enemy scouts and then concentrated on the Germans in the first assault waves. Their accurate fire first surprised, then confused, and finally halted the advance.[49]

> On 4 June 1918, in the first major battle for Marines during the war, the Germans advanced through a wheat field in front of Belleau Wood. Marine riflemen stopped the Germans as far out as 800 yards with aimed rifle fire. On that day, the skill of the Marines and accuracy of the M1903 Springfield became legendary.[50]

Hesketh-Prichard, writing in 1920,[51] suggested that the 'sniping war' had had four phases: the first, in which the Germans held the upper hand; a largely British and Canadian riposte, which gained momentum throughout 1916; the dominance of Allied sniping; and then the wholesale reduction of targets owing to the ever increasing fortress-like construction of the German lines.

Excepting Herbert McBride, whose 'sniper kills' were achieved with the Canadians before the USA became involved in the war, only two names are widely lauded – those of Alvin York and Herman Davis.

On 8 October 1918, little more than a month before the end of the First World War, seventeen men of 328th Infantry Regiment, 82nd Infantry Division, commanded by Sergeant Bernard Early, were sent to capture a German machine-gun position near Chatel-Chéhéry in the Meuse–Argonne theatre. Pinned down by hostile fire, the attack stalled until Corporal York elected to go forward alone. Caught in the open, trading shot for shot with his M1917 Enfield rifle, York not only silenced the machine gun but caused havoc among its protectors. Drawing his M1911 pistol, he then shot the six soldiers who attacked him with fixed bayonets in the belief that he had run out of ammunition. The Germans, 132 of them, then surrendered.

York's actions won him immediate promotion and the Distinguished Service Cross, subsequently upgraded to the Medal of Honor. Objections were made by Early and another of the survivors, claiming that York's version of events was not entirely truthful, but battlefield archaeology has largely supported the 1918 testimony.[52]

The story of Herman Davis was very different, as his sixty 'sniper kills', achieved with a standard open-sighted M1903 Springfield, were spread across time. Davis entered the US Army on 4 March 1918, then sailed for France on

Tennessee-born Alvin Cullum York (1887–1964), the subject of a 1941 film staring Gary Cooper, is usually credited with twenty-eight kills: an unremarkable figure, but only a part of his story. York was a man of contrasts. Previously violent and too fond of alcohol, he had joined a church with pacifist leanings in 1915. Once a self-declared conscientious objector, York was drafted in October 1917.

Herman Davis came from an impoverished rural background. Born in Manila, Arkansas, on 3 January 1888, Davis spent his formative years shooting ducks in woodland, and had become a hunting guide while still in his teens. Unlike York, who not only employed a publicist but also put his name to an untrustworthy autobiography, Davis's exploits achieved only local recognition.

15 June as part of the 113th Infantry Regiment, 29th Infantry Division. He was subsequently credited with eliminating a machine-gun position at Molleville Farm, having crawled to within fifty yards, and killing five German soldiers siting a machine gun at a range estimated by witnesses to be a thousand yards.

The recipient of awards including the Distinguished Service Cross and the Croix de Guerre, Davis was discharged on 29 May 1919. Returning to the USA, he went back to his hunter-guide life. But his health failed rapidly – perhaps due to the insidious effects of gas poisoning in France – and Davis died on 5 January 1923.

Neither York nor Davis used optical sights, relying simply on marksmanship honed by years of hunting. The failure of the M1913 Warner & Swasey sight in combat persuaded Frankford Arsenal to develop a conventional alternative simply by combining the best features of the 3.2× Goerz Certar that had performed best in pre-1914 trials with elements of the 5× Winchester A5 that

M1903 Springfield rifles of the First World War period. The upper gun has an M1913 Warner & Swasey sight and a cumbersome twenty-round magazine. The lower gun is one of the few to survive with a Winder 'sighting tube' built into the fore-end above the barrel. This was intended to allow sight to be taken through the smallest loophole, but suffered from a meagre field of view and was deemed unacceptable after US Army tests.
Courtesy of James D. Julia, auctioneers (www.jamesdjulia.com)

had been adopted by the US Marine Corps. Only three 2.6 × 14 mm Frankford sights were made in 1917, tested on suitably modified Springfield M1903 rifles, and provisionally approved for service issue as the 'Telescopic Musket Sight, Model of 1918'.

The new sight was to accompany the 'Telescopic-Sight Rifle, Caliber .30 Model of 1918', a half-stocked variant of the M1917 Enfield with the 'ears' or 'wings' on the receiver bridge ground away to accommodate a suitable mount. However, there is no evidence to show that M1918 rifles were made in quantity, and there may only ever have been a solitary prototype.

Tests undertaken in 1919 showed that the M1918 sights were far too fragile to withstand military service, and also that incorporating the windage and elevation adjustments in sporting-type mounts was mistaken. As there was deemed to be no role for the sniper in the post-war army, the sniper-rifle project was promptly abandoned.

Fighting on the Austro-Italian front, long and bloody, with well in excess of two million casualties on each side, rarely gains much attention in English-language sources. This is partly due to a preoccupation with affairs of the Western Front and, especially in Australasia, the disastrous Gallipoli campaign. Set-piece battles such as Caporetto (a disaster for the Italians) and prolonged fighting on the Isonzo are sometimes mentioned in passing, but little heed has been paid to the use of marksmen.

Most of the fighting took place in the north-eastern part of Italy, in the foothills of the Alps. The Austrians in the north were generally favoured with the high ground, and extensive use was made of tunnelling and mining to gain advantage. Mountain howitzers capable of lobbing shells over high ground were useful, but icy conditions were a very real handicap to troop movements.

Campaigns ground to a halt in the snows of winter, while Italians and Austro-Hungarians shot at each other from ridge to ridge or tried to make sporadic attacks through the mountain passes. The result was often stalemate, but stalemate that cost many men their lives.

The fighting was undertaken largely with conventional weapons, which to the Italians meant the Mo. 91 infantry rifle, the Mannlicher-Carcano; and to the Austro-Hungarians, the M95 straight-pull Mannlicher. Both guns were long and clumsy, ill-suited to mountain fighting, but there is no evidence to suggest that carbine or short rifle derivatives were widely preferred.

An Austro-Hungarian sniper takes aim with his M95 rifle while his colleague acts as spotter. The way in which all three of men visible in the picture are exposed to view suggests a training exercise somewhere in the Austrian Alps. The sight is most probably a 3 × Oigee or Goerz type, mounted too far to the rear of the rifle to be entirely effective. Though the ocular lens is fitted with a rubber eyecup, the sniper may still get struck in the eye socket when he fires. *Author's collection*

The Italians developed the twin-barrel 9 mm Villar Perosa, a machine gun in miniature, chambered for pistol ammunition, seeking a way of delivering higher volumes of fire without the penalties of excessive weight. The Austro-Hungarians then countered with a machine-pistol adaptation of the 9 mm Repetierpistole M11 ('Steyr-Hahn'), but neither of these innovations could shoot accurately beyond 200 yards.

The Austrians, in particular those of the Tirol, had a long tradition of marksmanship, as did some of the Italians who lived on the other side of the border that divided the countries. Skilled at tracking and comfortable in sub-zero conditions, they had the potential to be excellent snipers. However, there is no evidence that the Italians, even the Alpini units, took the idea very seriously.

This was probably at least partly due to the crudity of the Mannlicher-Carcano rifle, which was not regarded as accurate.[53] A few photographs show rifles fitted with optical sights offset to the left of the action; a handful of pre-1918 Mo. 91 rifles survive with mounts for this type of sight and some have clearly seen heavy use.[54]

The Austro-Hungarians faced a different problem. The M95 Mannlicher fired a cartridge that was ballistically poor, and service rifles, though well made and slickly operated when well-maintained and properly lubricated, were incapable of sustaining rapid fire. More than one test has shown that it is virtually impossible to fire ten rounds through an M95 before it jams. The same seems to have been true of use in freezing temperatures, when the bolt mechanism tended to lock.

The accuracy of the 8 mm-calibre Mannlicher rifles was regarded as acceptable, but not outstanding. A few guns were fitted with 3 × Goerz or Oigee 'Luxor' telescope sights, sometimes mounted centrally (but very high) and at other times offset to the left to allow the magazine to be loaded. However, it is assumed that they did not perform well enough in sub-zero conditions for widespread issue to be attempted. A few M95 short rifles (*Stutzen*) have also been found with optical sights, and it is assumed that they were intended specifically for use in the alpine highlands where cumbersome, full-length infantry rifles would have been a handicap.

When the authorities sought a sniper rifle, after exhausting supplies of the excellent turn-bolt Mannlicher and Mannlicher-Schönauer sporting guns, they turned to an unexpected but invaluable resource: thousands of Mausers which had been awaiting delivery to South America when the First World War had intervened.

Three essentially similar guns were issued as the 'M. 14' – the residue of an order placed in 1912 by Chile for 37,500 rifles and 5,600 carbines; about 5,000 Colombian Mo. 12 rifles left over from a March 1913 contract for 10,500; and 66,979 Mexican Mo. 12 guns, part of an order for 100,000 which had been kept in store since the summer of 1912. These rifles can be distinguished by the national arms above the chambers, and were all chambered for the 7×57 mm cartridge.

The 7 mm round has always had a good all-round reputation, allying flat trajectory to excellent accuracy. There were more than enough Mausers to spare a few for use as sniper's rifles. These were given sporting-type telescope sights, in two-piece mounts, generally offset to the left side of the breech to allow the magazine to be charger-loaded. However, too few survivors have been reported to be certain that all the guns had been treated in the same way.

Notes to Chapter Three

1. Viktor Amadeus, Herzog von Ratibor, Fürst von Corvey, Prinz zu Hohenlohe-Schillingfürst (1847–1923), was a high-ranking aristocrat and politician in Silesia. He had fought in the Franco-Prussian War, with the Potsdamer Leib-Garde-Husaren-Regiment and held the honorary rank of major-general in the Prussian Army. A proficient hunter, he was convinced of the value of marksmen and optical sights in military service.
2. It even became known in some circles as the *Judenflinte*, 'Jewish Rifle', after an allegation that it was all part of a plot to undermine the morale of the German Army (many of the principal manufacturers of guns, ammunition and propellant were Jewish-owned).
3. Hesketh Vernon Hesketh-Prichard, *Sniping in France, With Notes on the Scientific Training of Scouts, Observers and Snipers* (1920), p. 31.
4. Quoted in several English-language publications, including Daniel W. Kent, *German 7.9 mm – Military Ammunition 1888–1945* (Ann Arbor, 1973), p. 9.
5. The Adrian helmet was designed to resist a fragment weighing 11.35 gm (175 grains) travelling at no more than 120 m/sec (394 ft/sec); the first British Brodie-type helmets had to pass a test against a 171-grain (11.09 gm) shrapnel ball at 750 ft/sec (229 m/sec). The British requirement was reduced to 700 ft/sec when manganese steel was allowed to be used in November 1915, but the helmets still performed noticeably better than the French equivalents.
6. Viola Meynell, *Julian Grenfell . . . Reprinted from The Dublin Review* (London, 1917), p. 10.
7. Ibid., p. 12.
8. Martin Pegler, *Sniping in the Great War* (London, 2008), contains extracts from interviews with survivors.
9. It is sometimes claimed that he subsequently transferred to the Royal Flying Corps, but service records seem unclear: there were at least two officers named Walter Henry Fox, one serving in the South Staffordshire Regiment (confirmed in the rank of second lieutenant, 16 March 1915).
10. 'Sniper Sandy' lyrics: Pegler, *Sniping in the Great War*, p. 196. Written by Lieutenant E. A. Macintosh, killed during the Battle of Cambrai on 21 November 1917, it was sung to a popular 'tongue-twister' music-hall tune, 'Sister Susie's Sewing Shirts for Soldiers'.
11. Ibid., p. 59.
12. Among them was a 'Magazine Pattern Sporting Carbine' on the 'Lee Speed' action. Offered only in .303, the Pattern No. 1 'for Officers' use' was 40½ in. long, weighed 7 lb 6 oz empty, and had a 21 in. barrel. The detachable box magazine held five rounds; a leaf pattern back sight on the barrel was graduated to 2,000 yards. The military

style butt had the typical British pistol grip, and chequering on the butt and fore-end distinguished the No. 1 (which accepted a P/88 sword bayonet) from the other BSA sporters and many similar service-issue carbines.

13. 'Lock time' lasts from the point at which pressure on the trigger activates the sear to release the striker and the instant that the striker reaches and strikes the primer of a chambered round. The period is generally very short, but can allow aim to shift if the delay is too great. Figures of 8–10 milliseconds for some pre-1914 rifles compare very unfavourably with the 2–3 milliseconds expected of many modern bolt-action designs.

14. SMLE rifles issued after 1926 were reclassified as 'No. 1 Mk III' or 'No. 1 Mk III*'. All other patterns were declared obsolete and discarded.

15. Accidents involving 1905-type Ross actions had also been reported. These may have been due to faulty engagement of the locking lugs, or to the trigger releasing the striker before the breech was properly locked.

16. Scottish records confirm that Christie had been born in April 1867 to Joseph Christie and Jane Murdoch, who had married in Fifeshire in June 1865. It is believed that Jane Christie died and that Joseph re-married in Scotland before departing for Canada in 1885 or 1886 (sources disagree). Canadian censuses name his wife as 'Helen'. Eventually, the army realised that James Christie was older than he had claimed, and medical records leading to a judgment of 'Pemanently unfit for General Service' (27 July 1918) corrected the birth-date.

17. 'Treaty Indians' belong to the bands that signed any of eleven treaties agreed by the Canadian government and Native American (now 'First') nations from 1871 until 1922. 'Non-treaty Indians' existed outside these laws, and their status was not legally recognised.

18. It is usually assumed that the injury was sustained on the battlefield, from a bullet or shell fragment, but Ballendine's medical records reveal that 'Laxity of Crucial & Lateral Ligaments of right knee' had resulted from a kick received during an impromptu football match and then aggravated in the trenches by constantly twisting and turning while crawling into position.

19. Herbert W. McBride, *A Rifleman Went To War* (Endeavour Press edition, 2016), p. 241.

20. Johnson Paudash had married Florence Emma Johnson on 18 April 1914. Eight children of this union were born after his return from Europe at the end of the First World War. Paudash died in Lindsay, Ontario, on 26 October 1959.

21. Returning to Parry Island in 1919, and to the life of poverty and disillusionment that frustrated so many Native American soldiers, Pegahmagabow engaged in local politics. As chief of the Parry Island Band, he clashed repeatedly with the Indian Agent and those of his own nation who saw him as a troublemaker. Twice high chief of the Native Indian Government after the Second World War, Francis Pegahmagabow died on 5 August 1952. He was survived by a wife and six children.

22. Canadian military records suggest that Comego had been born on 1 January 1866, but the 1871 census gives his age as '7', the 1891 census as '30', and the 1911 census (perhaps the most reliable) notes that he was born in December 1866. The records of his marriage on 4 October 1906 to Eliza Jane Crowe gives his age as 38 ('born about 1868'). Comego claimed in his attestation to have been born in 1872, probably so that he would not be deemed too old to fight. He had been married at least twice, but was listed as a widower in 1911. His trade is given as 'railway section man' in 1891 and 'laborer, trapper' in 1911.

23. *A Rifleman Went To War* (2016 printing), p. 96.

24. According to the 1891 Canadian census, Joseph Patrick Riel was the third surviving son of Francis Xavier Riel and his wife Rose. [Joseph Guillaume] Francis Xavier may

have been born on 8 March 1845 to Maghin [sic] Riel and Géneviève Bejage, but the baptism records are very difficult to read. The paternity of Louis Riel (1844–85) is easier to establish, however, as an unbroken descent from 1704 is given by the *Genealogical Dictionary of Canadian Families, 1608–1890*. His father was Louis Riel *dit de l'Irlande* (1817–64, 'known as of Ireland' as the family came to Canada from Limerick) and his grandfather was Jean-Baptiste Riel (1785–1868). For Joseph Patrick and Louis David Riel to have been nephew and uncle, another of Louis Riel's sons would have to have been Joseph Patrick's father. Genealogical sources suggest that the only possible candidate, Charles Meunier Riel, had died two years before Joseph Patrick was born.

25. McBride's memoirs state that while he had wanted to fight in South Africa, and had been involved in recruiting for the Strathcona Horse, his offer to serve was rejected on the grounds that he was not a British citizen There are other questions about his life to answer: one genealogical source, unverifiable, says he was married to 'Marie Augusta Parry', though his death certificate gives his status as 'single'.

26. John Hamilton, *Gallipoli Sniper: The Remarkable Life of Billy Sing*, p. xii, gives the figures as 61,829 dead and 157,156 wounded, gassed, or prisoners-of-war of 324,000 Australians who had served overseas.

27. Hamilton, *Gallipoli Sniper*, pp. 102–3, quoting E. Ashmead-Bartlett, *The Uncensored Dardanelles* (London, 1926).

28. It has been claimed that Billy Sing used a Galilean sight (usually identified as a Lattey). However, this is not supported by eye-witness testimony even though others undoubtedly saw service in Gallipoli. However, Sing may have used a Martin aperture back sight at some point in his career, replacing the long-range back sight on the rear left side of the SMLE body; Martin also promoted a Galilean sight, which could explain the apparent confusion.

29. Billy Sing's personal life has been the subject of much debate. It has been suggested that, as his medical records reveal venereal disease, his marriage did not last. Birth records reveal that Gladys Elizabeth had two children by different men, Mary Millar Stewart (1919–2005), born in Edinburgh, 'father unknown', and Theo Malmborg [Stewart], born in Australia in 1924. Mary was born in September 1919, and so could not have been Billy Sing's daughter as sometimes claimed. He had left for France, never to return, in August of the previous year. Speculation that the father's surname was Millar may prove to be worthwhile.

30. It is possible that Alfred Hugh Dillon, born in Hastings in 1897, was a son of the well-known Otago politician Alfred Dillon (who came from 'rural stock'). Service records reveal that in 1916 he gave his trade as 'carpenter' and as next-of-kin as his mother, 'Mrs. H. Dillon'; the elder Alfred had died in 1915, which may not be entirely coincidental. However, much more research is needed before this can be confirmed.

31. William Neville Methven was born in St Ninians, Stirlingshire, Scotland, in 1884, to James Methven and Julia Allin. The 1901 Scottish census lists him as an 'insurance clerk', but he had emigrated to southern Africa by 1911. On 1 May 1917, he was promoted to 'Temporary Lieutenant (Supernumerary to Establishment)', and was awarded the Military Cross for gallant leadership on 1 June 1917. Returning to southern Africa after the war, he married Katharine 'Kate' Halsted and died in Bulawayo on 5 April 1974.

32. Pegler, *Sniping in the Great War*, pp. 66–7.

33. Langford Lloyd, a Past Master of the Royal Irish Rifles Lodge (No. 2,312), was co-proposer of Arduis Fidelis Lodge (No. 3,432) to 'cement and further the happy means of conciliating friendship among those who served in all branches of the Royal Army Medical Corps' with Captain A. R. Owst, FRCS. Lloyd was then a

captain and adjutant of the London School of Instruction.

34. Crum's father, William Graham Crum (1836–1928), an industrial chemist, owned one of the largest calico-printing factories in Britain. According to the 1891 Scottish census, it employed '802 men, 175 women, 127 boys and 66 girls'. William Crum, conscious of the unhealthy atmosphere in Thornliebank, Glasgow, moved his family to rural Cheshire. All of his children were born in Tarpley, and the family had no fewer than ten servants in 1891.

35. Hesketh-Prichard, *Sniping in France*, p. 21.

36. Ibid., p. 58.

37. Major F. M. Crum, *Memoirs of a Rifleman Scout* (2014 edition), p. 263.

38. The 'Sights, Optical (Lattey), Mark I' and 'Sights, Optical (Neill), Mark I', approved for issue on 28 September 1915, were introduced in L.o.C. §§ 17556 and 17557 of 1 December 1915. Oddly, the 'Sight, Optical (Martin), Mark I', which had been approved on 1 May 1915, was not announced until 1 June 1916 (§ 17757).

39. According to his enlistment papers of 24 September 1914, William Charles Bullock Beech was born in May 1878 in Wellington, Shropshire, England. He claimed to have served for five years in the Shropshire Yeomanry and two years with the field artillery volunteers during the South African War. Discharged from the Australian Army on medical grounds, he was eventually awarded £100 by the War Office for his ingenuity. Married to Isabella Grace l'Estrange (with a son and a daughter), he died in Condobolin, New South Wales, on 22 September 1929. However, no trace has been found of his birth in British records; he may have lied about his age. It has also been claimed (Hamilton, *Gallipoli Sniper*, p. 129) that Mauritius-born Private George Tostee (1887–1967, 'Toster' in the AIF Nominal Roll) of the 10th South Australia Battalion devised a periscope rifle, but it is not clear if he had simply made a prototype for Beech, and his claim to ingenuity, if indeed he had one, does not seem to have been pursued.

40. Hesketh-Prichard, *Sniping in France*, pp. 145–6.

41. British Patent 104,921, 'Improvements in or relating to Optical Instruments for Sighting Purposes', accepted on 22 March 1916 (but sought on 22 March 1915), protected a method of adjusting for elevation by displacing the reticle-collar within the sight tube.

42. Thousands of guns that survived in store (renamed 'Rifles No. 3' in 1926) were refurbished for use in the Second World War, though customarily restricted to home-defence units like the Home Guard. They served alongside .30 M1917 Enfields supplied under Lend-Lease.

43. Intriguingly, the British tried to develop a .303 high-velocity rimless cartridge for the P/14 in 1917. Trials dragged on for years, without results, and then the idea reappeared in 1935. One experimental .303/450 cartridge, loaded with the standard 174-grain bullet, gave a muzzle velocity of 3,372 ft/sec. The chamber pressure was estimated to have reached 50 tons/sq.in. – 112,000 lb/sq.in., which confirmed the great strength of the P/14 action even though the chambers were prone to stretch. Experiments were finally abandoned only when the Second World War loomed.

44. The finalised 'Cartridge, S.A. Ball, .303-inch, N.C., Mark VII.W' armour-piercing cartridge, with a steel core, was introduced to service on 30 July 1918 (§ 21354).

45. Colonel en Retraite Jean Martin, *Armes à Feu de l'Armée Française 1860 à 1940* (Paris, 1974), p. 383.

46. The Berthier retained a Lebel-like bolt, but had a clip-loaded box magazine instead of the under-barrel tube. Two rifles had been introduced prior to 1914: the reduced-scale Mle 1902, for the Tirailleurs Indo-chinois, and the full-size Mle 1907 for the Tirailleurs Sénégalais.

47. Problems with receiver failures were traced to poor heat treatment, and most of the guns made after the spring of 1918 had a special double-treatment that cured the problems. Some 1918-vintage Rock Island products had nickel-steel receivers identified by small 'NS' marks.

48. Hundreds of thousands of M1917 rifles were sent into store, to emerge for refurbishment when the Second World War began. Many were sent to Britain under the Lend-Lease scheme, where they equipped numerous Home Guard units. Most were given broad red bands around the butt and fore-end to remind firers that they chambered the American rimless cartridge instead of the rimmed British pattern, a potential confusion that caused many serious accidents before action was taken. Garands supplied under Lend-Lease were often marked in the same way.

49. Dick Camp, *The Devil Dogs at Belleau Wood: US Marines in World War I* (Minneapolis, 2008), p. 68.

50. Alfred V. Houde Jr., 'Every Marine a Rifleman: the M1903 Springfield Rifle', *Fortitudine: Bulletin of the Marine Corps Historical Program*, Vol. XXXI, No. 2 (2004).

51. Hesketh-Prichard, *Sniping in France*, p. 55.

52. Recent attempts to trace York's heroism, which have successfully moved the site from where it was once generally accepted to be, retrieved .30 rifle and .45 pistol cartridge cases. These have been matched forensically with York's guns.

53. Few experienced riflemen had much time for the Mannlicher-Carcano. Clifford Shore said simply, 'I have used an Italian service rifle on ranges once or twice but had a very poor opinion of them. I reckon that it is about the worst service rifle I have ever handled': *With British Snipers to the Reich* (1948), p. 190.

54. Ottavio Bottecchia (1894–1927), the first Italian winner of the Tour de France in 1924 and the leader from start to finish in 1925, is said to have been a champion sniper with as many as 200 kills on the Austro-Italian Front. Bottecchia, from humble background in San Martino de Colle Umberto, joined the Bersaglieri during the First World War. Few details of his service record have been found, but his principal job seems to have been ferrying messages with the aid of a special folding bike. Gassed in 1917 after the Battle of Caporetto, while part of the rearguard providing covering fire as the Italian forces withdrew, he was captured and almost immediately escaped. But there is currently no evidence to show that he was a crack shot or even an experienced rifleman.

CHAPTER FOUR

Redrawing the Map: Snipers Lose Favour Again

The period between the world wars was supposed to be one of 'Great Peace', inspired by the demilitarisation of Germany and the creation of the League of Nations. Tremendous quantities of war matériel were collected and destroyed, not only in vanquished Germany (where the Treaty of Versailles had limited the armed forces to 100,000 men and an emasculated pre-dreadnought navy) but also in the Allied states. There was no need of tens of thousands of machine guns, hundreds of thousands of rifles, squadrons of aircraft or huge battle fleets. Millions of men were returned to civilian life in which many struggled to adapt. Even HMS *Dreadnought* was sent to the breakers in 1921.

Frailties of human nature doomed the Great Peace to failure. There was huge resentment in Germany and, to a lesser extent, Austria, where the imperial dynasties had been swept away almost overnight. Political uncertainty led to an ugly struggle between the left of the Marxists and right-wing nationalists exemplified by the Freikorps, who put an end to a short-lived socialist republic proclaimed in Bavaria. The collapse of the German mark in the early 1920s added to the hardship, effectively creating the power vacuum from which the Nazi Party (NSDAP) would inexorably rise. The British economy came close to collapse immediately after the First World War ended; the French had suffered so badly from the war that their national spirit was cowed; the USA saw a chance to dominate world trade; and the post-Revolutionary USSR was regarded as a pariah by virtually everyone. In these troubled times military authorities tried to analyse the lessons of war, though progress was usually inhibited by political masters who were trying to avert their gaze from conflict.

The result of such introspection was the covert rise of German militarism. Even though the Dawes Plan promoted by the USA had stabilised the German economy after the collapse of the mark, the financial crisis of 1929, the 'Wall Street Crash', undermined all that had been achieved. Depression was felt not only in the USA; it fatally weakened the elevated position British industry believed it still held in the world.[1]

Germany's continuing militarism between the wars, especially when the ineffectual Weimar Republic gave way to the increasingly bellicose Third Reich, encouraged rearmament and a rise in the number of men under arms. *Author's collection*

In Germany, the Reichswehr, the post-1919 armed forces, paid no more than lip service to restrictions imposed by the Treaty of Versailles. Encouragement of sports associations and flying clubs, the use of armed postal and forestry services, and the rise of the Sturm Abteilung and comparable paramilitary groups promoted covert training. Weapons were easily stockpiled away from prying eyes, and the manufacture of new weapons was camouflaged by the introduction of coding systems in which legions of sub-contractors and 'feeder factories' were hidden. Development of new weapons was undertaken elsewhere: aircraft and artillery in the Netherlands; submarines in Finland, small arms in Switzerland; and even armoured vehicles in the USSR – diametrically opposed to Germany politically, but with the same cynical view of military domination.

Sniping fell victim to public opinion, which generally equated the sniper with cold-blooded murder. Optical-sighted rifles were withdrawn from British service in 1921, a process repeated in virtually every other army. The sights were sometimes removed to allow rifles to be returned to normal service, though some were simply put into store.

The value of sniping reduced towards the end of the First World War as the battlefields became more fluid, and the introduction of aircraft and then the first tanks changed the balance of power – or so their proponents had hoped. The value of marksmen working independently, particularly in defensive situations, was no longer seen.

As far as infantry rifles were concerned, four years of fighting had shown the virtues of some designs and the drawbacks of others. The desirability of semi-automatic infantry rifles had been suggested by the Browning Automatic Rifle and (less prominently) the French RSC, but many military experts fretted about a supposed need for greater power in rifle cartridges. Judged purely by flatness of trajectory, which minimised range-gauging errors, their views had some justification. However, high velocity was easiest to achieve by reducing calibre and this was something few wished to do. The US Army actually approved a .276 round for the Pedersen and the earliest Garand rifles, but the Ordnance Department and many senior officers who had served in Europe ensured that the well-established .30–06 was retained. Ironically, the arguments were repeated in the late 1950s when champions of .223 sought to replace the 7.62 mm (.308) service cartridge.

Before WWI the British had decided to replace the Lee-Enfield with a .276 Mauser-type rifle, the Pattern 1913. When the war began, however, the incomplete plans were abandoned and a .303 derivative, the Pattern 1914, was made in quantity in the USA. When the American Expeditionary Force faced a dire shortage of rifles, the P/14 was adapted for the .30–06 cartridge to become the 'Model 1917'. This was potentially the best sniper rifle of the First World War, as the .30 cartridge performed better at mid- and long-range than the British .303.

The US authorities had developed the 'Sniper Rifle M1918' simply by fitting a sight adapted by Frankford Arsenal from the Winchester pattern to a half-stocked derivative of the M1917. However, the sight, based on a design that was essentially suited only to target shooting, proved to be very poor. It is unclear if any M1918 rifles were ever made.[2]

The British, well aware of the success of the SMLE judged purely as a battle weapon, were reluctant to go back to the P/13 or standardise the P/14 (unlike the US Army, which had come within a hair's breadth of replacing the M1903 Springfield with the M1917 Enfield). Consequently, the SMLE was 'improved'.

The first result was the SMLE Mk V, adopted tentatively in 1922 but abandoned two years later after trials had suggested that much more radical improvements were needed. The Mark VI rifle, known after 1926 as the Rifle No. 1 Mk VI, incorporated a heavy barrel, a simplified body, and simpler fittings. The leaf-type back sight was mounted within distinctive protecting wings above the receiver bridge, and the nose cap was changed so that a few inches of the muzzle protruded. Finally, after extensive trials at home and abroad, the 'Rifle No. 4 Mk I' was approved in November 1939 and the creation of new production lines began. The sniper-rifle derivative, the No. 4 Mk I (T), is described in the next chapter.

The French undertook experiments with the old Lebel, which had proved to be much more robust than the Berthiers issued prior to 1918. Though the Lebels were retained for sniping and grenade-launching duties, the authorities developed a rimless 7.5×58 cartridge, introduced in 1924, which promised far better ballistics than the venerable 8×51R pattern. Prototypes were made in Tulle and Saint Étienne in the late 1920s, with charger-loaded magazines and socket or spike bayonets that could be reversed into the fore-end when not required. The most successful submission was Tulle's MAT 1932, with a two-piece stock and a simple bolt with the operating handle bent forward so that its ball lay immediately above the trigger guard. This was developed in Saint-Étienne into the MAS 34, the 'B1' version being accepted for service in 1935 as the 'Fusil MAS 36'.

The new rifle was easily distinguished by the massive forged-steel receiver, with a pistol grip butt separated from the fore-end and hand guard. A Mauser-type staggered-row magazine replaced the tube of the Lebel or the clip of the Berthier, but one of the most important changes concerned the locking system as the MAS 36, possibly influenced by the Lee-Enfield, had two lugs on the bolt body engaging seats in the receiver behind the magazine well. The MAS 36 lacked a manual safety catch, and had a tangent-type back sight on the receiver immediately ahead of the firer's eye. A spike type bayonet was carried permanently in a channel in the fore-end.

The first deliveries were made in 1937, but few MAS 36 rifles had reached the army when the Germans invaded France in 1940. Desultory experiments with optically sighted versions of the MAS 36 and its predecessors were undertaken before and then during the Second World War, but the optical performance of the sights, adapted from the APX of the old Lebels, was uninspiring.

The emergence during the First World War of *Sturmtruppen*, 'storm troops', had had an effect on tactical doctrines and weapons development. A need for short-range firepower had inspired the MP.18, the Bergmann submachine-gun chambering standard 9mm handgun ammunition, and the value of long-range shooting, so important in the earlier stages of the war, declined.

The Germans had been largely satisfied with the Gew. 98, but were well aware that it was too long and cumbersome to be used in a trench or similarly con-fined area. The Kar. 98k ('k', *Kurz*, 'short') was developed from the Kar. 98 AZ of 1908. Introduced in 1935, the new rifle retained the basic Mauser action, but a tangent-leaf back sight was fitted. There was a single barrel band, and the 'H' type nose cap was accompanied by a 4cm bayonet bar beneath the muzzle. The sling attachment consisted of a slot in the butt and a short fixed bar on the left side of the barrel band.

Though the earliest guns were made by Mauser-Werke, the rapid expansion of the German military and paramilitary formations was solved by recruiting additional contractors. Consequently, a variety of markings will be found on Kar. 98k receivers, beginning with numerical codes. These include '27', Fein-mechanische Werke; '42', Mauser-Werke, Oberndorf; '147', Sauer & Sohn; '237', Berlin Lübecker Maschinenfabriken; '243', Mauser-Werke, Berlin-Borsigwalde;

and '660', Steyr Daimler Puch. Most of these may be encountered as with an 'S/' prefix.[3]

Telescope-sighted rifles were not popular with the Army, which regarded the basic training given to recruits as good enough to remove the need for specialists. In the 1930s, surviving sniper rifles (two of which were still issued to each infantry company) were withdrawn. The sights were detached and sold, and the rifles went back into service without them.

Exceptions were made for the telescope-sighted Kar. 98k and adaptations of the Kar. 98 AZ and Kar. 98b used by the SS, which were retained. These guns usually had 4× *Zielvier* in special mounts, and have distinctive marks.

The Italians, who had made very little use of snipers, learned from the campaign in Ethiopia that the 6.5 mm cartridge and the 1891-type Mannlicher-Carcano rifles were poor, developing a 7.35 mm version in the late 1930s. Their experience was mirrored in the opening phases of the Sino-Japanese War, which began in 1937. These showed that the 6.5 mm cartridge chambered by the Arisaka rifles and modified Hotchkiss-type light machine guns did not have the range of the 7.9 × 57 customarily used by the Chinese. Abandoning the policy of adopting comparatively low-power cartridges suited to the small stature of the average conscript, the Japanese ordnance department hastily developed a 7.7 mm round and a variety of suitably chambered weapons.

Work on the 6.5 mm 38th Year Type ('M1905') Arisaka continued in Nagoya until 1941, and small quantities were made during the Second World War in Manchuria and China. A few are even said to have been made in Heijo (Jinsen) arsenal in occupied Korea in 1938–9. Approved for service in 1937, the 6.5 mm Type 97 sniping rifle had a 2.5 × optical sight held in the mount by a radial latch. The sight was offset to the left of the gun to allow the magazine to be loaded from chargers. Even if the sight and mount are missing, guns of this type can be identified by the elongated bolt handle, bent down towards the stock-side, and by the steel-rod monopod beneath the fore-end.

The 7.7 mm adaptation of the 6.5 mm 38th Year Type rifle, the 'Type 99' (1939 being the 2,699th year of the Japanese Empire, reckoned from its mythical foundation in 660 BCE), shared the bolt action of its predecessor, updated to accelerate production. The trigger guard was a sturdy bent-steel strip instead of a machined forging, and the stock fittings were greatly simplified. Most guns had a monopod beneath the fore-end, and the hand guard ran from the front of the back-sight base to the nose cap.

A small number of full-length rifles were made in 1939 to 'Nagoya Rifle Plan No. 1', but a competing short rifle was preferred. Huge quantities of the latter were made in various factories until the end of the Second World War, during which a sniper-rifle adaptation with an improved optical sight was introduced.

The greatest changes were made in the USSR, where several years of bloody civil war had followed hard on the heels of the October Revolution of 1917. Conflict had shown the Bolsheviks that their arms industry needed to be entirely reconstructed if the losses of equipment were to be made good. One result of the first Five Year Plan, therefore, was the construction of new manufacturing facilities even as efforts were being made to refurbish as many old tsarist weapons as possible.

Production of the 1891-type dragoon rifle or *dragunskaya vintovka*, 3 in. shorter than the infantry pattern, began again. Many leading Soviet designers, including Fedorov, had argued for a much more radical solution to the infantry's needs but were overruled by politicians who saw keeping production going as paramount; introducing an untried weapon would not only disrupt output, but would not necessarily be an improvement on the Mosin-Nagant.

Virtually every aspect of arms production needed to be re-examined. Ammunition was purchased from leading US manufacturers in the hope that analysing the propellant would enable production to begin in the USSR, and the development of an optical industry – vital to the issue of binoculars and telescope sights – was to benefit from German aid. An optical factory had been founded in St Petersburg in 1914, with largely French capital, but, though effectively nationalised after the Revolution, was never able to answer demands.

Co-operation with the Germans remained covert in the 1920s, largely because the Allied Control Commission was still active enough in Germany to prevent overt contact. Yet not only did the Russians create factories to make high-grade lenses, but attempts were soon made to introduce optical sights.

Almost alone in Europe, the Russians could still see the value of snipers. In the mid-1920s, while experiments were being undertaken in the Podolsk optical manufactory, telescope sights were acquired from Busch, Hensoldt, Voigt-länder, Zeiss and other German suppliers. These are said to have been fitted to 170 specially selected dragoon rifles, 'setting-up' being undertaken in Germany. The prime contractor has been variously identified as Simson & Companie or its successor Gustloff-Werke (which did not exist in 1929!), but was more probably the sporting-rifle division of Gustav Genschow & Co. Genschow's gunsmiths had perpetuated skills that had been largely lost in the Reichsheer.

The Russians rejected the traditional two-piece 'claw' mounts that had been used prior to 1918 on the S.-Gew. 98, favouring a one-piece system pioneered, it is said, by Ernst Busch of Rathenow. Production of the first Soviet telescope sights duly commenced in a factory owned by the NKVD and trials were undertaken by the 'Dinamo' organisation, which had been formed in 1923 to oversee training under the pretext of 'sport shooting'.

There were, apparently, three types of semi-experimental rifle sight. Evidence is still sketchy, but the first incorporated lenses supplied by Zeiss through a sham intermediary, the Netherlands Industrial Company ('Nedinsco'). These were superseded in 1931 by the first truly Russian-made sight, D-II, which was adjustable only for elevation; then came D-III with adjuster drums for elevation and windage, and this in turn became the PT or Obr. 1930g. sight, made in limited numbers before giving way to the perfected PE described below.

The sights were held in a two-ring monoblock, mating with a plate attached to the left side of the rifle body, where a combination of a peg-and-slot and a large-headed screw locked them in place. Credited to designer A. A. Smirnsky, the earliest mounts were plain and had a keeper-chain attached to the locking screw; later ones abandoned the chain, but had range tables for the Type L ball cartridge rolled into the mount-plate.

The perfected Obr. 1891/30g. infantry rifle, adopted in April 1930, had been designed with ease of production clearly in mind. Consequently, though the basic action remained unchanged and the chamber area of the body was still octagonal, swivels were replaced by sling-slots cut through the woodwork; the bands were held by springs instead of screws; and the leaf-type back sight was replaced by a tangent-leaf design. The front sight was given a protective hood, and the socket bayonet – an archaic survival – was improved by the substitution of a spring catch for the encircling ring of its predecessor.

The rifle was accompanied by the new PE sight ('Obr. 1931g.') in an over-chamber block. Held by six screws – three a side – and also often soldered, the sight base was machined to duplicate the octagonal body. The two-ring monoblock mount was retained by two large-headed screws which forced a triangular wedge against the base-rail. The monoblock was also cut away longitudinally to allow open sights to be used when necessary. The 4× PE, introduced to service in 1932, had separate elevation and windage adjusters; it was long, heavy, and had a focusing collar ahead of the ocular lens. Lens carriers were originally made of brass, but eventually became lightweight alloy.

The new Obr. 91/30g. sniping rifles (*Snayperskiy vintovky*) were tested extensively in the protracted, shambolic and bloody Spanish Civil War (1936–9), in which the pro-fascist Nationalists, with German and Italian backing, eventually overcame the pro-communist Republicans.

The new Soviet-made rifles passed scrutiny but the sights gave trouble. The annular collar could be jolted out of focus by recoil, while poor seals led to the ingress of moisture and excessive fogging. The immediate result was the development, attributed to a team led by Dmitry Kochetov, of the PEM ('Obr. 31/37g.' or 'Obr. 1937g.'). The new sight was essentially similar to the PE but the separate focus adjustment was abandoned

By this time, the body of the 91/30 rifle had become cylindrical, easier to make than the old octagonal form, and the optical-sight base had been modified accordingly. An attempt was made to reduce the six locking screws to four, but this was less secure and a return to six was made until a new side-rail system had been perfected.

The design of the side-rail resembled that of the old 'Dinamo' sights, with a peg-and-slot connection and a large-headed bolt with a tapered tip to give a self-centring effect. The 91/30 rifle with a PEM telescope in the side-rail mount weighed 10.5 lb, but shooting generally benefited from the additional weight.

The rifles were made in Tula and Izhevsk in amazingly large quantities, beginning with 749 assembled in 1932 and progressing to 19,545 in 1938, by which time output had totalled 54,160 at a time when the value of sniper rifles elsewhere was at its lowest ebb. Most of the rifles had 'C' and 'П' above the chamber, an abbreviated form of 'Snayper proveryat' or 'Sniper[-issue] Proved'. PE and PEM sights were made during the Great Patriotic War in huge quantities, by several contractors. Details will be found in the next chapter.

Guns were specially selected for accuracy at an early stage. The bolt handles were elongated and turned down against the stock, attention was paid to the tolerances and finish of the chamber and the bore, and the trigger was honed to be as smooth as possible. Soviet manuals reveal that the acceptance tests for the sniper rifles demanded the ability to place ten shots into a 3.5 cm circle at 100 metres, 7.5 cm at 200 metres, 18 cm at 400 metres and 35 cm at 600 metres. These equate to MOA figures ranging from 1.27 at 109 yards to 2.1 at 656 yards, and confirm the Germans' view that the Obr. 91/30g. could guarantee the 'head-kill' at 300 metres that the Zf.-Kar. 98k could not.

When the Winter War with Finland began in 1939, the 91/30 was the principal weapon of the Red Army snipers. Substantial numbers were captured by the Finns, though only a few of these show on official registers, and some of the optical sights were fitted to Finnish m/39 rifles. Some Finnish Mosin-Nagants will be found with prismatic sights, survivors of a few hundred which had been made experimentally in 1937.

By 1939, the Red Army was more concerned with the development of the semi-automatic Tokarev. Production of the Obr. 91/30g. sniper rifle stopped in 1940 and did not recommence until the Great Patriotic War had begun. Soviet authorities had been so keen to introduce an auto-loading rifle that trials had been undertaken, almost without a break, since the mid-1920s. A major problem lay in the comparative crudity of Russian industry. Even though great progress had been made by the Five Year Plans, much of the engineering industry was badly equipped and poorly manned.

A Degtyarev rifle had been tested in the late 1920s and provisionally adopted as the Obr. 1930g. in December 1931, but progress with a Simonov tried earlier in the same year led to the adoption of the Avtomaticheskaya Vintovka Simonova ('AVS') in 1936.

The AVS owed its introduction to a decree signed in March 1934, when it was adopted to replace the abortive Degtyarev, but production was very slow and constant minor changes were made even after the first series-made batches left Izhevsk in 1937. The AVS has a characteristic three-quarter-length one-piece stock, with a wooden guard and a sheet-steel ventilator above the barrel. Excessive distance from the pistol grip to the trigger, the cleaning rod set into

The author's drawing of the Obr. 91/30g. rifle and the PU sight introduced during the Second World War.

the right side of the stock, and the design of the muzzle brake/compensator aid identification. The curious bayonet clipped over a transverse bar beneath the muzzle, and was retained by a lug on the underside of the front-sight post.

Though series production was confined to just two years (1937–9), more than 60,000 AVS were made, including what may have been the first officially issued auto-loading sniper rifles; a few guns dating from 1936 survive with PE sights of similar date in special mounts on the left side of the body, suitably offset to the left to facilitate ejection. Field trials and combat experience in Manchuria soon showed that the rifles extracted harshly and that the open-top breech allowed dust and grit to jam the mechanism too easily. These weaknesses were exposed once again in the Winter War with Finland (1939–40), yet some Simonovs survived the German invasion of Russia in the summer of 1941 to serve until worn out. Photographs show them in use as late as 1943, occasionally with German units.

Throughout the summer of 1936, despite the official adoption of the AVS, trials continued with alternative rifles submitted by Tokarev, Rukavishnikov and others. By the beginning of September 1936, they had been resolved in favour of the Tokarev, which had taken part in most of the 1926–30 trials without ever encountering success. Tests undertaken late in 1937 ended with the Tokarev rifle beating the Simonov by a narrow margin, with the Rukavishnikov design a remote third. But the results were shelved until the failure of the AVS in service caused trials to be repeated in November 1938, with much the same result. Some of the features of the AVS were still greatly appreciated, but the Tokarev had proved to be more reliable.

On 26 February 1939, therefore, the Samozariadnya Vintovka Tokareva obrazets 1938 goda ('Tokarev self-loading rifle, model of the year 1938') was officially adopted. An improved Simonov appeared, but Fedor Tokarev was more acceptable politically than Sergey Simonov, and so, though a committee

The Simonov, the AVS-36 (*top*), was the first auto-loading rifle to serve the Red Army in substantial numbers. Unsuccessful, and soon replaced by the SVT-38 (*above*), survivors were withdrawn into store after the Winter War. Some of these were subsequently re-issued as sniper rifles, fitted with PEM sights in a side mount and restricted to semi-automatic fire. The SVT-38 soon gave way to the SVT-40, but unreliability and comparatively poor accuracy ensured that neither proved to be an ideal sniper's rifle.
Courtesy of James D. Julia, auctioneers (www.jamesdjulia.com)

investigating production-engineering problems reported in May 1939 that the Simonov was simpler and cheaper to make, mass-producing the SVT was approved. Tests were stopped in July 1939, therefore, and production of the Tokarev rifle commenced.

The 1938-type Tokarev was somewhat like the AVS externally, with charger guides above the breech and a box magazine projecting ahead of the trigger guard, but the receiver was noticeably taller and had a rounded top. The wooden pistol-grip stock was made in two pieces, with a full-length cleaning rod set in a channel on the right side. Ventilation slots were cut through the front part of the barrel guard, a short sheet-steel shroud protected the gas tube, and the muzzle brake had six narrow slots.

The Russians intended to make a minimum of 50,000 SVT in 1939, accelerating to 600,000 in 1940 and then to 2 million per annum by 1942: the first completed rifle was exhibited on 16 July 1939, series assembly began less than two weeks later, and by 1 October mass production was under way. But then the Red Army became embroiled in the Winter War with Finland, and the shortcomings of the SVT became obvious.

The worst problems made familiar reading: fragility, parts-breakages, extraction problems and persistent jamming. In addition, the Russians discovered their lubricants to be unsuitable in sub-zero conditions, and that, therefore, AVS and 1938-type Tokarev rifles often locked solid.

Production of the 1938-type Tokarev was immediately suspended while improvements were made in the basic design. The SVT obr. 1940g. was formally adopted on 13 April 1940 for NCOs of the artillery and infantry, as well as naval infantry and the Russian marines. Though many changes had been made, the most obvious concerned the position of the cleaning rod, previously carried in a channel on the outside of the stock (where it had proved to be vulnerable to damage and comparatively easily lost), which had moved to the conventional position beneath the barrel. It is difficult to judge how many 1938-type rifles had been made, though 50,000 seems feasible.

A US Army infantrymen armed with an M1 Garand hides behind a half-track during an exercise in Kentucky in 1942. *US Army*

In the USA, military production at Springfield Armory had virtually ceased by 1927, though .30-calibre rifles were made for target shooting and sale to members of the National Rifle Association. The improved M1903A1 was simply an M1903 with a modified 'Style C' pistol grip stock, which replaced the straight-wrist 'Style S' in March 1929. However, few M1903A1 guns were made and straight-wrist stocks were still being used in 1939.

By this time, the future was believe to lie with the semi-automatic Garand and the trusty Springfield was no longer deemed worthy of development. The genesis of the gas-operated Garand and its victory over the toggle-lock Pedersen rifle has been told in detail elsewhere.[4] However, though approved on 3 August 1933, the design of the M1 rifle was not really settled until trials of improved guns finished in October 1935. Only then was the M1 recommended for service, standardisation following on 9 January 1936.

Teething troubles with individual parts delayed the first deliveries of machine-made guns until September 1937, and a major redesign of the gas-cylinder assembly and barrel had to be authorised soon after sample rifles were badly received at the 1939 National Matches. When war began in Europe, therefore, the bolt-action M1903 Springfield was still the principal individual weapon of the US armed forces.

Combat experience showed the value of the Garand, which proved to be a more battleworthy weapon than the Pedersen would ever have been. Though the quirky eight-round en bloc ammunition clip prevented single rounds being loaded into the magazine in an emergency, production by VJ Day amounted to over 4 million; 600,000 or more were made immediately after the war, work began again during the Korean conflict and a licence was ultimately granted to Beretta. Production of the M1, superseded by the M14, finally ended on 17 May 1957.

Notes to Chapter Four

1. British prosperity had peaked at the end of the reign of Queen Victoria, but had declined thereafter until the country's 32.3% share of world trade in 1870 had dropped to just 13.9% by 1914. Germany and especially the USA had provided ever-increasing competition.

2. It has been suggested that no more than a single rifle, and that perhaps even a wooden mock-up, had been made. Work on the sight continued until 1921, when, realising that it was never going to perform to expectations, the US authorities simply abandoned the project.

3. The 'S' prefix was assigned to Simson & Co., later Berlin-Suhler Werke and then Gustloff-Werke, who according to the limitations of the Versailles Treaty were to be the sole supplier of Mauser-type rifles to the Reichswehr. The accompanying numbers identified 'sub-contractors' acting as part of the 'Sachsengruppe' to accelerate production covertly.

4. There have been several excellent books about the Garand, but those by Bruce Canfield, published by Collector Grade Publications in Canada, are particularly accessible. For a brief summary, see John Walter, *Rifles of the World* (2006), pp. 213–15.

CHAPTER FIVE

The Height of Fame: The Winter War to Korea

The Winter War or *Talvisota* was, as far as sniping was concerned, the prelude to the Second World War – and, in particular, to what Russians call the 'Great Patriotic War' consequent on the German invasion in June 1941.

The basis of conflict with the Finns was territorial: to safeguard what they saw as the essential integrity of Leningrad, which could by blockaded by the Finns and the Baltic states, the Russians demanded that the Finns cede territory on the Karelian Isthmus, lease the port of Hanko to the Soviets as a military base, and give the USSR access to the Rybachi Peninsula and Petsamo, the only ice-free Finnish harbour.

When the Finns refused to comply with all the demands, the Russians faked an 'artillery bombardment' of their own territory to force Finland into submission. They had reckoned without Finnish intransigence, but nonetheless tore up the Russo-Finnish non-aggression pact, broke off all diplomatic relations, and, at 9.15 a.m. on 30 November 1939, invaded Finland along a thousand-mile front:

> The invasion was camouflaged as a spur-of-the-moment incursion rather than, as it really was, a carefully preplanned strike by the Russian 7th, 8th, 9th, 13th and 14th armies ... perhaps numbering 350,000 men against the Finns' 250,000. The campaigns were initially a sorry catalogue of Russian misfortune ...
>
> The winter of 1939–40 was also one of the coldest for many years, with temperatures dropping as low as 50° below zero, and a great many Russians simply died of wounds that would not have been fatal elsewhere; blood plasma froze, and so, too, did the invaders' spirits. The 163rd and 44th divisions, each numbering some 18,000 men, were then slaughtered in quick succession at Suomussalmi.
>
> In the first month of the war, one source reported that the Russians had lost the staggering totals of 200 aircraft, 212 tanks and 100,000

soldiers killed and wounded: perhaps too high an estimate, yet, in light of some excessive post-war figures, dreadfully plausible.[1]

The Russians learned their lessons the hard way, but one of the immediate results was the rehabilitation of the few officers who had survived the purges. Timoshenko replaced Meretskov in command of the Karelian Front, and, eventually, the balance of power inevitably shifted. Peace was finally agreed on 12 March 1940.

Studies undertaken in Finland in 1940, on the basis of official casualty returns, gave Finnish deaths as 19,576 with an additional 3,273 men 'missing in action'; wounded numbered 43,557, substantial numbers of whom subsequently died. The currently accepted 'all causes' death-toll is 26,662, from 30 November 1939 to 31 December 1940. Soviet casualties were reassessed in the 1990s by a team led by Colonel-General G. F. Krivosheev.[2] Published in 1993, the report acknowledges 126,875 deaths (including a staggering 36,369 men 'lost without a trace') and 264,908 wounded.

One of the enduring stories of the Winter War is that of Simo Häyhä, who, by the end of hostilities, had been credited with 505 kills (some sources claim 542) as a sniper in a front-line career that did not even last for a hundred days. There is no doubt that the Russian soldiers were badly led, poorly trained and lacking clothing and equipment suitable for use in arctic conditions, often making themselves comparatively easy targets, but Häyhä's success was unparalleled.

Häyhä, son of a farmer, was born on 17 December 1905 in Rautjärvi, in what was then the Grand Duchy of Finland – part of the Russian Empire prior to the Bolshevik Revolution of 1917. In 1922, he joined the Protective Corps where his skill as a marksman was soon evident, and was conscripted at the age of eighteen into the army. He served until 1927, then returned to the family farm.

When the Winter War began, Häyhä, like all other reservists, was called back to service and posted to 34. Jääkärirykmentti operating in the forests to the north-east of Lake Ladoga along the line of the Kollaa river, where the outnumbered Finns had been able to halt the Russian advance. After serving as an infantryman, Häyhä demonstrated his prowess with the rifle by killing a sniper who had accounted for several officers. The Russian rose at dusk from his position, believing himself to be under no threat until Häyhä shot him in the head.

The toll steadily rose. Häyhä himself, quiet, reserved and exceptionally patient, had no qualms about killing Russians; he simply accepted it as part of his job, and set himself the goal of doing his job to the limits of his considerable ability. He was prepared to observe for hours hidden behind a snow-mound, with nothing but sugar cubes for sustenance, and took great care to camouflage his rifle and ensure that, by carefully tamping the snow in front of his position, no 'snow-cloud' would rise when he fired. The results could be spectacular:

Probably taken during the Winter War of 1939–40, or perhaps the 'Continuation War', when the Finns re-took territory lost to Russia, this shows that ex-Soviet weapons were pressed into service wherever possible. The centre man has an m/26 Lahti-Saloranta light machine gun (sometimes classed as an automatic rifle), but his companions have an AVS-36 (*left*) and an SVT-38 (*right*). *Sotamuseo, Helsinki*

in one three-day period in December 1939, he accounted for fifty-one Soviet soldiers; and later in that month scored his highest total for a single day – twenty-five confirmed kills.[3]

One of the most impressive of many statistics is that virtually all of his kills were achieved with an open-sighted rifle. His fieldcraft was impeccable, relying on effective camouflage and knowledge of the terrain to minimise engagement range, but 'The White Ghost', as he came to be called, was also an exceptional shot. Häyhä was also small, merely 1.60 m tall and slight of build. This and his habitual use of open sights lowered his profile appreciably. He is said to have eschewed a captured Obr. 91/30 sniper rifle, with an optical sight, because he would have to lift his head to look through the telescope and there was always the chance of light glinting off the objective lens to betray his position.

Pictures have been published purporting to show Simo Häyhä firing a 6.5 mm-calibre 1896-type Swedish Mauser rifle (instantly identifiable by the tube-hilted bayonet attached beneath the muzzle); however, Häyhä's memoirs confirm his rifle to have been Mosin-Nagant m/28-30 Sk.Y. No. 60,974, the standard pattern issued to the Protective Corps prior to the introduction of

Above left: Simo Häyhä pictured during the Winter War, when he had been presented with a new Mosin-Nagant rifle – which he never used in combat, preferring to retain his tried and tested m/28-30.   *Above right*: Häyhä after the end of the war and reconstructive surgery to repair his smashed jaw. He wears the uniform of a *vänrikki* (second lieutenant). *Finnish official photographs*

the perfected universal-issue m/39. One of the rogue pictures is actually of a Swedish volunteer rifleman serving in Finland.

The Mosin-Nagant was the principal infantry weapon of the Finnish armed forces until the adoption of the Kalashnikov in the 1960s, owing to the capture of large numbers of Russian rifles in Helsinki arsenal in 1917–18. Ironically, these pre-1918 guns provided the actions for all Finnish derivatives (even sniper rifles developed in the 1960s and 1970s), albeit rebuilt and refurbished many times. Consequently, the Finnish Mosin-Nagants can always be identified by their octagonal receivers, even if, in the case of the 1939 pattern, they present a much more modern appearance than the Soviet Obr. 1891/30.

The first Finnish variant, the full-length m/24, had new free-floating barrels made by Schweizerische Industrie-Gesellschaft (SIG), a front band that was pinned to the fore-end, and the original 'pace' (*arshin*) graduations on the Russian back-sight base replaced with metric near-equivalents. Next came the m/27 short rifle, made by Valtions Kivääritehdas ('VKT'), with a short barrel with a full-length guard, a hinged German-inspired nose cap with a bayonet lug, and an adapted Russian back sight re-graduated in metres. The m/28 and m/28-30 rifles were issued only to the Protective Corps, the Suojeluskuntain Ylieskunnen ('Sk.Y.'). They had short barrels, detail improvements in the trigger, and new stocks. The principal difference lay in the sights; original rifles had re-graduated Russian sights, whereas the later pattern had a new 2,000-metre tangent-leaf sight.

The m/27 (Army) and m/28 (Sk.Y.) Rifles were not interchangeable, and even their bayonets were separate patterns. Consequently, the defence ministry and the Sk.Y. leadership decided to standardise designs. The resulting 'm/1939' rifle retained the 2,000 m tangent-leaf back sight of the m/28-30 and the rifling of the lightweight army-type barrel. However, a modification of the two-piece Soviet 91/30-type cartridge interrupter was fitted, and the new pistol-grip stock was made of two pieces of interlocking birch. Production continued until after the Continuation War (1941–4), when the manufacturer, Sako, was sold to the Finnish Red Cross to keep the facilities out of Soviet hands.

Ironically, Simo Häyhä's career came to an abrupt end not at the hands of a Russian sniper, but while acting as leader of an assault party. Carrying a Suomi submachine-gun instead of his Mosin-Nagant, he was hit by a bullet from a rifle or machine gun that entered his lip to smash away part of the left side of his jaw. Carried from the battlefield unconscious, Häyhä did not emerge from his coma until 13 March 1940 – the day on which the Winter War officially ended.

He subsequently endured more than twenty operations to rebuild his jaw and palate, and was eventually discharged from hospital in May 1941. Häyhä returned to his farm, where, apart from unsuccessfully volunteering to fight in the Continuation War (1941–4), he remained for the rest of his life. Simo Häyhä died on 1 April 2002 in a veterans' home in Hamina, and was buried in Ruokolahti. He was probably the most effective sniper of all time.

The Russians made the greatest use of snipers of any of the forces participating in the Second World War. Not only did Soviet marksmen contribute much to the defeat of Nazi Germany, but markswomen also made contributions that were entirely unmatched elsewhere; indeed, Lyudmila Mikhaylovna Pavlichenko (1916–74, 309 accredited kills), ranks among the most effective snipers of all time. One modern Russian source names forty-five female snipers, each with more than sixty kills to her name.[4]

The constitution of the USSR conferred equality on all its population, regardless of gender, but this was largely ignored in the male-dominated armed forces. Women were expected to become medical orderlies, drive supply trucks, or undertake administrative tasks. However, when it became clear that the Germans were pushing ever deeper into Russian territory, any means of mobilising resistance was considered.

This included the establishment of sniper schools and determined efforts, not universally popular, were made to recruit men and women who had undergone specific training. Some had been involved in competitive shooting through

Pictured during the Great Patriotic War, in 1943, Russian sniper Ivan Petrovich Merkulov (144 kills) of the 203rd Rifle Division fires a 91/30 Mosin-Nagant fitted with a PU sight. *Soviet official photograph/Novosti Press Agency*

the Komsomol, the Soviet youth league; others came from remote areas where survival could depend on an ability to hunt efficiently. A decade previously, in 1931, the Communist Party hierarchy had ordered that basic military training should be mandatory for all boys and girls who entered elementary school. This naturally included marksmanship.

Credit for giving impetus to sniper training has been given to Feodosy Smalyachov (1923–42), who inspired a 'people's movement' in Leningrad in the autumn of 1941 once the Germans had elected to starve the defenders into submission. Smalyachov did not live to see the proliferation of marksmanship schools, initially created without the backing of the Soviet state, but there can be little doubt that these contributed greatly to the efforts of the Red Army to halt the German advance.

The earliest groups of snipers were drawn from the ranks of hunters and the 'Voroshilov Marksmen' of the Komsomol and other similar militaristically orientated organisations. Osoaviakhim, the Union of Societies of Assistance to the Defence and Aviation-Chemical Construction of the USSR, founded in January 1927 to provide trained reserves for the armed forces, not only claimed to have fourteen million members in 1941 but had well-equipped marksmanship schools in Kiev, Leningrad and Moscow. Training methods varied from group to group, but snipers graduated in surprisingly large numbers. Natalya Kovshova and Maria Polivanova, destined to die together in 1942, were among the instructors – training at least twenty-six women in the Moscow Communist Division before getting permission to go to the front themselves.

The performance of the Soviet snipers during the Great Patriotic War soon became legendary, and 'kill totals' in excess of 500 are claimed for several of the men. They include Mikhail Ilich Surkov of the 4th Rifle Division (702 kills), Vladimir Gavrilovich Salbiev of the 71st and 95th Guards Rifle Divisions (601 to 31 December 1944), Vasiliy Shalvovich Kvachantiradze of the 259th Rifle Regiment (1907–50, 534 kills), Akhat Adulkhakovich Akhmetyanov, 260th Rifle Regiment (502 to 15 January 1944) and Ivan Mikhaylovich Sidorenko of the 1122nd Rifle Regiment (1919–87, 500 kills).[5]

Sidorenko is still sometimes considered to be the top scorer, the implication being that those above him had their kills inflated either by involvement in submachine-gun assaults or by political masters intent on eclipsing the detested Finn, Häyhä, who had taken such a toll of Soviet troops during the Winter War.

The Russian snipers were unquestionably the subject of propaganda claims: Lyudmila Pavlichenko and Vladimir Pchelintsev[6] were dispatched on a tour of the USA and Britain in the autumn of 1942, and the exploits of others were glamorised in magazine articles, short stories and lurid novellas. But however much these may have based on truth and regardless of how greatly they appealed to the casual readership, the sniper instructors in Britain and elsewhere were singularly unimpressed. For example, Captain Clifford Shore wrote scathingly:

> I do not think there was any subject about which there was so much balderdash printed and published during the whole course of World War II than Russian sniping. If we are to believe every report we read about the terrific casualties inflicted on the Germans by the Russian snipers it was amazing that there were so many Germans left to face the Americans and British in N.W. Europe! I do not think I am guilty of exaggeration when I write that hardly a week passed in the British press that one did not read of '"—sky" the Russian sniper [who] has accounted for so many hundreds of Germans'! Figures of over 100 Germans per sniper were common. 'One bullet per man' was a widespread slogan and somewhere I read that one Russian had killed 127 Germans in 128 shots, another 187 Germans in 189 shots, and so on . . .
>
> After the war I questioned many Germans who had fought on the Eastern Front and asked them about sniping. I was told that there was actually little sniping on the Eastern Front as the Americans and British – and Germans – know it . . . I have met some Russians who had been Red Army men, and saw them shoot in the summer of 1945.

If their shooting prowess be taken as a criterion I think the printed Russian figures for sniper casualties should be divided by a hundred and the result taken as being something like the number of Germans accounted for by Russian snipers.[7]

And yet, oddly, Shore was prepared to accept without demur the 505 kills claimed on behalf of Simo Häyhä – achieved in less than a hundred days during the Winter War of 1939–40. If Häyhä could achieve so much so quickly, why shouldn't a Russian sniper who survived combat duty for three years return a higher score?

Shore came from the English middle class (according to the 1911 British census, his father had been a hotel owner), a group whose mistrust of 'Bolsheviks' was generally strong. Of course, the average Briton's view of Russia had been blighted by the Russo-German Pact of 1939, and Operation Barbarossa initially brought very little change. Many people resented what they saw as turn-coat Stalin's exhortations to open a 'Second Front' against Germany to relieve pressure on the Red Army, and even the heroic Arctic convoys could be viewed as a waste of resources that might be used much more effectively at home.

Gradually, however, an idea of the true horrors of the Eastern Front filtered through the fog of propaganda and disbelief. One of the problems that Shore faced in making his assessment was lack of information not only about the titanic scale of the battles, the defence of Leningrad and Stalingrad in particular, but also the terrible losses with which victory had been bought. Had he known that at least 9 million Soviet military personnel had been killed, with an additional 22 million wounded,[8] Shore would perhaps have been more charitable.

And had he known how many men the Germans had lost, and the ways in which these losses had been inflicted, he would have viewed the snipers in a different light. Shore accepted that among the Russians there were some outstanding hunters who excelled in the very skills he had been promoting; but he did not realise how many of these men there had been, the extent to which the Red Army trained snipers, or that many of the snipers were women. Surprisingly, he does not even mention Vasiliy Zaytsev or Lyudmila Pavlichenko, even though both had achieved iconic status long before 1945.

The Central Defence Ministry archives in Russia still hold many records, including combat diaries, which do much to allay the suspicion that the Soviet propaganda machine – which made good use of Pavlichenko and Zaytsev (and, to a lesser extent, of the spirited, glamorous but ill-fated Roza Shanina) – was playing a numbers game. Yet it seems quite likely that scores are overstated, owing to the difficulties of confirming a kill.

Roza Shanina claimed that, 'The way it worked was that you might shoot many times at targets when on the defensive, but it was impossible to tell

Russian snipers customarily worked in pairs, one shooting and one observing before the roles were reversed to prevent carelessness arising from fatigue. Here a team armed with Obr. 91/30 bolt-action rifles and PE sights seeks out a position on a river bank.
Soviet official photograph/Novosti Press Agency

whether you had a kill or just a wounding.' There was a particular problem with running targets, particularly if shots were taken from comparatively long range: 'they [her commanders] would sometimes put no entry at all [in her log-book], or give me the benefit of the doubt, or include something highly questionable'. This uncertainty sometimes inflated scores, but, equally, could fail to recognise legitimate kills. And successful shots undertaken while attacking, even with the sniper rifle, were very rarely included. Lidiya Bakieva, with an official tally of 78, reflected long after the war – according to Lyuba Vinogradova's *Avenging Angels* (2017) – that her 'real total' could have been 150 or even 200.

In addition, prior to 1943, there does not seem to have been approved 'scoring guidelines' and independent verification was erratic. This potentially affects the tallies of snipers operating on the Leningrad Front and even at Stalingrad. Attempts were undoubtedly made to record kills accurately, particularly by individuals, but elements of doubt must remain. Iosif Pilyushin's *Red Sniper on the Eastern Front* pictures the paperwork of his award of the Order of the Red Banner, dated 6 August 1943, which states his total to have been 55 (of, ultimately, 124). But it is not entirely clear when this had been computed. Similarly, on 22 February 1942, Yevgeni Nikolaev attended a Leningrad Front

'Snipers' Rally' awards ceremony where, according to *Russian Sniper* (2017), Vladimir Pchelintsev was being credited with 102 kills, Ivan Vezhlivtsev had 134, and Pyotr Golichenkov 140 of what ultimately became 225. At this point, largely on his own reckoning, Nikolaev had a score of 76.

In answer to questions posed in 2010, Klavdiya Kalugina confirmed that the trench commander or a suitable witness had to sign a record, and then the sniper returned with evidence to present to his or her superiors. Kalugina also agreed that it was impossible to tell when the 'kill' was actually a wound and that a 'kill' was generally recorded if the target 'fell down' when struck. In trench warfare or in the shattered industrial landscape of Stalingrad, for example, the results of each and every shot were not always easily observed.

There are several instances in which a sniper was hit, not fatally but in circumstances which could lead the firer to claim ultimate success. Pilyushin, Nina Isaeva and Bella Morozova all survived bullets that hit them in or near the eye, and a blow to the helmet which knocked the target backward could also give the appearance of a kill. Yet even if 10 per cent is subtracted from each sniper's total (which is probably ungenerous), there can be no doubt that the Soviet snipers set the standards to which others aspired.

When precise marksmanship was not required, snipers could be pressed into service as assault troops, but gained no credit for kills achieved in this particular role. Consequently, it has been suggested that Fedor Matveévich Okhlopkov (1908–68), serving successively in the 1243rd, 234th and 259th Rifle Regiments, reached a thousand kills if his performance as an assault submachine-gunner is added to his score as a sniper of 429.

Of course, sometimes the sniper who achieves a single kill contributes more to the progress of a battle or a campaign than one who scores a hundred. Killing a high-ranking commander can have a greater effect on the course of a battle than the deaths of a hundred low-ranking men, and attempts were usually made to eliminate officers. The 242 kills attributed to Liba Rugova are said to have included 118 officers. She recalled in an interview in 1980 that 'We were told that killing its leaders would destroy the German Army, so that's what we did.'

Many Russian snipers were drawn from the ranks of the hunters, who were used to operating with great stealth and patience in the harshest of conditions. They included Maksim Passar (1922–43), an ethic Nanai from Nizhny Katar in the Khabarovsk area of Siberia, who is one of the unsung heroes of Stalingrad. Most of his 236 or 237 kills were achieved there, though he was fatally wounded on 22 January 1943 during an assault on the village of Peshchanka. Passar is also credited with training snipers in the 117th Rifle Regiment, most of whom were native Siberians destined to be credited with 775 kills between them.

Semyon Danilovich Nomokonov was born in August 1900 into an Evenki family (an indigenous people of northern Asia), in the hamlet of Delyun in what is now the Zabaykalsky autonomous region of Russia. He learned to shoot at the age of seven and spent his formative years hunting wapiti and elk in the taiga. When war began, Nomokonov, who had trained as a carpenter, was conscripted to make crutches for the wounded. However, no sooner had Nomokonov begun work than he demonstrated his prowess with a rifle by getting the better of a German adversary simply by shooting first.

A rapid transfer to a sniper unit was arranged, followed by service as a sniper and sniper-instructor in Karelia, the Ukraine, East Prussia and Manchuria. His logbook is said to register 360 kills on the Eastern Front and seven or eight Japanese soldiers killed on the Transbaikal Front in Manchuria after the end of the war in Europe. Nomokonov died in the summer of 1973, not far from Nizhny Stan where he had begun work as a carpenter in 1928.

The best known of all the snipers is undoubtedly Vasiliy Grigorevich Zaytsev, author of *Notes of a Russian Sniper*, reprinted many times, and a central figure of William Craig's 1973 book *The Enemy at the Gates: The Battle for Stalingrad*.

Born in Yeleninskoye on 23 March 1915, he learned to shoot from an early age in the Ural mountains. When the Great Patriotic War began, Zaytsev, then a petty-officer clerk in the Navy, stationed in Vladivostok, immediately

Maksim Passar (1922–43, *above left*), one of the most valiant defenders of Stalingrad, and the high-scoring Ivan Sidorenko (1919–84), who became best known as a sniper-trainer. *Soviet official photographs/Novosti Press Agency*

Vasiliy Zaytsev (*left*) instructs fellow snipers 'Somewhere at the front'. That the two trainees have shouldered their rifles suggests that the picture was posed. Zaytsev holds a 91/30 Mosin-Nagant sniper rifle fitted with a PEM optical sight in the perfected side mount. *Soviet official photograph/Novosti Press Agency*

volunteered to serve in the front line. Transferring to the Army as a senior warrant officer, and after proving his capabilities as a marksman, Zaytsev was posted to the 1047th Rifle Regiment. On 17 September 1942, he and his colleagues became part of the 62nd Army defending Stalingrad.

Vasiliy Zaytsev's memoirs create a graphic impression of war among the ruined factories and wrecked industrial installations that characterised a battle in which German attackers and Soviet defenders were all too often within range of a thrown grenade, or, literally, a wall's width apart.

Zaytsev soon proved his value not merely as a sniper but as a teacher. He was particularly adept at making use of cover and became a leading advocate of changing position after each shot. Innovations included the use of 'sixes' – three two-man teams, one central and two flanking – to ensure that each target area was always covered from different angles. In addition, like many other Russian snipers, Zaytsev was always willing to work outside the Soviet defensive perimeter when the occasion demanded.

This use of snipers in an attacking role replicated that of the Native American snipers of the Canadian Army during the First World War (see Chapter Three) and contrasted with German marksmen, who tended to remain within lairs and strongpoints in their own territory.

Taking part in an assault on German lines in January 1943, Vasiliy Zaytsev was injured by a mortar-shell blast and temporarily lost his sight. He eventually recovered well enough to return to the front, and ended the war as a captain. He had been awarded the title 'Hero of the Soviet Union' in February 1943; four Orders of Lenin, two Orders of the Red Banner and a host of lesser medals eventually made him among the most highly decorated of the Soviet snipers. Settling in the Ukraine after the war, Zaytsev rose to become director of a textile factory and died in Kiev on 15 December 1991.

It is regrettable that so few memoirs survive, as Zaytsev's active career, comparatively short and centred largely around Stalingrad, does not typify everyone's experiences. However, Yevgeni Adrianovich Nikolaev wrote not only a memoir but also a history of sniping compiled largely by interviewing fellow snipers.

Yevgeni Nikolaev, born in Tambov on 1 September 1920, worked as a set designer in a theatre before being drafted in October 1940. Serving with the 154th Infantry Regiment of the NKVD in Karelia and then on the Leningrad Front, Nikolaev was subsequently transferred, with the remnants of his division, to the 14th Rifle Regiment of 21st (NKVD) Infantry Division of 42nd Army.

Among the first snipers to be deployed on the Leningrad Front, Nikolaev killed forty-three Germans from 25 October to 30 November 1941 (including eleven on 29 November alone); he was awarded the Order of the Red Banner, and was even given a new sniper rifle – possibly a 1940-type Tokarev with a PU telescope sight. By 1 May 1942, his tally had reached 124; and on 5 August 1942, *Pravda* reported that Nikolaev had killed 187 enemy soldiers, including '104 fascists in three days'. In December 1942, however, after sustaining yet another serious wound, Yevgeni Nikolaev was transferred to the counter-intelligence service, Smersh, and his 'sniper account' finished at 324 kills ('including a general').[9]

It is possible that Ivan Sidorenko is also due much more attention than he is generally given. Unusually among the snipers, who were often drawn from the rank-and-file, Sidorenko already ranked as a junior lieutenant when the Germans invaded the USSR. He proved to be not only an excellent shot, but also a first-class teacher and was charged with organising sniping courses at a time when many of the high-scoring snipers had hardly begun their service careers. Eventually, on 4 June 1944, after being wounded for the fourth time, Ivan Sidorenko was ordered from the front line to concentrate on training, and is credited with the development and introduction of techniques that are still widely used.

Snipers were only credited with kills if there were witnesses, and so, in theory, each success had to be verified by a colleague or their commander. Vasiliy Zaytsev testified that:

The rest of the guys had been keeping track of their kills – I wasn't the only one. We'd set up daily 'personal head counts'. Each sniper had to confirm his kill with an eye-witness's signature. It was true, I had quite a few more than anyone else, Sasha Gryazev's own signature attested to this.[10]

Among the snipers were some who specialised in counter-sniping, though their successes, hard earned, are now too often overshadowed by propaganda stories such as the duel between Vasiliy Zaytsev and the 'Major König' (or 'Konings') or between Lyudmila Pavlichenko and 'Otto von Singer'.[11] Counter-sniping proved to be a very real art. After describing his own meticulous preparations, which included lengthy periods of observation and even creation, if deemed necessary, of a dummy, Vasiliy Zaytsev observed that:

> Experienced Nazi snipers moved into their positions under cover of machine-gun fire, in the company of two or three assistants. Thereafter, they worked alone. When facing such a lone wolf, I would usually pretend to be a beginner, or even an ordinary infantryman. I would dull my opponent's vigilance, or simply play around with him a bit. I would set up a decoy to draw his fire. The Nazi would soon become accustomed to such a target and would stop noticing. As soon as something else distracted him, I would instantly take up the position of my decoy. For this I needed only a couple of seconds – I'd kick the decoy aside and catch the sniper's head in my crosshairs.[12]

Among the most successful of the Russian counter-snipers seem to have been Vasiliy Ivanovich Golosov of 81st Guards Rifle Regiment (70 snipers in a total of 422 kills gained by the end of July 1943), Lyudmila Pavlichenko of the 54th Rifle Regiment (36 in 309), Zhambyl Yeshevevich Tulaev of the 580th Rifle Regiment ('30 or more kills' in 313), and Yefim Pavlovich Kiryanov of the 81st Guards Rifle Regiment (30 in 200 kills achieved prior to September 1944).

The highest scorer among the women was Lyudmila Pavlichenko of the 54th Rifle Regiment with 309 kills, even though her service career was cut short by injury in 1942. Next came Liba Rugova (242 kills), but the rankings thereafter are debated. It is generally agreed that Olga Aleksandrovna Vasileva obtained 185 kills, followed by Ekaterina Zhdanova (155 kills), Inna Semenova Mudretsova (1918–2000, 143 kills), Genya Peretyatko (148 kills), Nadezhda 'Nina' Pavlovna Petrova (1893–1945, 122 kills) and Tatyana Ignatevna Kostrina (120 kills).[13]

Pavlichenko is the best-known. Born Lyudmila Belova in Bila Tsverka in the Ukraine on 12 July 1916, she fell pregnant while still at school and married Alexey Pavlichenko.[14] Expelled from school and something of an outcast in her native village, Lyudmila Pavlichenko soon left for Kiev. There she worked first as a labourer and then as a turner in the Arsenal manufactory, while studying at night to obtain a high-school diploma. She passed the entrance examination for Kiev University, and embarked in 1937 on a five-year history course

Pavlichenko joined an Osoaviakhim shooting club in her teens and had won many trophies with her 5.6 mm TOZ 3-8 rifle. She was among the first to volunteer for combat when the Germans invaded the USSR in the summer of 1941 – but only after having to persuade a sceptical recruiting officer that she was an experienced shot.[15]

Lyudmila Pavlichenko with her favoured SVT-40/PU sight combination, possibly pictured shortly after the official presentation of the gun in March 1942.
Soviet official photograph/Novosti Press Agency

Private Pavlichenko was assigned to the 54th Rifle Regiment, 25th Rifle Division, fighting initially to defend Odessa before moving on to Sevastopol. Her first hundred kills were achieved with a bolt-action Obr. 91/30 Mosin-Nagant. Her memoirs record how, in March 1942:

> The general looked at me calmly, even impassively. 'Comrade Sergeant', he said in a low husky voice, 'For success at the Front of this Command, I present you with this sniper rifle in my name. Beat the Fascists without pity.'
>
> The adjutant of the divisional commander gave me a brand new rifle 'SVT-40' with a telescope sight, 'PU', shorter and lighter than the sight 'PE'. On the metal tube could be seen, beautifully engraved, '100. For the first hundred kills by L. M. Pavlichenko from the [sniper]

Lyudmila Pavlichenko, seen in an official portrait taken shortly before she departed on her tour of the USA and Britain at the end of 1942.

teams. Major-General I. Ye. Petrov, 25th Div.' 'I serve Soviet Union!', I solemnly replied, then kissed the blued barrel.[16]

This confirms that Pavlichenko used both of the Red Army's standard sniper rifles, even though most surviving images (including some probably taken after she had been retired from front-line service) show only the Tokarev. However, photographs taken early in March 1942, included in post-war memoirs, show her carrying the Mosin-Nagant. By May 1942, Lieutenant Pavlichenko was being credited with 257 kills, but had been evacuated from the Crimea to fight elsewhere. In June 1942, a fragment from a mortar bomb badly injured her leg, and the authorities decided to withdraw her from combat to concentrate on her value as a propaganda tool.

Later in the year, Pavlichenko and another high-ranking sniper, Vladimir Nikolaevich Pchelintsev of the 11th Rifle Brigade, were sent on a tour to North America and then Britain to raise awareness of the desperate struggle on the Eastern Front. When she returned to the USSR, Pavlichenko was awarded the title 'Hero of the Soviet Union'. However, she was deemed to be too valuable to risk death in combat and, against her wishes, Lyudmila Pavlichenko became a sniper instructor. She ended the war as a major, finished her university course, became a research assistant to the supreme headquarters of the Navy (1947–53), and died on 10 October 1974. Decorated not only as a Hero of the Soviet Union, but also with two Orders of Lenin and several campaign medals. Pavlichenko contributed a brief account of her career to a Soviet historical publication shortly before her death, and her memoirs appeared posthumously.

'Lady Death' was unquestionably a very good shot and her score stands as an all-time record for a female sniper. Her performance has been questioned

Elizaveta Mironova, killed in 1944 after achieving about a hundred kills.
Her rifle is a standard sniper-type 91/30 Mosin-Nagant with a PE optical sight
in an over-chamber mount. *Soviet official photograph/Novosti Press Agency*

in print, implying that it owes more to propaganda than fact; unfortunately, Soviet personnel records, which could perhaps resolve the problem, remain inaccessible to Western researchers. Pavlichenko was not the only one to strike terror into the hearts of the enemies of Russia. The Red Army's women snipers were greatly feared by the Germans, gaining the nickname *Flintenweibe* ('women of steel').[17] As a group, they were young and extremely proficient – which may be partly due to admission criteria which included the mandatory completion of several grades of education. Many male snipers, by comparison, were so poorly educated that they could barely sign their names.

The renowned American cultural anthropologist Margaret Meade (1901–78) recorded in an interview that, 'There are no rules for the aroused female defending her young, her family or their territory. There's no built-in chivalry here; she'll fight to the attacker's death.'[18]

The women were drawn from all walks of life: when war began, Lyudmila Pavlichenko had been studying history; Klavdiya Panteleyeva, only fifteen years old, was working in a munitions factory; Aleksandra Medvedeva, still at school, had been working in a textile factory under an assumed name; and Nina Lobkovskaya, also still a scholar in June 1941, began to train as a medical orderly before transferring to sniper school.

Among the most interesting was Nina Petrova, who scarcely fitted the image of the twenty-year-old woman sniper that the propaganda machine was apt to conjure. Nadezhda 'Nina' Petrova had been born on 27 June 1893 in Oranienbaum, but moved with her family to St Petersburg. Her father died, leaving her mother to raise five children, and so Nadezhda enrolled in a trade school after ending her elementary education. After a short spell in Vladivostok, she returned to Reval (now Tallinn, Estonia), accompanied by a small daughter, to work in a shipyard.

An all-round sportswoman, she was also a good shot. Consequently, in the mid-1930s, Petrova became an instructor in the local Komsomol marksmanship school and in 1939 won the All-Union shooting prize at the age of forty-six. When war began, she volunteered for service on the Leningrad Front but was not initially allowed to act as a sniper owing to her age. By 1943, old enough to be mother to most of the snipers, Petrova finally overcame objections and opened her personal account; on one day alone, 16 January 1944, she achieved eleven kills.

Her score had risen to 122 by 1 May 1945, but she was killed in a vehicle accident next evening, 2 May. Petrova left a daughter and a granddaughter, and had written shortly before her death that:

> The fierce hatred for the enemy has led me, a fifty-year-old woman, to the sharp edge. I killed thirty-two Nazis and then, with my friend Konstantinova, began to train snipers. We have raised 774 master marksmen. Our students have destroyed more than 2,000 Germans.[19]

On 16 March 1942, the People's Commissariat of Defence decreed that every rifle section should have three snipers, making nine to a company and 87 in an infantry regiment. The potential shortfall was answered on 20 March by establishing the Central School of Sniper-Instructor Training in Vishnyaki, a short distance from Moscow. Work began in November to teach 1,000 instructors and 3,010 students (including 450 women), but no sooner had progress been made than an appeal made by Komsomol in January 1943 resulted in 490 women applying to be snipers. As a direct result, a Central Women's School of Sniper Training was established on 21 May under the direction of Nora Chegodaeva, a veteran of the Spanish Civil War. Of 1,885 female snipers/instructors who served the Red Army in 1942–5, 1,468 (including 407 instructors) were trained in this highly successful school in three large groups from June 1943 until April 1944.[20]

A sniper's badge was introduced on 21 May 1942 so that even comparatively minor successes could be recognised. Training was also given by armies in the field, unofficially at first, by Osoaviakhim and by Vsevobucha ('General In-Service Military Training'). As many as 400,000 are said to have qualified as marksmen.

To be accepted, the women needed to be 18–26 years old (overruled in a few specific cases), and physically fit with excellent eyesight; they had to have

Soviet women snipers of the Second World War: *from left*, Lyubov Makarova, Klavdiya Panteleyeva and Roza Shanina. *Soviet official photographs/Novosti Press Agency*

completed 'seventh year education', with no family problems, and should not have 'previous wounds'. Soviet records suggest that the 808 women who graduated with 'Outstanding Sniper' status achieved 9,994 kills.

An effort was made in the early twenty-first century to record interviews with surviving female sharpshooters, who, with the exception of Lyudmila Pavlichenko, whose publicity trips in 1942 made her an exception, had largely been airbrushed from history by the post-war male-dominated militocracy.

Their reminiscences, which are now readily accessible,[21] give a fascinating insight into the way in which Soviet snipers were trained. Not only did this begin in earnest well before most other nations – usually forgetting lessons learned in the earlier world war – started to take the value of sniping seriously, but it also developed on lines which are now commonplace.

For example, the validity of pairing snipers, alternately observing and shooting to maintain concentration levels and minimise the effects of fatigue, was conclusively demonstrated. Observation skills, use of camouflage and the importance of making the first shot count became paramount.

The life of a sniper was often short. Little more than five hundred women survived, approximately one in every four, and it is assumed (reliable statistics are lacking) that the casualty rate among the men was broadly similar. Margins between success and failure were fine. Klavdiya Panteleyeva (Kalugina) remembered how:

> We stood watches during the day, and soldiers watched during the nights; they slept in the daytime. And so Marusia [her partner Marusia Chikhvintseva, picked simply because they had stood next to each other on enrolment] and I placed our rifles at one embrasure and

watched the German defences from the breastwork. But the Germans also put a sniper to watch us. And so I was watching, observing during my shift (because the eyes would get tired), and Marusia said: 'Let me take the watch now.' She got up, it was a sunny day, and she apparently moved the lens. As soon as she got up, there was a shot, and she fell. Oh, how I cried! The German was 200 metres away from us. I screamed so loud it could be heard all over the trenches, soldiers ran out: 'Quiet, quiet, or they'll open mortar fire!' But how could I be quiet? She was my best friend.[22]

Nina Lobkovskaya, a highly successful sharpshooter who finished the war commanding a Third Shock Army sniper section recalled:

One day, quite unexpectedly, I looked into my sniper eyepiece and saw a German in a white shirt, a high-collar jacket draped over his shoulders and a peaked cap. I was shocked because he was so close! I sized up the distance and aimed. It was a hasty shot, so the bullet only smashed his cap badge. He understood that the shooting came from a sniper, shook his fist at me and disappeared behind his disguise. I remember him well enough: a young man, rather good-looking, I'd say. I couldn't sleep that night, my mind was racing as I was trying to recall every detail, every trifle. From then on, a hunting game began, each stalking the other. I felt his presence, and he felt mine, I could tell that.

The duel lasted for a week. One day I got careless and let him spot me. He fired right away. The bullet hit the metal setting of the back-sight, ricocheted into the helmet and singed my temple. As I saw blood, I thought of one thing only: to fire back. So I snatched my partner's rifle and began to watch him from another angle. I knew he was bound to reveal himself somehow. And he did, for I soon saw the line of his helmet above the bushes. I took aim below that line and shot. No more shots were fired from the spot. As it became clear afterwards, it was not a sniper but an officer practicing his skills by shooting at our soldiers. I was satisfied, but I will remember his hand clenched into a fist shaking at me for the rest of my life.[23]

Snipers were expected to act as infantry if there was no need for their specialist skills, exchanging their rifles for submachine-guns, and many of the women doubled as medical orderlies. Some of the latter are credited not only with kills, but also with saving many lives.

In addition to rifles, snipers were issued with two fragmentation grenades. These were to be used to repel close-quarter threats, but were unofficially regarded as suicide weapons. Lucky was the captured sniper to be simply shot; most were tortured before being killed in unimaginably horrific ways.

Women were especially vulnerable, and there are well-documented cases of mutilation prior to execution. There was always another option: on 14 August 1942 – at least according to a surviving witness – Natalya Kovshova, too badly wounded to flee to safety, pulled the pin on a grenade to kill not only herself and her partner Maria Polivanova, but also some Germans who had entered their trench.

Photographed in 1944, these snipers of the Second Baltic Front all carry the 91/30 Mosin-Nagant rifle and PU optical sight. The original caption names them as (*left to right*): Sofia Kutlomametova, Antonina Dyakova, Olga Maryomkina, Tatyana Kuzina, Antonina Komarova, Lidia Onyanova, Klavdiya Ivanova, Maria Aksionova, Raisa Skrytnikova and Olga Bykova. *Soviet official photograph/Novosti Press Agency*

The Soviet sniper's weapons were quite conventional. When the war began, most snipers, judging by published photographs, carried a variant of the bolt-action Obr. 1891/30 infantry rifle, the 'Three-Line' Mosin-Nagant developed in tsarist days. These guns had been specially selected for their accuracy, and were fitted with the PE and PEM sights described in Chapter Four.

By 1940, the Soviet authorities appeared to be convinced that the auto-loading Tokarev was an improvement on the manually operated Mosin-Nagant, and had even contemplated universal issue. Production gathered momentum in 1940, when the Snayperskaya Vintovka Tokareva ('Tokarev sniper rifle', SNT) was adopted at the beginning of October and production of the 1891/30 bolt-action sniper rifle was halted.

The SNT is little more than a standard SVT selected for its accuracy. There is evidence to suggest that the performance of standard Tokarevs can vary

This sniper fires a 1940-type Tokarev (SNT-40). The picture clearly shows the value of cloaking the rifle in netting! *Author's collection*

considerably – the German acceptance standard for captured guns was merely to place shots in a 12 cm circle at 100 metres – but no study has been made specifically of 'pre-invasion' examples, which are generally much better made than those dating from 1942 or 1943 and could, perhaps, be expected to perform better. Production of the SNT-40 amounted to 34,872 in 1941 and 14,210 in 1942.

Adapting the Tokarev for sniper use, however, posed a problem: the gun ejects up and to the right, and so, even though the box magazine was detachable, the PEM optical sight and the standard mounting rail of the 91/30 Mosin-Nagant could not be used. An answer was found in a new short-body 3.5× sight, the Pritsel Ukorochenniy ('PU'), which had been created in the Feliks Dzherzhinsky manufactory operated in Kharkov by the directorate of internal security forces (NKVD). Approved on 18 July 1940 specifically for NKVD units, to whom, presumably, some of the earliest Tokarev rifles had been issued, the PU was carried in a special cantilever mount with arms extending forward to engage grooves milled longitudinally in the receiver immediately above the stock. The mount was locked in place by screws.

The earliest sights had the adjuster drums on a separate collar, raised above the surface of the lens tube. The lens holders were brass and the tube was made of lightweight alloy. However, experience showed the sights to be not only too flimsy but also unnecessarily hard to make. A modified version with the adjuster drums mounted directly on a steel tube-body, with alloy or steel lens holders, was substituted rapidly enough for the original design to be rarely seen.

The success of the PU soon persuaded the authorities to extend issue to the Red Army, where the 1938-type Tokarev had been issued to artillery NCOs, and to the marines. Manufacture began in State Factory 357 in Leningrad, where about 15,000 sights had been completed by the end of 1940; the Dzherzhinsky factory contributed an additional 5,700.

Experience showed that the mount, though a clever engineering solution, was insufficiently rigid to maintain 'zero' reliably. Combined with operating problems and inaccuracy compared with the 91/30, this was sufficient to damage the reputation of the SNT-40 even though some individual snipers made good use of them (particularly women, owing to the softer recoil). The best-known is Lyudmila Pavlichenko: hardly a photograph of her in action in 1941–2 shows anything other than an SNT, and the same is true of snipers drawn from the marines for whom the Tokarev was standard issue.

Though the bolt-action Mosin-Nagant had soon been reinstated as the standard sniper rifle, issue of Tokarevs during the desperate defensive battles of the first phase of the war – Leningrad, the Crimea, Stalingrad – made good sense. The semi-automatic could fire a second and successive shots far quicker than a bolt-action rifle, and could, for example, eliminate a machine-gun crew in the blink of an eye. This could be the difference between life and death not only for the sniper but also his or her unit.

Though the Russians showed an ambivalent attitude to the Simonov and the Tokarev, which were deemed to be much too complicated and temperamental to suit the ill-educated Red Army infantryman, the Finns were only too pleased to turn guns captured during the Winter War against their erstwhile owners.

Germans fighting on the Russian Front, particularly in the early part of the war, were also keen to acquire Tokarev rifles on the grounds that they gave much better firepower than the Kar. 98k and had a much longer range than the Schmeisser machine pistols. Even when the Gew. 41 (W) was issued, the Russian rifles – Selbstladegewehre 258 [r] and 259 [r], 1938 and 1940 patterns respectively – were clearly much more combat-worthy. It is not unusual to identify Tokarev rifles in the hands of Wehrmacht personnel serving on the Eastern Front, and there are many testimonies to large-scale issue during the periods in which $7.62 \times 54R$ ammunition was readily available.

Major a. D. Hans Rudolf von Stein (1911–2000), a leading authority on the weapons of the Wehrmacht, recalled that he had procured 'more than thirty Tokarevs and a Simonov' to arm the cavalry squadron he was commanding in Russia in 1942 and also that they had been a valuable addition to his unit's firepower.[24] The Oberkommando des Heeres, the Army High Command, even published details of acceptance tests that should be applied to the Tokarevs before they entered German service.[25]

More than a million SVT were made in 1941, when production was gaining on the Mosin-Nagants (1,066,643 SVT and SNT compared with 1,292,475 Obr.

1891 rifles and Obr. 1938g. carbines). Immediately after the German invasion in the summer of 1941, however, the situation changed dramatically. With machine-time in short supply, owing to the occupation of much of the western USSR, and the enforced relocation of much of the industrial plant from threatened areas, the Tokarev was too complicated to justify production. In addition, severe operating problems were being reported from the front line and, as there was no time to resolve them, manufacture of the Tokarevs was radically curtailed. Only 278,358 were assembled in 1942, little more than a quarter of the previous year's efforts. Work concentrated instead on the Mosin-Nagant, which was simpler to make, easier to use and much more robust.

The Obr. 91/30 was put back into production in March 1942, and the SNT was officially withdrawn in October 1942. Reasons for withdrawal included too much muzzle flash, which could betray a sniper's position at night; too much mechanical noise during the firing cycle, which was a potential liability in silent conditions; and deteriorating accuracy at ranges beyond 800 metres.

No Tokarevs were made after 3 January 1945, and it is difficult to gauge how many had actually been made during the war. Clearly, there were at least 1.3

Said to have been taken in 1942, this purports to show Red Army soldiers defending their Motherland against enemy aircraft. In addition to the 91/30 Mosin-Nagant, with fixed bayonet, a PTRS anti-tank rifle can be seen. Vasiliy Zaytsev, among others, testified that 14.5 mm-calibre rifles were regularly used to support snipers when something more powerful than a Mosin-Nagant was needed. They were, perhaps, the inspiration for today's anti-matériel rifles. *Soviet official photograph/author's collection*

million of the 1940-type guns, but very few are to be found dated later than 1942 and it can be assumed that total production did not greatly exceed 1.5 million when assembly finally ceased. This was as much due to the dislocation of production facilities as to the gun's inherent weaknesses.

Known colloquially as *Sveta* ('Light' in Russian), the SVT is now widely regarded as a failure. However, its production exceeded all other wartime 'full-bore' automatic rifles excepting the US M1 Garand. It seems unlikely that production of the only other possible challenger, the German Gew. 43, approached half the Russian total. The Tokarev is somewhat complicated and difficult to field-strip compared with the Garand, but it has a better magazine and an efficient locking system that appears to have inspired the post-war Belgian ABL and FAL.

Trials have shown the SVT to be accurate and acceptably reliable as long as modern ammunition is used, and this illustrates its major problem – the rimmed 7.62×54 mm rifle cartridge, which contributed to rim-over-rim magazine jams and poor extraction.[26] Ammunition made in Russia in the period between 1917 and the end of the Second World War was, by and large, very poor owing to the parlous state of the contemporary propellant industry, and ammunition supplied from the USA under the Lend-Lease scheme contained a flash-suppressor which gave excessive carbon fouling. Fluctuations in pressure ensured that the self-loading weapons were prone to malfunction. In addition, the slower rifling twist did not stabilise heavyweight projectiles as well as the 91/30 Mosin-Nagant. Consequently, the SVT shot most accurately with Type L ammunition.

A few fully automatic derivatives of the standard SVT (the AVT Obr. 1940g.) were made during the winter of 1941 and the following spring, when the Red Army was desperately short of the machine guns necessary to stem the German advance. It has been claimed that these were developed specifically for snipers, but it is difficult to see why this would have been necessary unless the snipers, who were much better trained as a group than the rank-and-file, were seen as the best people to provide local fire support. Selective-fire guns could be distinguished by a much heavier stock, and may have introduced the two-port muzzle brake that was fitted to the semi-automatic guns thereafter.

Semi-automatic rifles seen in wartime photographs and propaganda films are almost always Tokarevs. However, shortly after the German invasion, when losses of weapons and the dislocation of production facilities forced virtually anything that could shoot to be pressed into service, the military authorities reconditioned the few AVS-36 that had been stored since the Winter War. At least some of these were re-issued to snipers in 1941, with a PEM optical sight in a special mount attached to the left side of the receiver.

The AVS had proved to be too light to fire fully automatically, and had a reputation for breaking parts. However, it was passably reliable and acceptably accurate, so the sniper version was simply restricted to semi-automatic fire. How many of these guns had survived is not known, but, as production of the AVS only amounted to about 65,800 (1934–40), there were probably no more than a few hundred of them.

The failure of the Tokarev sniper rifles under service conditions forced the Soviet military authorities to re-introduce the 91/30 Mosin-Nagant. The only major change was the standardisation of the short-body PU telescope sight that had been introduced with the SNT. Smaller, lighter and much easier to make than its immediate predecessor, the PEM, the PU performed adequately if unspectacularly. The worst feature was the absence of a focus ring, a problem shared with the PEM, which meant that eye-relief was critical (usually given as 115–120 mm) and only firers with good eyesight could make the best of it. In addition, the flat-fronted objective lens, without any protective shade or notable recessing, was prone to reflect light and probably cost the lives of many unwary snipers simply by drawing attention to their position.

A new mount was required to fit the PU to the Obr. 91/30g., so a baseplate was attached to the left side of the action body by two pegs, two screws and two set-screws. Credited to engineer D. M. Kochetov, the mount consisted of a two-ring monoblock attached to the baseplate rail by a ball-and-socket system at the front and a large knurled-head screw angled against the base at the rear. Set-screws in lugs protruding laterally at the rear of the baseplate allowed elevation adjustments to be made during the setting-up procedure.

The earliest mounts (now never seen) are said to have had two large weight-saving cut-outs in the vertical standard, but the perfected version had a small recess, cut vertically with a rotary miller, on each side of a central circular hole.

Very few changes were made to the rifles during the war, though finish and the standards of machining declined in 1942–3. Laminated stocks were introduced towards the end of 1943, but seem to have been confined to guns made in Tula. They are usually reckoned to have been at least equal in strength and torsional rigidity to the one-piece type. Most wartime Mosin-Nagants were made in Izhevsk. However, separating quantities of sniper rifles from standard production has proved to very difficult. It seems that about 54,160 were made prior to 1938/9 (when work stopped in favour of the SVT) with PE and PEM telescope sights, mostly set up in Tula; and that at least 275,000 were made in Izhevsk with PEM and PU sights in 1942–58. Tula contributed far fewer sniper rifles in 1943–5, almost all fitted with PU sights.

It is assumed that legions of sub-contractors were involved, but details are lacking. By 1945, optical sights had been made by at least five manufacturers. Distinctive markings on the sight bodies take the form of simplified lenses or prisms, the precise design identifying the originator. The marks are invariably accompanied by the hammer-and-sickle beneath a small five-point star.

The Obr. 91/30g. sniper rifle was a potent weapon. It was retained until the adoption of the SVD (Dragunov) auto-loader, which was issued for large-scale trials as early as 1958 – when the last batches of Mosin-Nagant sniper rifles were made in Izhevsk – and officially adopted in 1963. But many 91/30 rifles reappeared in Vietnam, where their performance was still good enough to force US military personnel to take care not to expose themselves unnecessarily.

Production that exceeded 250,000 units in 1941–5 alone makes the Obr. 91/30 numerically, if not necessarily technically, the most important sniper rifle of the two world wars.

The British Empire

The result of more than a decade of trials and tribulations, the Rifle No. 4 Mk I was approved on 15 November 1939 to replace the No. 1 Mk III* (the perfected form of SMLE). However, it was clear that the demands of war, which had been predicted since the German annexation of Austria in 1938 and the move into the Sudetenland area of Czechoslovakia in 1939, would not be satisfied by the output of Enfield (concentrating on Enfield revolvers and the Bren Gun) and BSA (making Browning machine guns, Boys anti-tank rifles and magazines, SMLE Mk III rifles, Bren parts and magazines in Small Heath, and Besa machine guns in Redditch).

The decision was taken not only to create new production lines in Britain, but also – in the dark days after the withdrawal from Dunkirk, where huge quantities of weapons had been lost – to recruit additional contractors in North America. This had been done successfully during the earlier war, when P/14 rifles had been made exclusively by Eddystone, Remington and Winchester, and it was anticipated that comparatively few problems would be encountered.

BSA built a factory in the Shirley district of Birmingham, and new Royal Ordnance Factories were erected in Fazakerley in Lancashire, close to Liverpool, and in the Yorkshire town of Maltby. The introduction to service of the No. 4 Mk I and a variety of No. 4 bayonets – Mks I (already 'Obsolescent'), I*, II and II* – was published on 13 February 1941 in L.o.C. § B4737.

ROF Maltby is said to have delivered the first series-made No. 4 Mk I rifle as early as August 1940, followed by the first 25 from BSA in June 1941 and then 300 from Fazakerley in July 1941. Production accelerated sharply in the autumn of 1941, however, and 12,749 rifles were delivered in December.

The delays had persuaded the British authorities to upgrade about 2,000 No 1 Mk VI Model B and No. 4 trials rifles to No. 4 Mk I standards. The No. 1 Mk VI can be distinguished from the standard No. 4 Mk I by the lower bolt-way wall on the left side of the action body and by minor differences in the stocking arrangements.

No sooner had the mass-produced No. 4 rifles reached service than complaints were being made about sharp edges, rough machining, and the

A left-handed British sniper takes aim with a Rifle No. 4 Mk I (T). Note the use of netting to replace a cap or helmet, and the use of a leafy bush as natural cover. However, no attempt has been made to break-up the silhouette of the rifle and telescope sight, which suggests that the photograph was taken during a training exercise. *Author's collection*

reduction in length that was due to the adoption of a 'spike' bayonet. Ironically, the new bayonets had a far greater effect on shot-strike than the Pattern 1907 sword bayonet had done. This was largely due to the way in which the bayonets were mounted. The No. 1 Mk III* (SMLE) had a heavyweight nosecap to which the bayonet attached, largely isolating its effect from the barrel; the No. 4 bayonet, though much lighter than the sword type, attached directly to the muzzle.[27]

Changes were made to individual components during the war in a successful attempt to simplify – and thus accelerate – production; there were several differing No. 4 back sights, and two-groove rifling, tested and approved in 1941, was successfully used until July 1945. Guns made during the later stages of the war could have butt plates of Mazak alloy, stocks of poor-quality wood, and swivels which were simply thick wire bent to shape. Clifford Shore was unimpressed:

> Many of the early rifles were poor and had many faults, at one stage opinion was dead against the model. Bolts needed a lot of work before they functioned decently; rear sights had considerable lateral play; magazines were faulty; the bores were poorly finished due to lack of machine operations, and there was great tolerance and allowance manifest when passing the bores for gauging ... Some of the rifles were terrible ... War production, in which time is a most vital factor, resulted in dispensing with a number of the peace-time gadgets which were incorporated in the original design of the rifle.[28]

> It is my opinion that the No. 4 did not stand up to the wear and tear
> of active service conditions as well as the S.M.L.E.

The No. 4 Mk I* rifle was approved on 14 June 1942 for manufacture in the USA and Canada, though introduction was not announced in the *Lists of Changes* until 1946. Alterations had been made to the action body, where the 'catch head, breech bolt' was omitted, and the trigger was pivoted on the underside of the body – far better than in the Mk I, where the pivot lay in the front part of the trigger guard/magazine floorplate assembly. Attaching the trigger to the body ensured that the relationship between the trigger and the sear was not affected by distortion in the stock or heat arising from rapid fire.

About 2.02 million Lee-Enfield rifles were completed by BSA and the Royal Ordnance Factories during the Second World War.[29] Ironically, more No. 4 rifles came from Canada and the USA than had been made in Britain: about 2.15 million of them, predominantly Mk I* though a few Mk I rifles were also made. Small Arms of Long Branch, Toronto, made approximately 911,000, and the Savage Arms Company of Chicopee Falls, Massachusetts, made 1.236 million (1,196,706 of which were dispatched through the government of the USA to the 'British Empire').

The need for snipers was acknowledged, in some quarters at least, long before Britain declared war on Germany on 3 September 1939. But there were far too few telescope-sighted rifles to fulfil the need. Consequently, the No. 3 Mk I* (T) rifles that had been in store were fitted with the P/1918 telescope sight, which had been adapted by ordnance technicians from the Periscopic Prism design of the First World War.

In addition, the surviving Aldis sights that had been taken from SMLE rifles in 1921 were fitted by Alexander Martin & Co. of Glasgow, in special non-detachable mounts, to about 400 No. 3 Mk I* W (F) rifles – Winchester made P/14 Mk I* with finely adjustable sights. Issued as 'Rifles No. 3 Mk I* (T) A[ldis]', they had mounts which were offset to the left side of the action to allow the guns to be charger-loaded.

Introduced on 12 February 1942 by L.o.C. § B6861, the No. 4 Mk I (T) sniping rifle was accompanied by 'Chest, Small Arms, No. 15 Mk I', 'Rest, Cheek, Telescope Rifle', a leather sling ('US loop pattern'), 'Telescope No. 32' with case ('No. 8 Mk I'), cap ends, an adjusting tool and a polishing cloth. Little more than a specially selected example of the standard No. 4, sniper rifles were all fitted with the Mk I or 'Singer' back sight with a screw-adjusted slide; the 'point-blank' aperture was removed from the base of the sight, as it caught on the telescope body, and the sight leaf was usually given a non-reflective Parkerised finish.

Production of the Telescope, Sighting, Straight, No. 32 ('recommended for adoption' in March 1940) began long before its introduction was announced

The breech of a standard No. 4 Mk I (T) rifle, with the No. 32 Mk III telescope sight.

alongside the No. 4 Mk I (T) on 12 February 1942. The 3×19 sight had been developed for the Bren Gun, with which it was never issued.

The No. 32 Mk I weighed 2 lb 3 oz, surprisingly heavy, but proved to be exceptionally robust once teething troubles had been overcome. Field of view was about nine degrees. An elevation drum on top of the sight, graduated from 100 to 1,000 yards, had 50-yard 'clicker' increments; the windage drum, on the left side, was adjustable in clicks of 2 MOA. A sliding brass eyepiece could be extended to shade the ocular lens, but experience showed this to be ineffectual and the eyepiece was abandoned after only a few Mk II sights had been made; many older sights were subsequently modified to comply.

Announced on 23 April 1943, the No. 32 Mk II sight was an improved form of the Mk I, with 1 MOA click adjustments on both drums. The No. 32 Mk III (7 October 1944) had an improved anti-backlash system to cure the occasional tendency of Mks I and II to 'lose zero', and the adjuster drums, previously slightly offset, lay on the centre line of the sight-tube. A change was made to the way in which adjustments were made, requiring a special tool to lock the drums – a backward step, as a third hand was needed to accomplish the task satisfactorily without putting the rifle down.

The sights are usually clearly marked with their designation and mark, but, if these have been defaced, the accompanying optical stores parts numbers can sometimes resolve confusion: 'O.S. 466A', Mark I; 'O.S. 1650A', Mark II; and 'O.S. 2039A' for the Mark III.

British No. 32 sights were made by well-known manufacturers of optical equipment including Cooke, Troughton & Sims (mark: a 'CTS' monogram), best known for theodolites and surveying equipment; the Houghton–Butcher Manufacturing Co. ('H.B.M.Co.'); A. Kershaw & Sons ('A.K. & S.'); Taylor, Hobson & Co. ('T.H. & CO.'); Vickers Instruments ('V.I.L.'); and Watson & Sons ('W'). The cast-iron mounts were supplied by several contractors, their origins usually hidden by monograms, initials such as 'JG' and 'KD', or regional identifiers. The 'Northern Region' code 'N92', often found on the mounts, was allocated to John Dalglish & Sons of Avenue Ironworks, Pollokshaws, Glasgow.

Canadian No. 4 Mk I and I* sniper rifles may be encountered with the 3.5 × 24 C No. 32 Mk 4 sight, made by Research Enterprises in Sherbrooke, Ontario, which was a straight-tube design with the windage adjuster on the right side instead of the left. Most firers regarded this to be an improvement on the Mks I–III, as adjustments could be made without moving the supporting hand.

A few Canadian guns were fitted with an American-made Weaver sight in a Griffin & Howe rail-type mount, attached to the left side of the action body with two screws and two pegs, but they were never issued in quantity.

No. 4 Mk I (T) rifles were selected from those that had shown excellent accuracy during acceptance tests. They were duly stripped and refinished to the highest standards, with special attention paid to the fit of the fore-end in which the barrel could float freely. Pads were precision-machined on the left side of the action body to receive the cast iron telescope-sight mount, which attached with two large thumbscrews running laterally.

The No. 32 sights were fitted individually, but the gun/sight combination was accepted only if it met the approved standards: seven shots of seven in a four-inch circle at 200 yards, six out of seven in a ten-inch circle at 400 yards. The rifle and the sight were then numbered to each other.[30] Some individual guns, of course, were capable of bettering the accuracy requirements by a considerable margin. These were easily capable of one-shot 'head kills' at ranges as great as 400 yards.

The earliest No. 4 Mk I (T), 1,403 of which were eventually adapted from trials rifles, were set up in Enfield in May and June 1940 but work was subsequently passed to London gunmakers Holland & Holland. Beginning in September 1942, Holland & Holland delivered 23,177 No. 4 sniper rifles against orders totalling 26,442 by the time the war ended in the Far East in August 1945.[31] Guns supplied from the BSA Shirley factory were preferred, as Maltby was principally an assembly point for components made by the ever-growing legion of sub-contractors and 'feeder factories', and the output of Fazakerley was generally regarded as the poorest.

The principal problem was quality control, and there is evidence to suggest that tolerances of the 'go/no-go' bore-diameter gauges were relaxed in 1943 to .303 ± 0.002. A tight bore would have a diameter of .301, therefore, and a loose bore could measure .305. These margins are broad enough to have a considerable effect on the accuracy of individual guns.

A few hundred Stevens-Savage No. 4 Mk I* (T) rifles were supplied to the British Army in 1942, but survivors of this group are rarely encountered; in addition, Small Arms of Long Branch made 1,141 No. 4 Mk I* (T) rifles in 1944–5 for Canadian service, and about 950 similar guns for Britain (though several hundred of these may have been lost in transit in the Atlantic).

When the No. 4 Mk I (T) rifle was introduced, though the likelihood of a German invasion had receded, Britain was in the grip of aerial assault – the Blitz – and the Army was engaged abroad only in North Africa. But British marksmen did not prove to be particularly effective in the Western Desert, where neither the Germans nor the Italians made much use of snipers.

Sandscape offered very little cover, and even the occasional patch of desert scrub was poor camouflage. Captain Clifford Shore, writing in *With British Snipers to the Reich*, drew attention to a less obvious problem:

> The standard of shooting in hot climates was found . . . to be lower than in temperate fighting zones, and one of the reasons put forward for this was that it was impossible to produce a high standard of shooting when men had to shoot in shirt sleeves, the causes being: – one's elbows must have a comfortable position on the ground; with only a shirt covering the elbows when crawling and shooting on hard ground life, for the sniper, was not much fun; the rifle must have a firm bedding in the shoulder, and this can never be obtained with any rifle when there is only the thickness of a shirt between a steel rifle butt and the shoulder bone . . . with a sniper's rifle it is even more difficult since due to the height of the 'scope sight the firer's head is forced into a position high on the cheek rest and consequently the end of the butt is forced down into the shoulder that amount; this makes it difficult to bed the rifle securely in the shoulder with even a sufficiency of clothing . . .

After the invasion of Italy in 1943, however, snipers were finally able to make a meaningful contribution to progress. This was greatly helped by much more helpful topography, from grassland to wooded hills and mountain foothills, and to the cover provided by deserted villages and shattered monasteries.

The same could be said of operations after D-Day, the Allied invasion of Normandy on 6 June 1944, and in the protracted advance from the *bocage normand* across France and the Low Countries to and then beyond the Rhine. However, German snipers, often hardened by service on the Eastern Front, took a toll of British, Americans and Canadians. Much of the effect was undoubtedly psychological, as snipers were often deployed simply in support of machine guns and light artillery, but there was little doubt that they were feared by the Allied rank-and-file, however much some commentators have suggested that the Germans were neither well trained nor particularly good shots.

Pictured on 6 October 1944, Sergeant Harold A. Marshall (1918–2013) of the sniper section of the Calgary Highlanders carries a No. 4 Mk I* (T) rifle fitted with the No. 32 sight; a 36M grenade ('Mills Bomb') and the handle of a Gurkha *kukri* can also be seen. The *kukri* not only made a good entrenching tool, but was also a first-rate close-quarters weapon – much better than the regulation bayonets! *Library and Archives Canada collection, PA-140408*

The British took their post D-Day sniping seriously enough to create special training establishments: permanent facilities in Britain, especially the school established in Llanberis, and more mobile groups to accompany individual armies as they advanced. The most experienced sniper-trainers saw proximity to the front line to be essential to efficient training:

> It is clearly apparent that to do its utmost good a sniping wing should be entirely on its own, as was the B.L.A. Sniping School in N.W. Europe in 1944–1945, not far behind the line and in touch with the forward area units all the time. After Cassino, the school [part of the Mediterranean Training Centre] ... was 300 miles or more behind the line and ... completely out of touch. And all the efforts to get the staff sergeants up to the line for a shot or two were useless ... Every unit in the line was crying out for assistance.[32]

Tactics and techniques had been improved, even though an unforgivable lack of overall control sometimes allowed unsympathetic officers to block progress within their own units. This was particularly true of those, and there were many, who either saw sniping as 'Bad Form' or were fearful of immediate retaliation.

The Germans, often in the blink of an eye, could hit suspected sniper positions with a short-lived (if frantic) barrage of mortar bombs and artillery shells – known colloquially as a 'stonk'. However, German soldiers were usually just as fearful of British snipers, who were said to favour head-shots and had proved to be very effective almost as soon as they had landed in France.

Though more than 20,000 No. 4 sniper rifles had been made by 1945, they were often in short supply in the front line and it is probable that many were still in Britain when the war ended. Denis Edwards, a scout-sniper of the 2nd Battalion, Oxfordshire & Buckinghamshire Light Infantry, of 6th Air-Landing Brigade, was among those dropped by glider on the night of 5/6 June 1944 to capture Pegasus Bridge over the River Orne at Bénouville. His memoir, *The Devil's Own Luck*, describes his successes and occasional failures as a sniper. On 25 July 1944, after more than six weeks in France, he noted in his diary:

> To my surprise and delight, I was issued with a brand new sniper rifle [No. IV Mk I (T)] straight out from Ordnance. It was covered with a thick layer of grease and wrapped in greaseproof paper. I spent much of the day taking it apart and cleaning it. Until this time my rifle had been a completely standard issue Lee-Enfield .303 Mk. IV, with no telescopic sights, and this new weapon was much more suitable.[33]

It is interesting to note that a marksman involved in such a strategically vital task in the prelude to D-Day should have had to use a standard No. 4, particularly as the official scales of issue of sniper rifles for British paratroops and air-landing battalions were considerably greater than those of the infantry-battalion sniper sections.[34]

It is interesting, too, to consider the complete absence of information concerning the activities of British snipers of the Second World War. There can be little doubt that some men recorded tens of kills, if not the hundreds of the earlier conflict (or the Russians of 1941–5) and that their contribution to individual campaigns would have been notable. Yet they remain anonymous. Perhaps this simply reflects the opinion, widely held by those who were not directly involved in combat (and even by some of those who were), that militarism in general and sniping in particular were necessary evils.[35]

Clifford Shore observed in *With British Snipers to the Reich* that 'in Italy, one private soldier had over 60 certified kills to his own rifle. A South African officer had more than forty kills and got the Military Cross for his sniping, but was eventually killed himself.' But even Shore names neither man: his book was published in 1948, at a time when the individual deeds of war, still raw, often passed unacknowledged.

Resisting the temptation to experiment with small-arms design, shared with the Russians, was one of the great strengths of British ordnance affairs during the Second World War. It was the polar opposite of the Germans' tendency not only to accept needless 'refinements' but also to impose manufacturing discipline only reluctantly and allow each branch of the Wehrmacht (as well as the SS) to meddle with procurement.

Consequently, the British were able to develop what was, perhaps, the most effective sniper rifle of the Second World War; tests undertaken in 1945 suggested that the No. 4 Mk I* (T) would usually outshoot its rivals, often by a considerable margin. The combination of gun and sight was undeniably heavy at 11.2–11.7 lb, but the No. 32 telescope was unusually robust and the additional weight provided a steady platform.

The Germans

The utility of the optically sighted sniper rifle had been questioned as early as 1918. Those that had been used in the Reichsheer, usually short Kar. 98AZ or full-length Kar. 98b, were withdrawn in the early 1930s (see Chapter Four).

The rise of large-scale training, camouflaged by the formation of many 'sporting clubs', was accompanied by the manufacture by Mauser of the so-called 'Deutsche Reichs-Post' or 'postal services' rifle. Many of these were supplied to paramilitary formations such as the SA and the SS (the latter sometimes with optical sights in a special mount), but the military view was that marksmanship training rendered the telescope sight superfluous. It was assumed that *all* soldiers would be able to use open sights effectively at normal combat ranges.

Mass-production of the Kar. 98k had been largely hidden by the use of numerical codes, but these gave way to alphabetical marks in 1940–1.[36] Manufacturing standards fell as the fighting progressed, creating the *Kriegsmodell* ('war model') of 1942. Stamped nose caps, barrel bands and butt plates replaced high-quality forgings and castings; crudely finished strip-metal trigger guards were fitted; and stocks were thinly varnished. Many guns had laminated stocks, the result of trials stretching throughout the 1930s. Plywood laminates resisted warping better than conventional one-piece stocks, did not require lengthy maturing, and were less wasteful not only of raw material but also of machine time.[37]

When the Second World War began and the Kar. 98k was being made in large numbers, views had changed and the issue of limited numbers of optically sighted *Zielfernrohrgewehre* was sanctioned. The *Zielvier*, with 4× magnification, was duly standardised. However, individual sights were simply acquired on the commercial market, and therefore came in a profusion of shapes and sizes.

They were usually mounted in a side rail, held to the left side of the action body with three screws. The sight was usually clamped in place. Unfortunately, the considerable power of the 7.9 mm rifle cartridge rapidly loosened the

A typical Zf.-Kar. 98k (*top*), fitted with a *Zielvier* in a side-rail mount secured with a clamping lever. Note the extension protruding above the cocking piece, which makes the safety catch easier to apply. A Zf.-Gew. 43 (*above*), with the special Zf. 4 in the standard clamping mount. *Courtesy of James D. Julia, auctioneers (www.jamesdjulia.com)*

screws, and many remedies were tried. First came three additional lock-screws, then two locating pins, and finally even an additional centrally placed vertical clamp. None of the changes proved to be as effective as hoped, as nothing had been done to cure the basic problem: the sight rail was too short and insufficiently rigid.

A better solution was badly needed, and a surprising variety of mounts appeared in service. There were post-and-claw types, acquired from commercial sources, and then the first turrets. These were made in 'high' and 'low' form, depending on the diameter of the ocular-lens tube. The sights were attached to the rifle by inserting the foot of the front clamp-ring post into the front mount at right angles to the axis of the bore, then twisting the sight-barrel round until the rear post could be clamped onto the back base-block.

Turret-mount Zf.-Kar. 98k seem to have been made only by Mauser-Werke and J. P. Sauer & Sohn. High mounts were often cut away to allow open sights to be used, but the additional height of the optical sight above the bore was a potential liability in combat, as it exposed more of the sniper's head.

The invasion of Russia in the summer of 1941, initially very successful, exposed the German Army to Russian snipers, men and women, who exacted a dreadful toll on their opponents. The Obr. 1891/30 Mosin-Nagant sniper rifle proved to be much more accurate than the Zf.-Kar. 98k, something even the Germans admitted, and the SNT (Tokarev), while incapable of matching the bolt-action 91/30 for accuracy, could sustain a greater rate of fire. One immediate result was that the Germans pressed captured rifles into service, to arm not only infantrymen but also Wehrmacht snipers. Sepp Allerberger achieved his earliest kills with a Mosin-Nagant he had retrieved from an arms-dump, before moving to a standard open-sighted Kar. 98k, a Zf.-Kar. 98k and finally, apparently, a Zf.-Gew. 43.

An attempt to improve the shooting capabilities of the Kar. 98k was made by introducing an extraordinary small-magnification optical sight, the Zf. 40 (soon upgraded to 'Zf. 41'), which could supplement the standard tangent-leaf back sight. The sight consisted of a small tube in a one-piece mount which slid into a dovetailed slot cut in the left side of the back-sight base.

Made only in the Mauser factories in Oberndorf and Berlin-Borsigwalde, and by Berlin-Lübecker Maschinenfabrik, the rifles were specially selected for accuracy but were otherwise standard issue. The idea was simply to improve the chances of infantrymen hitting targets such as vision slits in tanks and pill boxes, which were difficult to engage over open sights if light levels were low.

There is little doubt that some of the sights were initially appropriated by marksmen, particularly in the days before anything better was available on the Eastern Front, but their optical performance was very poor. Magnification was only 1.5×. The narrow field of view was supposedly counteracted by the ability of the firer to simply look over the top of the Zf. 41, which had an eye relief of about 20 cm compared with only 5 cm for a *Zielvier*.

By 1944, the OKH had effectively abandoned the Zf. 41. Yet tens of thousands of the little sights had been made by at least twelve primary contractors,[38] and attempts had even been made to provide an adaptor allowing any Kar. 98k to mount the sight.

The sight leaf was detached, the tangent-curve part of the back sight was removed – leaving the original bed – and the adaptor was substituted before the sight leaf was reattached. The Zf. 41 could then slide onto the side-rail when required. Unfortunately, this plan, which made sense on paper, had not considered that the external surfaces of the back-sight bed were not necessarily machined precisely in line with the bore axis. Consequently, the optical-sight adaptor was often too far out of alignment to allow even the extreme windage correction built into the sight drum to bring shots onto the target. Tests

A *Gebirgsjäger* examines his Zf.-Kar. 98k, which is fitted with a 4× optical sight in what appear to be high-turret mounts. The mountainous areas of southern German and annexed Austria provided some of the very finest Wehrmacht snipers.

suggested that only one in five Kar. 98k could be altered satisfactorily, and so the adaptor was abandoned.

By 1944, German snipers had become as efficient as their Soviet rivals. Increasingly forced onto the defensive, an environment in which their skills could be put to the greatest use, individual marksmen recorded kill after kill. Yet, apart from a few well-known names, most of the Germans have disappeared into history. Albrecht Wacker relates in *Sniper on the Eastern Front* how while 'Russian and Allied marksmen are honoured as heroes, German marksmen – even in their own country – are considered as wicked killers.' In the original German language edition of the book, published in 2005 while Allerberger was still alive, the names were all changed: the sniper masqueraded as 'Franz Karner'.

The success of snipers is often difficult to establish, owing partly to a paucity of records but also to the establishment of the *Scharfschutzenabzeichen* (marksman's badge) on 24 November 1944. The badge could be awarded in three grades, but only for kills validated after 1 September 1944. There can be no doubt that the scores of many snipers had already reached respectability.

Discounting a few probably apocryphal candidates, the leading snipers were Matthäus Hetzenauer (1924–2004), with a score of 345 or possibly 346; Joseph 'Sepp' Allerberger (1924–2010, 257); Bruno Sutkus (1924–2003, 209); Friedrich

Pein (1915–75, 'at least 200'); and little-known Jakob Hechl (121). Excepting the infantryman Sutkus, born in East Prussia to a Lithuanian father, they all came from the Tirol and served in the *Gebirgsjäger* or mountain troops, among the hardiest and most trustworthy in the Wehrmacht. Allerberger and Hetzenauer both saw service with 144. Gebirgsjäger-Regiment, and Pein, initially at least, with 143. GJR.

Hetzenauer and Sutkus were already experienced shots when drafted, but Allerberger, an apprentice carpenter, simply had an untapped ability which became obvious only when, after an unexpected display of marksmanship, he was allowed to take an optically sighted Obr. 91/30g. Mosin-Nagant from a dump of captured Soviet weapons. And all three men were young, born in the same year, and so still in their teens when their service careers began.

Throughout the period in which the snipers made their names, often against the odds, equipment was in short supply. Wacker observes that, in 1942:

> With the onset of more static warfare and defensive actions the problem became more obvious and urgent. The lack of telescopic sights in the German Army was now critical. The introduction of a telescopic sight with a magnification of only 1.5× was utterly inadequate for long-range shots ... Until regular production of more powerful telescopic sights could begin ways had to be found to improvise. Captured weapons such as Sepp's [Allerberger's Obr. 91/30] were pressed into service, and hunting weapons with telescopic sights were gathered up in Germany and sent to the front. The few marksmen's rifles available from barracks back home and in the possession of the police were brought together and provided the first equipment of the Wehrmacht's marksmen.[39]

This explains why many post-1941 photographs show German soldiers with old full-length SG. 98 and Mosin-Nagants, and, in many ways, supply problems still existed in 1945. The Russians stuck rigidly to weapons which could be made almost anywhere in huge quantities, but the Germans rarely took a similar path.

Manufacturers were allowed too much freedom of design, and the ultimate result was chaos. Virtually nothing interchanged satisfactorily, leather cases made by individual contractors even differed from one another, and the ability to keep weapons in service merely by changing sights and mounts was lost. Another problem lay in priorities. The armed forces rarely agreed among themselves, the SS and the other paramilitary formations generally procured weapons independently, and, at the end of the day, Hitler could (and sometimes did) overrule everyone.

Attempts were made to impose order as the war ran its course, but too few and too late. The difficulties could all be seen in the development of auto-loading rifles. The Gew. 41 (M[auser]) and Gew. 41 (W[alther]) shared gas operation and a conical restrictor on the muzzle. As the bullet emerged from the muzzle-cone, gas pressure, unable to dissipate into the atmosphere, momentarily built up behind its base. Kinetic energy was transmitted to an annular piston behind the muzzle, and then by way of an actuating rod to open the breech.

Trials in Russia showed the Gew. 41 (M) to be unreliable. Essentially an old design locked by allowing an internal cam-and-track system to rotate lugs on a two-part bolt into engagement as the operating rod beneath the barrel pushed the bolt-carrier open, it was abandoned after only about 6,700 had been made.

The Gew. 41 (W), with the operating rod above the barrel, relied on the retraction of locking flaps into the bolt as the action opened. It performed much better than the Mauser and was adopted in December 1942. About 123,000 were made, including some Zf.-Gew. 41 with *Zielvier* offset to the left of the breech to allow ejection to take place efficiently, but the muzzle-cone system was susceptible to jamming and the effects of corrosion.

Eventually, the flap-lock breech of the cumbersome Gew. 41 (W) – which was notably muzzle heavy – was matched with the gas-piston system of the Tokarev, and a detachable box magazine replaced the fixed charger-loaded case that had restricted the rate of fire. Adopted in April 1943, the Gewehr 43 was potentially excellent. But series production had hardly got under way when Germans were pushed back from the Soviet Union and penned into the rapidly diminishing 'Greater Germany'.

The Zf. 4 sight and mount issued with the Zf.-Gew. 43, showing the clamp lever in the 'free' (LOSE) position. Designed with ease of manufacture in mind, it was inspired by the Soviet PU. *Courtesy of James D. Julia, auctioneers (www.jamesdjulia.com)*

Critical raw materials began to run short, machine time had to be selectively allocated to maximise production, and the incorporation of castings and stamped parts (techniques still then in their infancy) effectively reduced the life of each gun. Breakages and extraction problems seem to have dogged what could have been the best semi-automatic rifle of the Second World War.

The near-simultaneous development of the FG.42 and the MP.43 proved to be a burden that the German arms industry could not bear, even though more than 425,000 Gew. 43 were made by the three principal manufacturers and a legion of sub-contractors. About 50,000 of them had been completed as Zf.-Gewehre 43, with a new 4× sight (originally developed for the FG.42) based on the Russian PU pattern introduced with the Tokarevs and then used with the 91/30 Mosin-Nagant. It was carried in a one-piece mount clamped to a rail on the left side of the receiver. However, declining manufacturing standards were not compatible with extreme accuracy, and the Zf.-Kar. 98 was preferred.

The USA

The Japanese attack on Pearl Harbor, in the morning of 7 December 1941, caught the USA largely unaware. But the declaration of hostilities also found the US armed forces desperately short of key weapons and essential skills. This was particular true of the snipers, who had lost whatever recognition they had gained in 1917–18.

The failure of the M1918 sniper rifle and the accompanying optical sight, which had been abandoned in the early 1920s, was not seen as a problem. Even though marksmanship was fostered by inter-service rivalry, such improvements as could be made were confined largely to target shooting.

The entry of the USA into the Second World War forced the military authorities to act. Production of war matériel was accelerated as fast as industry could adapt, and largely successful attempts were made to introduce new weapons such as the M1 Carbine. However, though the M1 Garand had been approved for universal issue even before the war in Europe began, a combination of teething troubles and limited manufacturing capacity not only ensured that the bolt-action M1903 Springfield remained in service in large numbers, but promoted other solutions – not all of which were realistic.

The ferocity of the hand-to-hand fighting, and the casualties ascribed to Japanese snipers, who were not nearly as effective as initially perceived, soon persuaded the US Army and the Marine Corps to deploy snipers of their own. Headquarters Army Ground Force had already decided to introduce a variant of the Garand with an optical sight, but development would clearly take time. This was partly due to the design of the Garand magazine, in which the 'en bloc' clip was an essential component. Garands could not be loaded with loose rounds, which was a major disadvantage not only to the rank-and-file, but also

A US Army sniper checks his M1903A4 rifle shortly after Allied landings in Italy in 1943.
His colleagues are armed with M1903 Springfields and M1 Garands. *Author's collection*

to snipers whose sights would have to be offset from the centreline.

National Match and similar M1903 derivatives were hastily impressed, fitted with various sights of commercial origins, and sporting rifles were also acquired in quantity.

Springfield Armory was committed to Garand production, and so there was no immediate source of M1903 Springfield rifles. However, prior to the Japanese attack, the US gunmaking industry had been supplying weapons to Britain and it had been suggested – a provisional contract had even been signed with Remington Arms – that a variant of the M1903 chambered for the British .303 round could be made. This project proceeded as far as mock-up models when the British called a halt, realising that their rifle-making facilities could supply more than enough No. 4 Lee-Enfields (with help from Canada and the USA).

By this time, the US War Department and Remington had considered re-starting production of the M1903 rifle, using machinery that had been stored since work ceased in Rock Island Arsenal shortly after the end of the First World War. Feasibility studies suggested that work could begin as soon as the machinery had been overhauled and missing tools had been replaced.

On 17 September 1941, therefore, Remington was given a contract for 134,000 'M1903 Modified' rifles (very few changes had been made at this time) and the first 1,273 rifles were delivered in November. The rapid enlargement of the US armed forces after the attack on Pearl Harbor led to additional orders, which by March 1942 amounted to 508,000.

Remington had simplified the M1903 Modified rifle (which was essentially the same as the regulation M1903A1), and the M1903A3 was duly adopted on 21 May 1942 with a straight-grip stock.[40] The first new guns were delivered at the end of 1942, with an aperture backsight on the bridge ahead of the bolt handle, a variety of stamped parts, and changes to individual components to reduce machining time.

Once the supply of rifles had been assured, thoughts turned to snipers' weapons. Protracted trials undertaken by the Infantry Board and the Ordnance Department led to a recommendation that the Model 330C optical sight made by the W. R. Weaver Company of El Paso, Texas, and the 'Junior' mount made by the Redfield Gun Sight Company of Denver, Colorado, should be adopted. The US Rifle, Caliber .30, M1903A4 (Snipers) was duly approved on 14 January 1943, and, four days afterward, a 20,000-gun contract was passed to Remington.

Numbered from 3407088 to 3427087, the guns were taken from M1903A3 production, selected for accuracy, and honed to give the smoothest possible action. The bolt-handle shank was given a concave 'outcurve' to lie closer to the stock, in which a special recess had been cut, and allow the bolt handle to clear

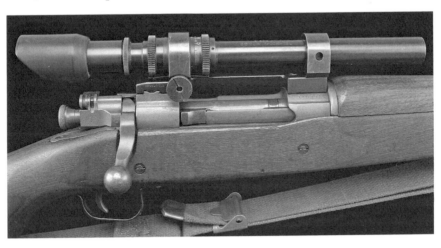

The M1903A4 sniper rifle. This gun, probably reconditioned after the end of the Second World War for service in Korea, has a variant of the M73B2 telescope sight made in France by Optique et Précision de Levallois (OPL) instead of the more common Weaver type. *Courtesy of James D. Julia, auctioneers (www.jamesdjulia.com)*

Four iconic sniper rifles: *from left*, an M1903 with a Winchester sight; the M1903A4 with a Weaver M73B1 sight; the Winchester Model 70 with an Unertl target sight; and an M1D Garand with Lyman M84. *Courtesy of James D. Julia, auctioneers (www.jamesdjulia.com)*

the sight-tube during the loading stroke. The guns lacked open sights, which proved to be detrimental if multiple targets had to be engaged at short range or if the optical sight was damaged.

The earliest M1903A4 rifles had Type C pistol-grip stocks, blanks being supplied from Springfield Armory and by the Keystone Co. Later guns had a 'scant stock' with the merest hint of a pistol grip; supplied largely by Keystone, the first batches were converted from blanks that had been intended for the abortive '.303 Springfield', with the characteristically hooked British butt-form.

The M1903A4 proved to be broadly acceptable, though there were many who saw it as a poor answer to an important question. Some commentators were scathing in their condemnation. Roy Dunlap, author of several well-respected gunsmithing textbooks and an armourer during the Second World War, was among the most vocal critics of the M1903A4.

Airborne and infantry units involved in Operation Torch, the comparatively bloodless landings made in North Africa early in November 1942, were the first to use the M1903A4 in combat. But the rifles were to prove their worth only after the invasion of Italy in 1943, where fighting was far fiercer than in North Africa, and then island-hopping in the Pacific. Complaints were soon being heard: the optical sight was weakly made, the adjuster drums often broke, the seals were ineffective – a particular problem in the tropics, where the lenses fogged, encouraging mould – and the rifles often 'lost zero'.

The ordnance authorities had always considered the commercially available 330C sight and its militarised variant, the M73B1, to be an expedient pending acquisition of M73 Lyman Alaskan all-weather sights. However, very few Lyman sights were fitted to Springfields prior to 1945; use of M81 and M82 optical sights was restricted largely to M1C and M1D Garands.

The 2.5× Weaver 330C/M73B1 sight was about 10⅞ in. long and could be focused from 25 ft to infinity. But it had a very slender tube, merely ¾ in. diameter, and its light gathering capabilities were poor. There is no doubt that the Japanese snipers were better equipped than the Americans in this particular respect. The standard reticle was a crosshair, though the M8 sight, an improved form of the M73B1, usually had a tapered post and single crosshair.

The Weaver sights were attached to the Redfield base by hooking a lug into the front mount, turning the telescope through 90 degrees until the tube aligned with the bore, and then locking the rear mount onto its lateral dovetail with two large-headed screws.

Sniper rifles were bore-sighted for 100 yards, the use of shims being permitted if alignment was otherwise impossible. Any guns that failed test-firing reverted to M1903A3 standard and returned to regular service. A particular problem seems to have arisen from changes to rifling.[41]

Remington received another contract for 1903A4 sniper rifles on 20 June 1943, but it is not clear how many had been made by the time the last 31 of them were accepted in June 1944. It has been claimed that the contract amounted to 8,365, but this may have been the quantity actually delivered against a demand for 52,923 of them.[42]

Nearly 30,000 M1903A4 rifles were delivered to US forces engaged in the Second World War, but they were far from the best that could have been had. A desire for quantity led to a deterioration in manufacturing quantity, and the low-power Weaver sights were notably ineffective. Yet many, reduced to 'Limited Standard' status, survived to serve in the Korean War. A few even made it to Vietnam, fitted with Lyman sights.

Ironically, Remington had suggested at the outset that the Model 720 bolt-action sporter would have made a better sniper rifle. The subsequent success of the Winchester Model 70 and Remington Model 40 (derivatives of which are still serving in large numbers) shows Remington to have been right, but the Ordnance Department was wary of losing the virtues of commonality of parts that lay in adapting the Springfield.

Marked 'M1903A3' on the action-body, M1903A4 rifles also have Ordnance Department 'crossed cannons' on the stock, with 'R.A.' for 'Remington Arms' and the 'FJA' mark of Colonel Frank Atwood, chief of the local Ordnance District inspectorate. Blued finish eventually gave way to Parkerisation, but many guns were delivered with a mixture of parts and others have simply been refinished. The sights were numbered, almost always with an electric pencil, but simply to control inventory; the numbers do not match those of the guns.

Marine Corps snipers fared better than the US Army men. Traditions of independence allowed a blind eye to be offered to 'army standard'; the M1903 was sometimes replaced by the Winchester Model 70, and the Weaver sight was usually abandoned in favour of high-magnification Unertl target patterns.

The use of the M70 by the Marines and Army snipers has been the subject of controversy. There can be no doubt that they were used in Korea, as eyewitness testimony proves, but use in the Pacific campaign is often questioned. However, 373 Model 70 rifles, supposedly to be used for training, were purchased from Winchester by the USMC in May 1942. Numbered in the 41000–50000 range, they were fitted with 8× Unertl sights – subsequently deemed by many observers to be too flimsy for arduous service – and pressed into service as sniper rifles in the period before the M1903A4 was standardised. The same seems to have been true of the very few guns that found their way into Army hands. Most M70/Unertl-sight combinations were replaced by the M1903A4 and Weaver sight as campaigns rolled on.

The Model 70, approved for manufacture on 29 December 1934, had been introduced in the 1937 Winchester catalogue. Derived from the successful Model 54 of 1927, which was itself derived from the M1903 Springfield, the Model 70 used a two-lug Mauser-type bolt with a special patented bolt-guide rib to smooth the bolt stroke. An important advance in design had been made during the production life of the M54, when the introduction in 1932 of the 'Speed Lock' had reduced the lock time from .0055 to .0033 seconds.

Any reduction in the delay between pressing the trigger to release the sear and the strike on the primer of a chambered round was universally beneficial to a sniper. Those few milliseconds could be the difference between life and death. The Winchester M70 was greeted with acclaim, but production stopped at the end of 1941 in favour of war work, and did not recommence until 1946.

The development of the Garand, which had been adopted to replace the M1903 Springfield in 1936, is sketched in Chapter Four. When the Second World War began, the worst of severe teething troubles had been overcome and mass production in Springfield Armory was under way. Yet there were far too few auto-loaders to equip the US Army, particularly as recruitment was accelerating rapidly. The USMC was forced to procure the Johnson, designed by a maverick ex-marine, but manufacturing output was small and the recoil-operated rifle was much more delicate than the gas-operated Garand.

The intention had always been to introduce a Garand sniper rifle, and the M1C (developed as 'M1E7') was standardised in July 1944. The preferred M73 (Lyman Alaskan) or less effectual M73B1 (Weaver 330) telescope sight was carried in a Griffin & Howe mount on the left side of the body. This was due to the quirky clip that was essential to the Garand design, ejecting automatically when the last round had been fired. The optical sight and its mount had to be offset to allow the rifle to be reloaded; unlike the bolt-action Springfield, inserting single rounds into the magazine or the chamber was not an option.

M1C rifles, often issued with the conical T37 (later 'M2') flash hider, weighed 11.2–11.5 lb with their sights. The M1D ('M1E8') was introduced in September 1944, but few had been made when fighting in the Pacific ceased in August 1945. The M1D was developed to allow Garands in need of repair to be transformed into sniper rifles; it has even been suggested that, to avoid shipping weapons back to the USA after the war, work was done in Tokyo arsenal, where costs were low and expertise was high. M1C and M1D rifles were fitted with the M81 or M82 optical sight, the former offering a crosshair reticle while the M82 had a post.

The limitations of the Garand rifle, the clip-loaded magazine in particular, led to the 7.62mm (.308) M14 rifle, accepted in 1957. Essentially a 'product improved' M1 with a detachable twenty-round box magazine, this rapidly supplanted the older rifles. Even after the .223 AR-15/M16 rifle had been accepted for universal service, optically sighted M14 sniper rifles and, eventually, the M21 derivative gave US Army and USMC snipers good service.

Snipers lost their status in the US Army and the USMC almost as soon as fighting in the Pacific stopped. When war began in Korea in 1950, lessons that had been learned from bitter experience had to be re-learned. Suitable rifles were in short supply, and so M1903 Springfields and M1 Garands were fitted with optical sights. By this time, the old M73 Lyman and M73B1 Weavers had been largely replaced by derivations of the 'all-weather' Lyman Alaskan design: M81, M82 and M84. A few hundred USMC M70 Winchesters held in store, apparently mostly in Japan, were also reissued.

The Winchester rifles performed well enough to be refurbished from 1956 onward with Douglas barrels and new stocks, work being undertaken in the USMC repair facilities in Albany, New York. Many ended their days in Vietnam. Unfortunately, changes made in the new Model 70 action introduced in 1964, with a view to cut costs, were largely detrimental.

The Ordnance Department objected particularly to the replacement of the Mauser collar-type extractor with a small hooked bar set into the bolt-face. The new bolt shrouded the head of a chambered round, improving safety should a case-head rupture occur, but the extractor engaged the case rim only as the bolt closed; the collar design engaged the rim as the cartridge was travelling forward from the magazine. Consequently – though largely unnoticed commercially – the rifles were more likely to misfeed, particularly if they were canted or fired upside down.

The Winchester Model 70 lost official support, and was replaced by the Remington Model 40, which had been introduced in 1961 on the basis of the proven M700 action (which was itself derived from the pre-war Model 30, and, ultimately from the British P/14 by way of the US M1917 Enfield). The Remington soon showed structural weaknesses, and was not brought up to an acceptable standard until the much-improved M40A1 appeared in 1989. This in turn led to the M24, which, in its current M2010 adaptation, remains in limited front-line service.

Experience in Korea, where open ground increased engagement range, showed that there was little to choose between the M1903 and M1. Though 700 yards was widely regarded as optimum engagement range, both rifles could hit man-size targets 1,000 yards away if conditions were right, if the sniper was experienced, and if good-quality ammunition was used. The Springfield held a slight edge at long range, but the ability of the Garand to fire follow-up shots rapidly could be beneficial if multiple targets were to be engaged.

Many sniper-trainers went as far as to suggest 300 yards to be the minimum engagement range, as closer shots significantly increased the chances of detection in conditions (such as Korea) where natural cover was scant. In addition, as few hits could be guaranteed beyond 1,200 yards – admittedly, there were noteworthy exceptions – use of M2 .50-calibre Browning machine guns fitted with optical sights, first tried in Korea, gained popularity when the Vietnam War began. Set to fire single shots, the Browning was capable of effective hits at twice, perhaps even three times as far as the .30 rifles.

What of the men of the Second World War and Korea? They were often very young – Private Daniel Cass Jr was just nineteen when he achieved his first kill in the Pacific in 1945 – and even Bert Kemp, a highly experienced shot, was only twenty-four when he reached Europe after D-Day. Chet Hamilton and Ernest Fish were twenty when they reached Korea; unusually, both were experienced target shots.

Comparatively few men came from the cities; instead, wild and often lawless places such as the Appalachian mountains or the wilds of Texas had given them the ease with which they handled weapons, and the skills (especially patience and hand–eye co-ordination) without which no sniper could succeed. Yet the casualty rates were horribly high; in Italy, it has been estimated that six snipers of every ten were killed in action.

This was partly due to the absence of co-ordinated training. Though many individual units created 'sniper schools' of their own, the absence of service-wide standards reflected a widespread reluctance to acknowledge the effect a few snipers could have in the field. An idea of the improvised nature of pre-1953 training can be gained from the recollections of USMC Corporal Ernest Fish:

> We placed the white barrels at staggered intervals. There was one at 100 yards and so on up to 500 yards. The gunny issued us standard M-1s [probably M1D Garands] to which armorers had attached 2½× Weaver scopes. We felt a little slighted when we heard Army snipers were using bolt-action Springfields with 10× scopes, but we were told we couldn't get any of those. Big as those barrels were, it was still just plain Missouri luck if you hit them at 500 yards . . .
>
> We all laid down on the ground and fired at the barrels, while the gunny stood and looked through field glasses. After the clips pinged empty, we grinned confidently at each other and . . . sauntered out to count the holes in the barrels. The grins turned sheepish as we approached the most distant barrels and found maybe only one or two holes in them.[43]

The Far East

In addition to the US Army and the USMC, the Australians and the Japanese made good use of snipers in the Pacific jungles. However, almost none of them have ever been identified individually, even though Clifford Shore noted:

> In Timor, one of these kangaroo hunters had a great sporting time, and whether right or wrong, he played sportingly with the Japs and never used his telescopic sight . . . [at] less than 300 yards range! He was credited with 47 Japanese killed but with characteristic modesty claimed only 25 certainties . . .
>
> Another 'roo hunter presented with a mass target of automaton Japs – the target which comes once only in the life of one sniper in a hundred – got twelve of the yellow-men in twelve shots in fifteen minutes.[44]

A sniper, probably serving in Europe during the Second World War, aims his No. 1
Mk III* SMLE. Heavy-barrelled rifles of this type were regulation issue in the
Australian Army, as the Lithgow factory did not make the No. 4. *Author's collection*

The anonymity of individual Japanese marksmen is still so complete that
Tom McKenney, contributing 'Japanese Sniper: A Long Walk to a Short Life' to
The Sniper Anthology in 2012, had to invent 'Corporal Taki Nakamura' to allow
an undeniably eclectic 'personal story' to be told.

Though they were initially successful, attaining an almost god-like
veneration from Allied soldiers, increasingly efficient counter-sniper fire and
the inability of most Japanese snipers to improvise – even when their lives
were threatened – changed the balance. Among the Australians were many
excellent shots, notably kangaroo hunters trained never to spoil a skin with a
misplaced bullet, who took a dreadful toll.

Japanese snipers favoured the treetops. Their boots often had climbing
spikes, and they lashed themselves in place to ensure that they could wait,
even doze, without falling to the ground. They were masters of camouflage
and exceptionally patient. Yet they rarely fired at parties of more than two, and
were often killed simply by raking the treetops with a Bren Gun from hundreds
of yards away. In addition, Japanese ground troops, intent on obeying orders
without question, often advanced directly over a man who had fallen victim to
a sniper . . . to become the sniper's next victim. One report tells of more than
twenty Japanese soldiers falling to a single Australian, who had simply to keep
his position in a way most sniper-trainers would have abhorred.

The No. 4 Lee-Enfield rifle was not made in Australia, where the Lithgow small-arms factory was still making the No. 1 Mk III* SMLE when war in the Pacific began. Consequently, the standard sniper rifle was a special heavy-barrelled version of the No. 1 Mk III* with an Aldis (or similar) telescope sight mounted on the centreline, high enough to allow cartridges to be loaded singly. The Japanese snipers had the pre-war 6.5 mm Type 97, and also the 7.7 mm Type 99 approved in 1942. The newer gun, fitted with a monopod beneath the fore-end, usually had a 4× Type 99 optical sight offset to the left in a one-piece mount. Like its predecessors, this sight usually had neither windage nor elevation adjusters; Japanese snipers were taught to 'aim off' as required, even though this limited their effectiveness to comparatively short ranges.

As the war in the Pacific ran its course, the quality of the Arisaka declined markedly. Additional contractors were recruited from private industry, and a 'Substitute Type 99' (sometimes known as the 'Type 99 Model 2') was approved in 1943 in an attempt to conserve raw material. The bolt cover was abandoned, chrome-plating the bore and the bolt-face was abandoned, and sapwood was often used for stocks. This process continued until the last guns to be made were terrible, but such rapid deterioration does not seem to have affected sniper rifles; very few of them were made after 1944.

Notes to Chapter Five

1. John Walter, 'David and Goliath: Sako and the Winter War', in *Shooter's Bible*, 1983, p. 47.
2. Grigory Krivosheev (editor): *Grif sekretnosti snyat: Poteri vooruzhennykh sil SSSR v voĭnakh, boevykh deĭstviyakh i voennykh konfliktakh* ('Tragic secrecy lifted: the USSR Armed Forces' losses in wars, hostilities and military conflicts'), 1993. Published in English in 1997 as *Soviet Casualties and Combat Losses in the Twentieth Century*.
3. Roger Moorhouse, 'White Death', in *The Sniper Anthology: Snipers of the Second World War* (2012), p. 9.
4. Some are included in the general list available from http://airaces.narod.ru/snipers; A. I. Begunova, *Angely Smerti: Zhenshchni-snaĭpery, 1941–1945* ('Angels of Death: Women snipers, 1941–1945'), 2014, is the best source of reliable information about the weapons, equipment and career of the principal women snipers. Lyuba Vinogradova's *Avenging Angels* (2017), though eminently readable and wider-ranging, makes controversial claims which often seem to be based more on personal reminiscences than official records.
5. Snipers' names have been transliterated in *Snipers at War* according to British Standard 2979:1958. Unfortunately, there are several such systems – Russian passports have used three since 1997 – and names of people and places can vary sufficiently to compromise identification. *Snipers at War* has tried to avoid unnecessary diacritical marks and tie-characters, but BGN/PCGN, the system that is easiest for Anglo-American readers to understand (and comes closest to actual pronunciation), gives unfamiliar results. Consequently, somewhat reluctantly, the British Standard has been preferred.
6. Vladimir Nikolayevich Pchelintsev, born in 1920, served with the 11th Rifle Brigade and is now usually credited with a total of 456 kills.

7. Captain C. Shore, *With British Snipers to the Reich* (1948), pp. 95–7.
8. Actual losses have been (and are still) disputed. The current official Russian estimate, the result of the investigation undertaken under the chairmanship of Col.-Gen. Krivosheev, dates from 1993. It states 'all causes' military deaths to be (at least) 8,806,453 and wounded to number 22,610,148. These are staggering totals, but were almost immediately attacked for being too conservative. A study undertaken in 2008 in the Central Defence Ministry archives claims the total of dead and missing to be 14,241,000, and some seek to raise even this figure. Civilian losses are more difficult to analyse satisfactorily but it is generally believed that the 1993 estimate of 26.6 million deaths (military and civilian) is also much too low. Among the limiting factors are difficulties of assessing loss of life in the Stalinist purges, inherent inaccuracy in censuses – which are not always the reliable base for predictions that is usually assumed – and the inability of fighting units surrounded by the Germans to submit casualty reports. Losses in captivity and to forced labour (where Russian claims and German admissions are poles apart) are also impossible to reconcile.
9. Nikolaev went through the war, reaching the Berlin Reichstag with the 96th Howitzer Artillery Brigade of 23rd Artillery Division. Ending the war with the rank of captain, Yevgeni Nikolaev worked for the newspaper *Tambov Trud* as a journalist, and then in the Pushkin regional library. He died on 22 February 2002.
10. Vassili Zaitsev [Vasiliy Zaytsev], *Notes of a Russian Sniper* (2015 edition), p. 91.
11. There is no doubt that both these stories have a basis in truth, as it would make sound tactical sense to detail the most experienced sniper available to eliminate an enemy sniper who had been identified as a potent threat. But that sniper would most probably be recruited locally and not imported specially from Germany. Zaytsev's memoir makes convincing reading, but the hand of the propagandist can probably be seen and the accepted version of the story has been questioned by Anthony Beevor in *Stalingrad: The Fateful Siege, 1942–1943* (1998). Many similar stories have been told, including the supposed duel between the Australian Billy Sing and the Turk 'Abdul the Great' at Gallipoli in 1915.
12. Zaytsev, *Notes of a Russian Sniper*, p. 201.
13. Other claimants include Tari Vyutchinnik (155), and the team of Natalya Kovshova and Maria Polivanova (killed together in 1942, said to have had 317 kills between them). David Truby, in his article 'Soviet Women Snipers of World War II' published in *Small Arms Review*, named Ziba Ganieva ('nearly 300 kills to her credit') and Vrna Zworykin [sic], who 'retired with the rank of Colonel in 1974. From 1941 to 1943, though, she was one of their army's top snipers . . . According to Soviet records, [she] killed 85 German officers in the summer of 1942 alone . . . Another Soviet female officer behind sniper's crosshairs was Tari Vucinich [sic], who operated in the Ukraine . . . She was credited with 155 kills before dying in a Luftwaffe attack on her unit in 1944.' Unfortunately, Vyutchinnik excepted, there is no readily accessible confirmation of the standing of any of Truby's claimants. Russian sources reveal that Junior Lieutenant Ziba Pasha qizi Ganieva, born on 20 August 1923, part-Azerbaijani part-Uzbek, served in the Great Patriotic War as a sniper attached to the 3rd Moscow Rifle Division. She gained 21 kills, sometimes also acting as a radio operator and cross-lines scout, but was seriously wounded in 1942 and her active career seems to have come to an end. She died in Moscow in 2010, apparently claiming '129 kills' but it is not clear if this total includes successes achieved with a submachine-gun.
14. Almost nothing is known about Alexey Pavlichenko, who was 'many' years older than Lyudmila; he is not mentioned in her service record, now kept in the central archives of

the Ministry of Defence, and may have been killed during the war. Her son, Rostislav Alexeevich Pavlichenko, was born in 1932. He graduated from Moscow University with a law degree and spent sixteen years teaching in the Higher School of the KGB. Ill-health eventually forced him to retire with the rank of colonel, and he died in 2007.

15. Pavlichenko, well dressed and immaculately groomed, was only able to persuade the sceptical recruiter that she should serve in a combatant role by producing her shooting certificates: *Ya – snaĭper* (2015), p. 31.

16. Ibid., p. 89.

17. The term has also been translated as 'Gun Women', because *Flinte*, used as a suffix, can refer to a shotgun or sporting gun.

18. Quoted by David Truby, 'Soviet Women Snipers of WWII', in *Small Arms Review*.

19. Quoted from 'Which of legendary snipers was called "mother Nina" and why?' – http://rusarticlesjournal.com/1/19906.

20. Mike Markowitz, 'Women with Guns: the Red Army Female Snipers of World War II', published on 19 November 2013, gives the total as '1,885 snipers and instructors', but this is likely to include those who had been trained in the Central Sniping School before the specialist women's training centre was created.

21. See, for example, A. I. Begunova, *Angely Smerti: Zhenshchni-snaĭpery, 1941–1945* ('Angels of Death. Women snipers, 1941–1945'), 2014.

22. From a transcript of an interview of Klavdiya Kalugina (née Panteleyeva) by Artem Drabkin, published on 21 September 2010.

23. 'Woman sniper' by Olga Troshina, taken from Lobkovskaya's reminiscences.

24. Letter to the author from Hans-Rudolf von Stein, 12 June 1976.

25. *Allgemeine Heeresmitteilungen*, 9. Jahrgang, 12. Ausgabe, 7 May 1942: '405. Anschiessen der russischen Selbstladegewehre „Tokarew" (Mod. 1940)'.

26. Many of the SVT rifles made in 1942 have grooved chambers to facilitate extraction and ejection.

27. Attaching a bayonet to the barrel can have a surprising effect on shot-strike. Trials undertaken in Britain in July 1908 by the School of Musketry showed that fixing a bayonet to the Russian Obr. 1891 Mosin-Nagant rifle moved the point of impact 24 in. down and about 5 in. to the left at 200 yards.

28. The Mk III* sight was a considerable improvement on the Mk III, from which it differed principally in the design of the slide-latch and more robust construction. However, No. 4 Mk I (T) rifles were always fitted with Mk I sights from which the 'battle sight' had been removed; Shore's original text suggests that the Mk I was actually an improved form of the Mk III, which is clearly not so.

29. BSA delivered about 665,000 guns, 737,000 came from Maltby, and Fazakerley contributed 619,913.

30. The number of the gun was customarily marked on the sight and sometimes also the sight mount; the number of the sight was added to the gun-butt, and both numbers were displayed prominently in the chest-lid.

31. Skennerton & Laidler: *The British Sniper Rifle*, p. 123.

32. Shore, *With British Snipers to the Reich*, p. 286.

33. Denis Edwards, *The Devil's Own Luck: From Pegasus Bridge to The Baltic* (1999), pp. 127–8.

34. A parachute battalion had 32 snipers, an air landing brigade had 38: Dan Mills, 'Fighting Irishman', in *The Sniper Anthology: Snipers of the Second World War* (2012), p. 173.

35. There are many references to the low esteem in which snipers were held almost everywhere except in the USSR. 'Seventy years ago, snipers were often condemned as

cold-hearted killers who took lives without risking their own, an attitude that extends at least back to the Napoleonic era. Second World War sniper veterans often concealed their wartime duties, reluctant to expose their experiences to a misunderstanding in public': Major John L. Plaster in his introduction to *The Sniper Anthology*. In addition most memoirs contain references to commanders who were reluctant to employ snipers

36. The codes included 'ar' for Mauser-Werke, Berlin-Borsigwalde; 'ax', Feinmechanische Werke; 'bcd', Gustloff-Werke; 'bnz', Steyr Daimler Puch; 'byf', Mauser-Werke, Oberndorf; 'ce', Sauer & Sohn; 'ch', Fabrique Nationale d'Armes de Guerre; 'dot', Waffenwerk Brünn; and 'svw', Mauser-Werke, Oberndorf.

37. The *Kriegsmodell* inspired development of the Volkskarabiner 98 of 1945, which, though usually built on a magazine-feed action, had simple fixed sights, a rudimentary half-stock, and very little attention to finish. The goal was quantity instead of quality at a time when raw materials had become scarce and industrial production was at its lowest.

38. Makers of the Zf. 40/41 series included D. Swarovski of Wattens/Tirol (code, 'cag'): Dr. F. A. Wöhler of Kassel (code, 'clb'); Emil Busch of Rathenow ('cxn'); 'Oculus'–Spezialfabrik für ophthalmologischen Instrumente of Berlin ('ddv'); Opticotechna of Prerau ('dow'); Runge & Kaulfuss of Rathenow ('dym'); G. Rodenstock of Munich ('eso'); Spindler & Hoyer of Göttingen ('fvs'); Feinmechanik of Kassel ('fzg'); Ruf & Co. vorm. Carl Schütz of Kassel ('gkp'); Ernst Ludwig of Weixdorf/Sachsen ('jve'); Établissements Barbier, Benard et Turenne, Paris ('kov'); and Seidenweberei Berga, C. W. Crous & Co., of Berga/Elster ('mow').

39. Albrecht Wacker, *Sniper on the Eastern Front* (2015), p. 86.

40. The pistol-grip 'Stock, Rifle, D1836', or 'Type C', had been approved on 15 March 1928; the M1903A1 rifle, appropriately fitted, replaced the straight-grip stock M1903 on 5 December 1929.

41. The first guns had four-groove rifling, cut conventionally. Based on British experience, Remington then adopted two-groove rifling to accelerate production. However, accuracy seems to have deteriorated far enough for four-groove drawn rifling to be introduced.

42. The widely accepted total of 28,365 M1903A4 rifles was contested by Lt.-Col. William Brophy, who proposed 29,964 on the basis of serial numbers. It seems possible that the former total refers only to the guns delivered prior to June 1944, and that a few hundred were subsequently assembled from 'overrun' components. More research is clearly needed.

43. Charles Sasser & Craig Roberts, *One Shot – One Kill*, p. 93.

44. Shore, *With British Snipers to the Reich*, p. 117.

CHAPTER SIX

Snipers of Today:
Better Rifles, Better Sights

When the war in Vietnam began, most armies were still equipped with the guns that had served them throughout the 1950s, when the bolt-action infantry rifle of the Second World War had given way to the first generation of autoloaders. The British (and many others) had adopted a variant of the Belgian-developed FN FAL, the USSR had mass-produced the Kalashnikov, and the US Army had developed the 7.62 × 51 (.308) M14, adopted in 1957, from the Garand.

The British had replaced the .303 No. 4 Mk I* (T) with the L42A1, but this was little more than the original rifle with a new 7.62 mm barrel. US Army and USMC snipers continued to use Remington and even old Winchester rifles while the M14 was adapted to become the XM21/M21 series; and the Soviet Army, and most of its satellites, initially retained the Mosin-Nagant. Rebellions, incursions and territorial conflicts rumbled on against the spectre of a world annihilated in nuclear conflagration. And yet, oddly, despite the Cuban Missile Crisis of 1962, there had been no large-scale wars since Korea.

Small arms continued to develop, even though many high-ranking officers still saw war as something to be contested by massed troops at comparatively long range. The exception was provided by the Russians, who relied on a comparatively weak 7.62 × 39 round even though this was too feeble to be an effective long-distance fire-support tool. Though the Soviet infantrymen were equipped with the AK (replaced by the AKM from 1959) and light support weapons such as the RPD, the old 7.62 × 53 rimmed cartridge was retained for machine guns and a new sniper rifle, the SVD.

However, the 'small-calibre revolution' – from 7.62 mm (.308) to 5.56 mm (.223) – moved rapidly with the approval of the ArmaLite AR-15 rifle for US service as the 'M16'. Among the advantages claimed for the new rifle were its light weight and high velocity. In addition, the reduction in the weight and size of an individual cartridge allowed soldiers to carry much more ammunition.

Introduction of the M16 did not proceed smoothly. So many problems were reported that the authorities came close to abandoning the weapon altogether.

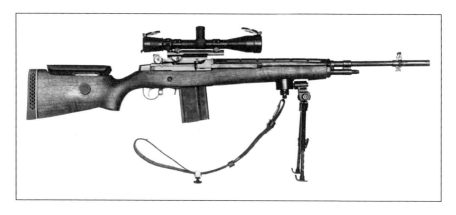

The 7.62mm M21 was a dedicated sniper rifle derived from the M14, which was an upgrade of the old M1 Garand.

Eventually, what had been touted as a 'self-cleaning gun' was improved, and problems with the propellant were largely overcome. A far bigger problem concerned tactics. The ability of US industry to produce war matériel in vast quantities allowed 'fire-density' and 'quick kill' to become buzzwords. The goal was to ensure that soldiers laid down a curtain of fire when, for example, disembarking from helicopters: shooting magazine after magazine into surrounding cover, even if there was no hostile fire.

One important effect, apart from a stupendous waste of ammunition,[1] was to reduce the value of individual marksmanship almost to nothing. There was no shortage of recruits who could shoot, but their abilities were not initially appreciated. Lessons of two world wars, and the war in Korea which had ended scarcely a decade previously, had to be learned once again.

Unfortunately, US losses grew far too quickly. The M16, though surprisingly accurate at short range, was ineffectual if the distance extended much beyond 300 yards and so, once again, optically sighted Springfields and Garands were brought out of storage to be refurbished and upgraded. Attempts were also made to purchase sporting rifles, though by this time the Winchester Model 70 had been largely discredited and the Remington Model 700 was preferred.

US Army and USMC snipers showed their effectiveness on the battlefield, but there was a serious shortage of equipment and no appreciation of the value of training. Major John Foster of the 101st Airborne Division, trying hard to overcome apathy, took thirty experienced soldiers to the firing range:

> Each man stepped to the firing line with his M16 and two fully loaded thirty-round magazines. The target was a single man-sized silhouette placed fifty meters downrange. The shooter was instructed to fire as many rounds as he could within a one-minute time limit. His position

and rate of fire were left up to him. Most of the soldiers shot from a standing position on full automatic. Many didn't even bother to use their rear sights. The group averaged four to six hits *total* out of approximately eighteen hundred rounds.[2]

This appalling performance was highlighted by the worst shot of the group, who had been taken aside to learn the rudiments of controlled fire. Firing with newly acquired knowledge, the novice always scored many more hits from his sixty rounds than the group had done simply by blasting away at the target.

Foster, and like-minded colleagues who could see the value of accurate shooting, persisted with criticism until the authorities finally took heed. But it had taken time, and could be measured in hundred of lives needlessly lost.

Success was still limited by equipment, particularly as shots were often taken at such long distances that hits could not be guaranteed. The skills of even the finest marksman were largely wasted if his rifle could not perform to the same levels.

These problems were appreciated in virtually every army in which sniping was considered to be valuable, and, increasingly, in police and counter-terrorist agencies to which collateral casualties were an ever-present inhibitor. Everyone agreed that a first-round kill was desirable, and that there was a pressing need to extend range far beyond the capabilities of a modified service rifle.

One immediate result was the Anti-Matériel Rifle (AMR), with a pedigree stretching back to the big-game rifles tried in the early stages of the First World War by the British against the Germans' iron trench-shields. Increased demand and the ever-increasing influence of police Special Weapons and Tactics (SWAT) squads also created a new generation of sniper rifles, exemplified by the bolt-action Steyr-Mannlicher SSG69, special variants of commercially available sporting rifles such as the Savage 110 or the Ruger M77, heavy-barrel AR-15 / M16 derivatives and, ultimately, sophisticated designs such as the Heckler & Koch PSG-1.

Perhaps more importantly, ammunition underwent a transformation. Even the finest rifle cannot perform with poor-quality cartridges, and snipers had always been careful to select from specific batches. This can be essential in wartime, when an emphasis on quantity inevitably leads to a decline in manufacturing standards.

There has been a tendency, particularly in recent years, to procure equipment from only a handful of manufacturers. This trend towards universality, redolent of the way in which the Mauser rifle-making cartel ensured that its products were distributed worldwide prior to 1914, now transcends borders; no longer does chauvinism play a decisive role in military trials. Ineffectual weapons were once accepted for service simply because they were 'own make'. The French, Austro-Hungarians and Italians were particularly vulnerable to this; oddly, the British often tried to adopt the best gun for the task regardless of its origin.

Special Forces, and their snipers, are usually allowed to procure weapons required for specific tasks. An occasional exception to this rule, such as the enforced withdrawal of the HK416 from some US special operations units serving in Afghanistan in favour of the M4A1, simply draws attention to the trend towards commonality.

But, equally, cost-consciousness acts as a brake on development. The price of new specialised equipment can be staggering – the Tracking Point Smart Scope, for example, costs about $25,000 – and military authorities are readily persuaded instead to upgrade existing weaponry purely on fiscal grounds. The ageing M14, the box-magazine Garand derivative adopted in 1957, is still serving as the US Army's M14EBR-RI Designated Marksman Rifle and the essentially similar USMC Model 39 Enhanced Marksman Rifle.

The story of the Remington bolt-action rifles of the US Army and USMC is particularly interesting. Based on the long-action Model 700 sporting rifle, the militarised M40A1 developed into the M24 Sniper Weapon System, introduced to service in 1988 with a 10× Leupold optical sight. Supposedly superseded by the M110, a refinement of the Knight's Armament Company SR-25 (combining the best features of the AR-10 and AR-15/M16 series), M24 procurement did not cease until the beginning of 2010. By this time, acknowledging that the bolt-action Remingtons still had value, a decision to upgrade the entire inventory to M2010 Enhanced Sniper Rifle status had been taken. The first adaptations reached service in March 2011 and, by the time the last had been delivered in April 2014, 2,558 had been made. Chambered for the .300 Winchester Magnum cartridge, fitted with the M5A2 (Leupold Mk 4) 6.5–20×50 sight, the M2010 is said to return sub-MOA accuracy at 800 yards or more.

The US Special Forces Operational Command (SOCOM) has taken its own course as far as weapons are concerned. In addition, the use of Navy-type 'Mark/Model' descriptors at the expense of the Army's 'Model' can obscure links with service weapons. In 2014, SOCOM finally approved the Mk 21 PSR ('Precision Sniper Rifle'), based on the Remington Modular Sniper Rifle that was readily available on the commercial market. Fitted with an AAC sound-suppressor, 5,150 PSRs have been ordered. Owing to readily exchangeable barrels, they can chamber the 7.62×51, .300 Winchester Magnum or .338 Lapua Magnum cartridges. Accuracy is claimed, perhaps optimistically, to be 1 MOA at distances as great as 1,500 yards. But whether this will actually ensure commonality of equipment across all SOCOM units is still an open question.

Beretta, Colt, FN Herstal, Glock and Heckler & Koch, among others, have supplied weapons to countless military, paramilitary and police forces. Details can be found in my *Guns of the Elite Forces* (reprinted in 2016) and in Leigh Neville's authoritative *Guns of the Special Forces 2000–2015* (2016).

The AW ('Arctic Warfare') sniper rifle, developed successfully for trials in Sweden in the early 1990s, presaged the British Army L115 series. *Accuracy International*

As the Glock pistol has proved, success can grow from the least promising background. The Accuracy International rifle resulted from the enthusiasm of two Scots engineers, David Walls and David Caig of C&W Products, and the participation of the renowned target-shooter Malcolm Cooper (1947–2001).

Caig and Walls built the prototype target rifle with which Cooper took a silver medal in the 1978 world championships. The design was duly refined, and the third gun, fitted with a thumbhole stock and a unique flash suppressor/recoil compensator with spiral ports, developed into the Precision Marksman Rifle (PM). A flat-bottom action body bonded onto an innovative aluminium-alloy chassis gave an unusually rigid platform.

An early PM rifle was evaluated in 1984 by the British forces, leading to purchases of eight for the Special Boat Service and thirty-two for the SAS. The highly successful 'Green Meanie', so called because of its colour, then won the competition to replace the L42A1 in British service against rivals which included the Heckler & Koch PSG-1, the SIG-Sauer SSG 2000, the Remington 700, and the Walther WA2000.[3] Finally, on 11 March 1985, AI was given a contract for 1,212 L96A1 rifles; success was assured, to be reinforced by adoptions world-wide as the 7.62 × 51mm L96 evolved into the Arctic Warfare Rifle (L118), accepted in Sweden in 1991, and then, in 1995, into the L115 series chambered for the .338 round developed jointly by AI, Sako and Lapua. A variant of the L115 chambering the .300 Winchester Magnum round was adopted by the Bundeswehr as the G22, the first of the series to feature the Caig folding stock.

Another unexpected success has been that of Blaser Jagdwaffen, whose R93 sporter, based on patents granted to Gerhard Blenk and Meinrad Zeh,[4] has been adapted for snipers so effectively that it has been adopted by, among others, the German federal police. The Blaser has a straight-pull bolt, a class that has a pedigree – Mannlichers, the Ross rifle – but rarely finds favour in combat. These action are slick, but apt to jam in adverse conditions. Primary extraction is generally poor. Snipers, however, are rarely called to fire continuously and there is no good reason why a straight-pull rifle should not be advantageous.

The Blaser R93 Tactical-2, with the unique self-centring multi-point locking system, is capable of exceptional accuracy. Published figures suggest 0.25 MOA can be obtained at 300 metres when match-grade ammunition is used.

Some guns are ubiquitous: the Kalashnikov is still the principal weapon of the Taliban, Isis/Daesh, terrorist cells and unnumbered 'freedom fighters'. The sniper equivalent is the Samozariadniya Vintovka Dragunova (SVD), issued for field trials in 1958 and approved for service in 1963. Guns of this type feature prominently in the Middle East, where the exploits of female Peshmerga snipers (and Scotsman Alan Duncan) have often made headlines.

Opinions of the SVD and the perfected PSO-1-M optical sight differ. Many in the West are quick to disparage their capabilities, damning them, often with faint praise, as typically unsophisticated products of the old Soviet bloc. Others laud the rifle's accuracy and reliability, even though the action is basically an adaptation of the Kalashnikov and the rimmed cartridge originated in the nineteenth century.

The current SVD has synthetic furniture, an adjustable butt plate, and a selection of greatly improved sights which includes the 3–9×42 Minuta telescope and an effective image-intensifier. The SVDS has a short barrel, an improved flash-suppressor, and an adjustable bipod.

Russian sources claim the Dragunov to be more accurate than the Mosin-Nagants that equipped many Communist snipers active in Korea and then in Vietnam. The claims include a 100 per cent dispersion circle of 395 mm at 600 metres compared with 440 mm for the Obr. 91/30g. Analysis of the 25-shot sequence shown in Target 11 suggests that the Russian figures err on the side

The Dragunov or SVD, with NSPU-3 intensifying and PSO-1M2 optical sights. *Izhmash*

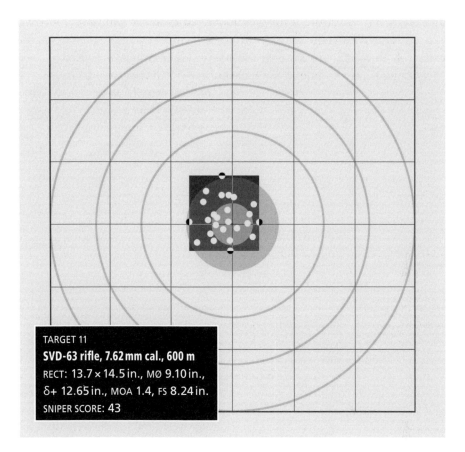

TARGET 11
SVD-63 rifle, 7.62 mm cal., 600 m
RECT: 13.7 × 14.5 in., MØ 9.10 in.,
δ+ 12.65 in., MOA 1.4, FS 8.24 in.
SNIPER SCORE: 43

of pessimism, though the average SVD probably does not outrank many Designated Marksman Rifles. The Steyr-Mannlicher SSG-69 has been credited with a 100 per cent dispersion circle of 40 cm at 800 metres, markedly better than the SVD, and the best current Western sniper rifles – such as the bolt-action L115A3 and even the auto-loading SR-25 – will out-perform a Dragunov.

However, out to at least 600 yards, an SVD firing special match-grade 7N1 ammunition is to be feared. There have also been suggestions that, unusually, accuracy of the Dragunov improves by fitting the bayonet (even though mean point of impact will change). The additional weight may have a beneficial effect on barrel harmonics, perhaps by limiting muzzle-whip.

Yet even the Russians seem to have recognised limitations in the potential of the SVD and have developed bolt-action sniper rifles including the SV-98, in 7.62 × 54R or 7.62 × 51, which is a militarised version of the well-proven Rekord-CISM target rifle fitted with a 7× PKS-7 optical sight. The SV-99, designed by Vladimir Susloparov, is a compact design with a straight-pull bolt adapted from a biathlon rifle; merely a metre long, it chambers the .22LR rimfire cartridge and is usually fitted with an integral silencer.

The .50 Browning Barrett Light Fifty was one of the first large-calibre anti-matériel rifles (AMRs), and also one of the most successful. Straight-line construction and an efficient muzzle brake reduced 'felt recoil' by about 30 per cent compared with bolt-action rifles of the same chambering. From a brochure for the Model 82, c. 1985. *Author's collection*

The improvement in the accuracy of sniper rifles and the introduction of more powerful ammunition has increased engagement range; where once this was expressed in hundreds of yards, snipers now talk in thousands. Carlos 'White Feather' Hathcock (1942–99) recorded a kill at more than 2,000 yards in Vietnam in 1967, with an optically sighted .50 Browning M2 machine gun, and the advent of the AMR has raised public interest in the 'longest shot'.

The record was extended when a Canadian sniper, supporting Operation Anaconda, achieved a 2,430-metre kill with a McMillan TAC-50 in 2002; then taken by Briton Craig Harrison, whose 2,475-metre success in Helmand province of Afghanistan in 2009 was achieved with a .338 L115A3. This is still the longest-distance kill with a small-calibre sniper rifle, though an unidentified Australian, operating in Helmand with a .50-calibre Barrett M82A1 in 2012, has been credited with a kill at a scarcely believable 2,815 metres.

The use of large-calibre rifles in sniping has a lengthy pedigree, dating back to the Soviet 14.5 mm PTRD and PTRS anti-tank rifles of the Second World War and to optically-sighted .50 Browning machine guns in Korea. However, though the .50 bullet is exceptionally destructive, capable of inflicting fatal trauma even if it strikes outside what would normally be regarded as the kill-zone, it is not as accurate as the .300 Winchester Magnum and .338 Lapua cartridges. Chris Kyle, of *American Sniper* fame, remarked that his 'longest kill', achieved in Iraq in 2009, was basically a lucky strike. And Craig Harrison, 'Maverick 41', confirms in *The Longest Kill* (2015) not only that these long-distance hits can be achieved with appropriate preparation but also that several misses may be needed to gauge the adjustments necessary to compensate for bullet-drift or crosswinds.

Sights and associated equipment

It could be argued that some of today's rifles are yesterday's designs, such as those based on the Remington Model 40 which was based on the Remington 700, derived from the Model 30 which was a variant of the Model 1917 Enfield and, ultimately, the British P/14 rifle of the First World War. This is undoubtedly true. However, there have been game-changing advances in accessories.

The sniper of 1917 would have been more than happy to use a modern Remington rifle, but he would have been baffled by its accessories. Even the best optical sights available in 1918 would be poor tools compared with image intensifiers, laser designators, target illuminators, electronic data-gatherers, and the computerised audio/video recorders that see through the sights precisely as does the sniper. The advent of sophisticated aids has revolutionised the role of the sniper by increasing hit-probability, extending engagement range, and improving the chances of a second or successive hit.

The first attempts to improve weapon sights relied simply on magnifying the image so that the target could be seen more clearly. Yet targets could not be engaged in darkness however much the image was enlarged. Some of the best-made telescope sights were capable of improving performance in poor light, it was true, but the gains were limited by the performance of the firer's own eye. Variable or 'zoom' magnification, wide-angle lenses to broaden field-view, computer-designed multi-element lenses and non-reflective coatings to reduce reflections or improve transmission of light, and development of graticles to facilitate range-gauging have all had positive effects on performance.

Collimator sights appeared in the 1970s, making use of a well-known optical illusion by combining an illuminated aiming mark within the sight body with the ability of the firer's binocular vision to accommodate the reflected aiming mark and a view of the target simultaneously. The first successful sight of this type, 'Singlepoint', appeared to project a red dot on to the target and undoubtedly facilitated rapid fire. However, few of these sights (even those with powered reticles) have proved to be of much use in low-light conditions.

AimPoint 7000 collimator sights were used by Randy Shugart and Gary Gordon of the US Army when they took the courageous decision to help the crew of a downed Black Hawk helicopter in Somalia in 1993. The two men saved the lives of four crew members, but at the cost of their own. The red-dot sights undoubtedly allowed Shugart and Gordon, outgunned by the insurgents, to sell their lives dearly.

Trijicon ACOG reflex sights, tested by the US Army for compatibility with the Close Combat Soldier Enhancement Program of the 1980s, not only found a niche with Special Forces but are also, in an improved form, still in front-line service. The intensity of the amber-colour aiming dot, powered by a combination of ambient light and a tritium lamp, adjusts automatically to prevailing light.

The light-gathering properties of large-objective telescope sights can give the impression of being able to see in twilight. However, 'light' is no more than a name for the visible portion of the electromagnetic-radiation spectrum. The human eye is capable only of resolving electromagnetic radiation with a wavelength of roughly 390–750 nm. Animals often have better night vision than humans, which can be due to better light-gathering but also to abilities to see into near-ultraviolet (short wave) and near-infrared (long wave) spectra.

The infrared portion of the spectrum is particularly useful, as it extends from 700 nm to 10^6 nm. Only near-infrared, 750–900 nm, is customarily used in television controls and infra-red film, but anything up to 2,000–2,400 nm can be accommodated by special cameras. Even longer wavelengths – emitted by 'hot' objects – can be used to create a thermal image (see below) merely by recording emissions of electromagnetic radiation from the surface of an object.

Infrared radiation is simply 'light' that is invisible to our eyes; it can be focused and reflected and can be polarised in the same way as visible light. The underlying principles have been known for many years. Fahrzeug-Gerät 1229, developed in the late 1930s by the Forchungsanstalt der Deutschen Reichspost in collusion with Leitz of Wetzlar, relied on a primitive converter to present the human eye with an otherwise unseen image, achieved by transforming 'near-infrared' radiation to lie within the visible spectrum. The primary goal was to allow vehicles safe access to unlit roads without headlights betraying their presence to a potential enemy.

The night sky contains more than enough near-infrared radiation to create an image, but the inability of the earliest converters to magnify this significantly gave very poor results. Using an infrared lamp to flood the target area greatly improved the converted image by providing more reflected light, but this, the 'active' system, has a major drawback. Any suitably equipped enemy can detect the presence of the infrared lamp and, by using a passive detector relying simply on reflected light, do so without being seen.

A soldier practises with an M110 (Knight's SR-25) rifle, fitted with a Leupold optical sight and a sound-suppressor to mask the origins of the shot. *US Army photograph*

Early in the Second World War, the Germans adapted their driving aid to create the Zielgerät 1229 Vampir. A 13 cm-diameter transmitting lamp, the converter and a magnifying-eyepiece lens in a telescope-like tube were mounted on a Kar. 98k, a Gew. 43, or an MP. 43/Stg. 44. A separate electrical supply was provided in the form of a bulky battery. Vampir sights could locate targets in poor light, but only at the expense of excessive dimensions; the power pack alone weighed 15 kg (33 lb).

By 1945, the US Army had created the Sniperscope on the same principle as the Vampir, but improvements in technology had discarded the infrared floodlight and the battery compartment was small enough to be mounted on the sight body itself. Though the passive-mode Sniperscope was heavy, awkward and difficult to handle, it worked surprisingly well. Perfection made possible by solid-state electronics was little more than a step away.

Thermal imagers reconstruct an image from differences in the heat signature of individual parts of an object. The night sky contains considerably more electromagnetic radiation in the near-infrared spectrum than in the limited window in which the human eye performs best, but phosphor present in most targets emits photons at wavelengths that can be accepted by a dark-adapted or 'scotopic' eye.

Poor performance of the photocathode in the earliest converters required the areas under observation to be flooded with infrared light. This was acceptable as long as infra-red sights were rarely encountered or, alternatively, were denied to an enemy. However, infrared sights can act as passive detectors without using the floodlight; once sights of this type are widely distributed, therefore, utility is largely lost. Non-detectable thermal imaging systems clearly have an advantage.

However, the thermal emissions from potential targets are usually absorbed or scattered by the atmosphere before reaching an observer. Fortunately, there are two narrow bands (3,000–5,000 nm and 8,000–13,000 nm) where thermal emissions penetrate the atmosphere efficiently. Sights adapted to work in these bands performed better than those relying on infrared, though they needed to be cooled continuously – often with liquid nitrogen – and were bulky. In addition, reconstructed images lacked definition and could be bizarrely coloured.

The Lasergage LWS 1060 night-sight of the 1980s could engage man-size targets effectively in low-light conditions up to 300 metres away. Overall length was 10.4 in., objective-lens diameter was 2.95 in., and the sight weighed 31.6 oz.

Compact image-intensifying systems have been among the most remarkable advances in sighting equipment, and reduction in cost has considerably broadened distribution. The greatest single improvement concerns the converter, which rapidly developed to a point where the image could be enhanced electronically, greatly improving resolution and extending engagement range dramatically.

Intensifier sights follow the same general pattern, though individual details vary greatly. The degree of amplification, releasing as many electrons as possible for each initial photon strike, depends on the design of the 'cascade tube'. Even first-generation sights, though bulky, were capable of 60,000-fold gains.

Light radiation from an image is focused onto the photocathode[5] window by conventional lenses, causing electrons to be emitted from the light-sensitive layer. These electrons are accelerated by an electrical current to strike a luminescent screen forming the inner surface of the rear window of the tube. There the phosphor coating converts electrons back to light radiation and forms an image corresponding to the original but incomparably brighter. The reconstructed images are almost always inverted, but erector lenses ensure they are upright when entering the eye.

The diodes of the 1960s gave excellent image resolution but only a moderate gain in image intensity, usually 200–500 times. They could process images in which there was considerable contrast between light and dark, and, therefore, had a 'wide dynamic range'. They were also comparatively free of interference ('low noise'). The tubes were focused either by reducing the distance between photocathode and phosphor screen to a minimum ('proximity diode') or by allowing an electron-lens to focus the electrons emerging from the photocathode before they reached the luminescent screen. These 'inverter diodes' returned the image to its upright form without requiring an additional ocular lens, but were considerably more clumsy than the proximity type. Among the advantages of proximity diodes were the absence of geometric distortion,

The AN/PVS-14, one of the most widely distributed of all intensifier sights.

high resolution over the entire area of the photocathode, and true 1:1 or 'same size' image transfer. They were also largely immune to electrical and electromagnetic interference.

The major drawbacks of first-generation intensifier sights were expense, excessive size – the Rank Pullin Individual Weapon Sight SS20 was 18.8 in. long and weighed 6 lb 2 oz – and the delicacy of the converter unit. Improvements in converters, which soon changed from cascade-type photocathodes to fibre-optic micro-channel plates, resulted in surprisingly compact intensifier sights. These could identify a man-size target under starlight conditions to a range of 300 m, and act as a passive infrared detector at 500–600 m.

Second-generation systems also took proximity- or inverter-diode form. The major change lay in the introduction in 1973 of a micro-channel plate (MCP), placed between the photocathode and the luminescent screen. This not only enhanced the energy of individual electrons but also allowed them to multiply. Electrons from the photocathode passed through tiny holes, rarely more than 10^{-7} mm in diameter, in a conductive glass plate. Secondary emissions occurred as each electron struck the sides of the hole; for each electron that entered, 10,000 could emerge. Image resolution and dynamic range suffered by comparison with first-generation diodes, but luminous gain was incomparably improved: as much as 10 million times for two MCPs in stacked configuration.

Third-generation diodes were proximity-focus patterns with micro-channel plates and special gallium arsenide photocathodes. These gave better luminous sensitivity (1,200 μA/lm instead of 300 μA/lm) than the bi-, tri- and multi-alkali photocathodes that had been used previously. They were intended specifically for use in infrared and near-infrared spectra and were unsuited to ultraviolet. Sensitivity made them susceptible to interference from heat ('thermal noise').

New fourth-generation converters, introduced in the USA in the 1990s by ITT and Litton Industries, improved manufacturing techniques in a quest for more gain and greater resolution. For example, the MX10160B had an 0.7 in.-diameter inversion cathode with a gallium arsenide 'filmless' photocathode bonded to glass; micro-channel plate current amplification; and an inverting fibre-optic phosphor screen. The converter was 1.448 in. long, with an external diameter of 1.225 in., and weighed a few ounces.

Intensifier sights can 'black out' under even 1-lux[6] conditions unless suitable filters are used, as they are designed to perform under less favourable conditions. Some sights can vary filtering level simply by adjusting the operating voltage automatically, but others must be adjusted manually. Excessive filtering can be needed in sunlight, in which case conventional optical sights can be preferable: intensifiers perform at their best in conditions from overcast starlight to full moonlight with light cloud-cover. Woodland, overcast starlight or unbroken cloud cover may reduce even a third-generation intensifier to impotence.

The AN/PVS-14 night sight, made by Litton Industries (now L-3 Warrior Systems), has sold in large numbers throughout the world. Non-magnifying, but with a field of view of 40° and an adjustable-diopter eyepiece, it can be fitted to a variety of weapons. The sight is 4.5 in. long and weighs about 12.4 oz. The third-generation intensifier tube allows a standing-man target to be engaged under starlight (0.001 lux) out to 300 yards.

Attempts have been made to combine thermal imagers with intensifiers, seeking to use the strengths of one to cloak the other's weakness. A thermal imaging/image-intensifying sight offered by Officine Galileo in the 1980s could even superimpose infrared and thermal-emission images to improve clarity.

The name 'laser' is an acronym of 'light amplified by stimulated emission of radiation' – widely touted as the 'death ray' beloved by science fiction. A much more obvious benefit has been the development of continuous monoline projectors used for alignment and surveying. Lasers of this type inspired the development of aiming projectors.

However, though the principle of the laser has been known for many years, the first commercially practicable system, based on rods of ruby, was not perfected until the early 1960s. 'Visible lasers' now customarily rely on excited helium/neon mixtures, whereas infrared designs may rely on carbon monoxide or even hydrogen cyanide.

Laser designators operating in the visible spectrum project a beam which can be seen by the firer and the target at all times, often in itself a great deterrent, but those that operate in the infrared bands require an additional detector. The LA-5 APTIAL ('Advanced Target Pointer Illuminator Aiming Laser') combines a visible spectrum red-dot laser with an infra-red beam visible only through suitably receptive goggles.

Laser rangers, which remove a lot of guesswork, can be used in conjunction with optical sights or image-intensifiers. They can also improve shooting skills dramatically; as complexity and sophistication increase, paradoxically, shot-resolution becomes easier. As Leroy Thompson has observed:

> The sniper still needs to know how to estimate distance using Mil Dots or other ranking reticles on his 'scope in case the laser range-finder becomes inoperative. But the Barrett Optical Ranging System (BORS) can take various data in conjunction with range, wind, temperature, etc., to 'dope the shot' and will even dial the proper elevation on the scope. Current programs such as those from GeoBallistics allow the sniper to use a cell phone or an iPad to get a firing solution. This can be as easy as placing a pointer on his position and a pointer on the position of the target on the map – with the weather meter attached to the sight getting temperature and wind, the solution for the shot will appear on the screen.[7]

What the future holds only the prophet knows!

Notes to Chapter Six

1. It has been estimated that for each 'confirmed kill' in Vietnam, 200,000 shots had been fired – nine times poorer than the US Army had achieved in the Second World War.
2. Charles Sasser & Craig Roberts, *One Shot – One Kill*, p. 137.
3. The Parker-Hale M85, with a conventional 1898-type Mauser action, was recommended as 'fit for service', probably as a safeguard against the failure of Accuracy International to translate the trials-submission rifle into series-production reality.
4. See for example, US Patent 5,458,046 of 17 October 1995. An application had been filed in Germany in 1993.
5. The photocathode, derived from the photocells and photomultiplier tubes (PMTs) developed for medical and scientific research, is essentially the inside of a window formed as part of a vacuum tube. The first PMT was developed by the French Atomic Energy Commission in 1953, then improved by the Philips LEP laboratories in Paris (1956). Production began in the late 1950s in Brive, in the factory now owned by Photonis SA.
6. The lux is defined as the amount of illumination produced when 1 lumen is distributed evenly over an area of 1 square metre; or, alternatively, the illuminance on any point of a surface 1 metre from a point source of 1 candela (1 international candle power). In practical terms, this is generally comparable with the level of illumination achieved by good street lighting, or by a full moon on snow or desert sand. By comparison, dusk, with the true colours still largely visible, will increase illuminance by a factor of ten; full sunlight can increase it by 100,000 times or even a million-fold.
7. Letter to the author, 20 February 2017.

What Makes a Sniper?

Even the most cursory investigation of the strengths and weaknesses of snipers confirms that, almost without regard to era, the 'core values' remain much the same. To avoid needless repetition, therefore, relevant material from the periods discussed in the foregoing chapters is presented here as an overview.

Successful kills have usually resulted from all or most of these phases: background; training; target selection; the approach; observation; the shot; evasion and withdrawal; and, lastly, reflection. Of course, every individual instructor (and now every author) has views on each of these that differ in detail, but there is a general consensus.

Background

The 'official views' of sniping, often hostile, have altered as the years have rolled past. It is probably fair to record that, the USSR/Russia excepted, the sniper has never enjoyed public esteem in a way accorded to men whose heroism was lionised. In the words of Charles Sasser and Craig Roberts:

> Although each war of this [twentieth] century has produced a need for snipers, a need that was temporarily filled at places like Salerno and Normandy and Pork Chop Hill and Chu Lai and Beirut, the standard reaction after the need ended was to cork the snipers back in their bottles, as though they never really existed. It was as though we were ashamed of them and what they had done, as though sniping was morally wrong and unfit for a role in the . . . armed forces.[1]

When World War I began, heedless of lessons learned in conflicts from the US Civil War to South Africa, the establishment view was often simply that a good target shot would make an ideal sniper (if no alternative to 'unsporting' sniping was to be found) or that men from army marksmanship schools would simply achieve success on the front line. Hesketh-Prichard explained:

> In the [British] Army there has always been in certain quarters a prejudice against very accurate shooting, a prejudice which is quite

understandable when one considers the aims and ends of musketry. While sniping is the opportunism of the rifle, musketry is its routine. It would obviously never do to diminish the depth of your beaten zone by excess of accuracy. But this war, which, whatever may be said to the contrary – and much was said to the contrary – was largely a war of specialists, changed many things, and among them the accurate shot or sniper was destined to prove his extraordinary value.[2]

By the spring of 1915, the notion that sniping had no value was being steadily disabused. Yet, in the absence of Army-wide direction (which was to happen, grudgingly, in 1917), the British achieved only individual laurels – usually gained by junior officers acting on their own initiative. The Germans still saw sniping as a 'temporary job' for qualified marksmen to whom optically sighted rifles were issued by a NCO.

The lack of an effective riposte to the German snipers cost the British dearly. An early, and largely misplaced belief in the value of established target-shooters soon gave way to the notion that training could only be effective if snipers were drawn from those who were practised at stalking, hunting or wing-shooting.

Target-shots such as Lieutenant George Gray of the Scottish Rifles, winner of several leading competitions and an excellent sniper-instructor, often met these criteria. But Clifford Shore, who had a poor opinion of the military value of target shooters, wrote at a later date:

> During the training of snipers there were many instances of men being excellent target shots but failing in the role of sniper shots. When I mentioned this matter ... with an officer, a very well-known Bisley shot and a sniping instructor in England and Italy during the [Second World] war, I was taken to task and my opinions rebuked in similar sentences to the following: – 'In my experience, the lad who has shot at school, or at Bisley has been streets ahead of anyone else, because he's a better shot.' Frankly, I should have thought that anyone who had sniped and worked as a sniper trainer would have found instances qualifying my contention that many range and competition shots did not make the grade ... This was a point mentioned to us at the sniping school in Holland [that] some men who were very good target shots ... found that shooting under service conditions was a very different proposition.
>
> Psychologically, the difference between resting on a placid firing point and shooting in war, even if one is in a comparatively safe position oneself, is the greatest possible contrast.[3]

Much more productive was recruitment from the fieldsmen – Scottish ghillies, beaters and even poachers; Canadian Native American trackers, hunters and guides; Australian kangaroo hunters; big-game chasers from

India and southern Africa (and also, of course, the British aristocracy and officer classes).

Men of the Lovat Scouts, drawn from northern Scottish great estates (see Chapter Two) had shown their worth in the South African War and were to be of great service in Gallipoli and on the Western Front. Canadians from the trackless expanses of plain and forest took a dreadful toll of Germans from 1915 onward, and Anzac volunteers such as Billy Sing proved to be highly effective marksmen in Gallipoli. The skills were needed again throughout the Second World War, in Korea and Vietnam, in Iraq, and in today's struggle against ever-increasing terrorism.

One of the strangest features was the regularity with which sniping attracted disapproval, usually because it was believed that problems would be overcome simply by training ordinary soldiers to be marksmen. The British withdrew all their sniper rifles, telescope sights and associated equipment in the early 1920s, and the Germans followed the lead in the 1930s. Only in the USSR was large-scale marksmanship training encouraged, a view which was to be vindicated by the success of Russian snipers, men and women alike, on the Eastern Front.

Many years later, Marine Captain Edward James 'Jim' Land was to reflect on his experiences in Vietnam. Land was an excellent marksman, who in 1960 had created a sniping/scouting school in Hawaii – the only one of its type in the US armed forces at the time. In 1965, Captain Robert Russell of the 1st Marine Division started a sniping school in Vietnam; a year later, Land was charged by his commanding general, Herman Nickerson, with the creation of a 3rd Marine Division school whose graduates would outperform their rivals in the 1st Division. Inter-unit rivalries of this type almost always improve results by adding an element of competition. But Land got off to a difficult start:

> [He] had to cajole and finagle equipment and weapons from reluctant and misunderstanding superiors whose war aims were directed else-where. He finally managed to acquire a supply of Winchester 30–06 Model 70 hunting rifles that the division kept in special services for his troops to check out for deer hunting. Then . . . there were no ordnance funds available with which to purchase telescopic sights. Men [were sent] to the post exchange in Okinawa to buy scopes along with volley-ball nets and basketballs.[4]

Very little had changed since the First World War!

The psychology of the sniper is the most contentious of questions. A popular view of snipers merely as cold, calculating, emotionless killers is misleading. There are undoubtedly several notable snipers who fell into the sociopathic category and it is true that many were 'loners'. Yet there were many other explanations: settling scores with the enemy, fighting for cause or country, political considerations, or simply obeying orders.

Pictured in 1915, this German soldier carries a Scharfschützen-Gewehr 98 fitted with an optical sight. Training at this time was rudimentary, but the British, in particular, had made great strides by 1918. *From the collection of the late J. Anthony Carter*

Even some of the finest sniper-instructors have struggled to explain motivation, but virtually all are agreed that a desire for revenge, unless carefully channelled, almost always leads to poor practice.

One of the most important requirements is, of course, perfect eyesight. Eyes are rarely the same, but the differences in an individual can be great enough to affect binocular vision and, therefore, the ability to gauge range effectively. This is easily tested, but peripheral vision, hand–eye co-ordination, reaction time and innate spatial ability may not be so obvious. Some men with excellent vision have proved to be poor performers when speed of decision is involved: shooting at disappearing targets, for example, or wing shooting. However, the best snipers have almost always had extraordinary abilities.[5]

Studies have often been done in an attempt to assess *visual acuity*, the ability of the eye to resolve small details at distance. A 'normal' eye is reckoned to be capable of resolving one minute-of-angle ('1 MOA') at a distance of 100 yards, but tests undertaken on more than a hundred Soviet marksmen in 1965–6 showed their eyesight to be significantly better: binocular vision averaged .572 MOA.[6] There was a perceptible difference between right eye (.595 MOA) and left eye (.610), but this was largely due to the fact that most of the individuals – like almost all snipers – were right-handed.

Experience also had a part to play. Marksmen who had been shooting for only two years recorded .704 MOA (monocular vision, dominant eye) while those with ten years' experience recorded .602. This discrepancy, and the universal better-than-average visual acuity of the group, explain why the best snipers have always had a reputation for seeing things that their colleagues do not. Other experiments suggest that filters can sometimes be advantageous. On a cloudy day, a blue filter improved 'naked eye' performance markedly, from .625 to .582 MOA, but even the best of the filters – yellow – had little significance in bright sunlight, improving .585 only to .578 MOA.

Conclusions drawn from these trials have particular significance for snipers. Visual acuity can be improved by practice, even if the firer's vision is better than average; periods of rest after intensive practice have an unpredictable effect, as visual acuity recovers better in some people than others; and rapid fire invariably reduces visual acuity sharply. Similarly, holding aim without blinking or shifting focus reduces the firer's accuracy by as much as 50 per cent. A wise sniper has to be aware of all these limitations.

Physical fitness is a prerequisite, especially if long periods of observational inactivity are necessary. Consequently, most sniper-training courses include self-defence classes, football and similar sports to ensure that their trainees reach (and stay) in peak condition. Today, awareness of individual physiological assessments and even nutritional requirements have to be considered during the recruitment process.

The ability to sustain minor wounds without affecting performance is also valuable. Snipers have almost always suffered heavy casualties, yet many managed to shrug off injuries from bullets and shell fragments. Among the most unusual case is that of Iosif Pilyushin, author of *Red Sniper on the Eastern Front*, whose service career in and around Leningrad seemed to have been terminated by the loss of his right eye. Amazingly, Pilyushin then taught himself to fire left-handed and was able to resume his combat career. But there were unexpected hurdles to overcome:

> I began to train vigorously: I dragged cases with cartridges and grenades to the front and hauled beams for constructing pillboxes and bunkers. Each morning I spent time crawling on my belly and leaping ditches. When leaping I couldn't always judge the distance correctly [with one eye] and I often didn't reach the other side of the ditch . . .
>
> [Eventually] I took a lot of shots and successfully so: the shots were all on target, although the bullet holes were scattered. Despite all the difficulties of firing with my left eye, I had achieved my main objective: I could defend myself in combat. With each shot, the bullet pattern

in the target became tighter and tighter. But it still required a lot to calculate each shot accurately from any distance.[7]

Many of the best-known Soviet snipers were eventually reduced to training roles when wounds reduced their effectiveness. Iosif Pilyushin, Ivan Sidorenko, Yevgeni Nikolaev and Lyudmila Pavlichenko were among them.

One useful trait in a potential sniper has always been calmness under pressure, allowing breathing to be regulated, and even the heartbeat to be slowed. Shooting with precision cannot be done if natural tremors affect aim. However, instructors have often found that many otherwise excellent candidates panic when required to take a difficult shot or, alternatively, are so excited that any natural caution they possessed is overriden in the heat of the moment.

Something that can rarely be predicted, however, is the reaction of a sniper when faced with his first human target. For many, the natural reaction is to hold fire. Looking into a man's eyes, knowing your shot will kill him, does not always come easily. Countless soldiers have surrendered their lives simply because they 'froze', hesitating just long enough for the intended victim to shoot first. Even experienced snipers – Henry Norwest, for one (see page 136) – were killed because their shot was a fraction of a second too late.

Norwest's shot, fired while the fatal bullet was in flight, is said to have killed his German opponent. There have been other well-documented mutually fatal encounters, and others where the same result was avoided only by good fortune. One of the most extraordinary stories concerns US Marine Sergeant Carlos Hathcock (1942–99), one of the most effective snipers of all time:

> The Cobra sniper's arms and legs flailed and thrashed. His body repeatedly propelled itself into the air, blood spurting, as it cleared a red-smeared nest in the bushes. It arched its torso against the sky; the entire body began trembling desperately until life was simply gone and the thing that had been a man minutes ago collapsed to earth.
>
> Burke was the first one to him. He picked up the dead sniper's rifle ... Both lenses of the scope, front and back, were shattered ... My bullet [had] smashed through his scope and into his right eye. At the moment I shot him, the Cobra had his scope trained on me and was a hairsbreadth away from claiming the bounty on my head. I just happened to get on the trigger first.[8]

This 'through the sight' story has been questioned. Indeed, a recently attempted duplication was judged to be a failure: the bullet deflected within the sight tube and failed to pass along its axis. But the testers had simply bought a standard SWAT-type rifle and optical sight, paying no heed to limitations that would have been conferred by the original equipment. When another trial was undertaken with .30-calibre US military ammunition and a Soviet PU sight, much simpler internally than most modern multi-element designs, the shot *did* pass straight through the sight body.

It is important, too, that the sniper and his observer are attuned to the conditions; someone brought up in the Arctic may not cope with the heat of the desert. Australians fighting in Gallipoli in 1915 coped well enough with the daytime heat, but the cold nights gave rise to hypothermia and even mild frostbite. There was (and still is) no point in expecting a man from the northern wastes to perform as well in the tropics without lengthy acclimatisation.

Snipers are also often required to maintain a lengthy vigil with very little sustenance. Simo Häyhä sucked sugar cubes to maintain his metabolism; US snipers in Vietnam preferred jelly beans and other compact high-calorie snacks. Dehydration can be a serious problem, even in temperate zones, and a well-filled water bottle is essential to any lengthy foray.

Training

One of the problems with sniping was that, especially in the early days, few men had enough knowledge to make use of the tools they were given. This even applied to the rifle, the most basic of infantry weapons. Training as an officer cadet, Clifford Shore was one of those unfortunate enough to be issued with:

> S.M.L.E.s and told these rifles were no use as regards shooting! This surprising statement was amply borne out when we went on to the ranges for the first of our very limited visits. My first three shots at 100 yards were wash-outs, and according to the signal very low and much to the left ... For my fifth and last shot, with 600 yards on the sights, I took an aim half-way up the flagpole which was standing at the extreme right-hand side of the butts, and lo and behold, up came a disc to signal a bull.[9]

Among the most important elements of rifle-shooting is 'zeroing': ensuring that the shot goes where the sights have predicted. This has been an age-old problem, beginning as soon as mass-production of rifles began.

Ever-increasing output, and demands made by military authorities, ensured that quantity was the byword. No time could be spared to regulate each gun individually, and so, at best, random batches were tested to ensure that there was no underlying long-term problem. The result of this short-sighted policy was that the Lee-Enfield and Lee-Metford rifles issued during the South African War shot exceptionally badly, and were sighted in such a way that adjustments could not be made in the field. Some individual marksmen learned to compensate for their poorly sighted rifles, but the rank-and-file generally blazed away regardless of where their bullets were striking. And, in the confusion of a large-scale engagement, it was all but impossible to follow

Naturally left-handed snipers are rarely encountered, especially in the period in which bolt-action rifles were commonplace. An exception to the rule was Lyubov Makarova, who ended the Great Patriotic War with eighty-nine kills. The forward-placed Mosin-Nagant bolt is among the easiest to use by reaching over the action body, though the sight-base attached to the left side provides something of an obstruction. *Author's collection*

the flight of an individual bullet. Deficiencies in musketry training were still obvious when the First World War began.

Once a rifle had been zeroed satisfactorily, it was essential to check the settings periodically. The slightest knock could move the sights by the fraction that meant inches at a hundred yards and feet at a thousand yards: the difference between hit and miss, or life and death.

One of the most important aspects of sniper training was to provide enough variety to prepare individuals for the challenges of the battlefield. Shooting at static targets helped at the beginning, but few real targets were static. Disappearing targets were useful, still better if they could move as well as fall. Shooting at replicated 'enemy trenches' taught the value of observation as well as accuracy; 'crawl, observe, consider, shoot' was wise practice.

Work needed to be undertaken at realistic distances, from all angles and under widely differing conditions. It was well worth exploring ways of using ranging or reference marks to prepare for a shot, or even as aim-points for the shot if, for example, the position of a loophole clearly observed with the spotting telescope was not as clearly defined in the less powerful optical sight.

The sniper – indeed, any rifleman – had to learn to fire from many positions, but there was often no consensus. Many instructors, such as Hesketh-Prichard,

McBride and Shore, were critical of the kneeling position on the grounds that it was not especially stable; standing, unless cover could be used beneficially, was rarely an option. Some championed the traditional back position, others favoured the Hawkins position. Special ways were even devised of dragging rifles into position when crawling.

The Germans had, perhaps, the most regimented system of training prior to 1945, and were widely believed to follow 'scientific principles'. Traits such as shooting not only with the rifle fore-end on a rest but also with the left hand cupping the under-edge of the butt were evident in training manuals, and trees were recommended either as a vantage point or to support the rifle. Yet the Germans produced few world-ranking snipers, excepting a group from the Tirol and men like Bruno Sutkus, who had grown up in the wilds of Lithuania.

The Wehrmacht, the German armed forces, acquired surprisingly large numbers of *Zielfernrohrgewehre*, but this merely means 'telescope-[sighted] rifle' and does not refer specifically either to marksmen or snipers. Indeed, there is no directly equivalent German word for 'sniper', even though they produced many highly competent snipers in the Second World War. The goal, which modern experience shows to have been ahead of its time, was to improve the overall performance of infantrymen until 'everyone was a marksman'. Consequently, the hopelessly ineffectual Zf. 41 appeared as a 'sighting aid' masquerading as a sniper sight, and 50,000 Gew. 43 were fitted with a Zf. 4 that had simply been copied from the Russian PU.

Several Allied commentators were critical of lack of individualism among the German snipers, and there is little doubt that training was effectively 'by rote', with little of the maverick elements that could sometimes be encountered elsewhere. Such regimentation was a far cry from the inspiration that created, for example, Hesketh-Prichard's papier-mâché heads. Sometimes apparently smoking a cigarette, these had proved an effective 'sniper lure' in the trenches of the Western Front. But there was method in such apparent eccentricity; not only did the three-dimensional heads prove to be invaluable training aids, but they were also highly effective 'sniper locators' when used in conjunction with a sliding elevator and a trench periscope.[10]

The use of trees (sometimes artificial) for cover or as vantage points, which the French and Germans had used regularly during the First World War, was widely decried by British and US sniper instructors. In their view, not only was the shot-angle rarely level, with difficulties of assessing the shot trajectory, but the trees gave very poor protection once the position had been detected. Many snipers were killed simply by raking the tree tops with a machine gun or by calling up a mortar team or artillery.

Nor was the use of a rest recommended, though trials sometimes showed that sandbags, provided that they had a reasonable degree of resilience, could be very useful. Current thinking encourages the use of anything that can help to steady a shot or improve observation skills: bipods and monopods fitted to the rifle, ski-poles tied together, or rudimentary tripods made of branches. Though a resilient pad can negate the worst effects, a rigid rest such as a concrete lintel or window-ledge will invariably have a disruptive effect on barrel harmonics if the point of contact – usually the underside of the fore-end – bears on the barrel by way of (for example) encircling bands or the nose cap. Only in the case of a free-floating barrel, or one that is precisely bedded for its length, will there be negligible disruption.

Textbooks were written in many languages to regularise training, but what a sniper often needed in particular circumstances came not from any printed page but from personal experience and even instinct. Some men had a 'sixth sense' to warn them of unseen danger; others ventured blindly onward.

Instruction has varied greatly from army to army. The Russians trained men for a week, then sent them to the Front in the charge of an experienced sniper. However, it has been claimed, even by Vasiliy Zaytsev, that 80 per cent of these *Malenkiy zaychik* ('little hares') lasted less than two weeks. Effectively, if mistakenly, relying on the survival of the fittest to give good results gave way to the establishment of the first specialised sniper schools in 1942–3. The British and Americans initially paid only cursory attention to sniping, until the fortunes of war dictated otherwise. For the British, the period that followed the Dunkirk evacuation was the nadir:

> After Dunkirk, sniping was thought to be completely out and the sole cry was for more dive bombers, tanks, anti-tank guns, artillery and fire-power generally. And so, although snipers equipped with the P/14 rifle and the 1918 pattern 'scope continued to exist on paper, there was no sniper training carried out at all. After Tunisia and Sicily, however, the position was totally changed, and one of the dominant cries was 'Sniping' . . .[11]

In the USA, it was the post-Pearl Harbor Japanese advance across the Pacific that showed how badly equipped the Army and Marine Corps had become.

If problems with the rifles were bad, those of the telescope sight were worse. Hesketh-Prichard gave a good illustration of their scale in *Sniping in France*. After discussing the way in which optical sights were issued in the German Army in the early stages of the First World War – to experienced NCOs who ensured that the rifle/sight combinations were properly regulated before handing them to individual snipers – he observed that this method gave far better results than:

> Upon our side, where very often . . . the man using the telescopic sight knew nothing about it. On one occasion I had gone down on duty to

The lessons of the First World War had been lost by the time war began again in 1939, and had to be re-learned. These Canadian soldiers practise with optically sighted P/14 rifles, their 'service issue' until replaced in 1944 by the No. 4 Mk I (T). *Author's collection*

a certain stretch of trench and there found a puzzled-looking private with a beautiful new rifle fitted with an Evans telescopic sight.

'That is a nice sight,' said I.

'Yessir.' [The private replied]

I examined the elevating drum, and saw that it was set for one hundred yards. 'Look here,' I said 'you have got the sight set for a hundred. The Hun trenches are four hundred yards away.'

The private looked puzzled.

'Have you ever shot with that rifle?' I asked.

'No, sir . . .'[12]

The adjustment of optical sights was problematical, and mounts were often insufficiently rigid to ensure that the telescope remained in place from shot to shot. Indeed, some of the target sights pressed into military service were spring-loaded; consequently, the Winchesters of the First World War and the Unertls of the Second World War often had to be pulled back to position before each shot.

There has always been disagreement about telescope sights: should they remain on the rifle as long as it is being used, or detached as soon as possible into a protective case? During the Second World War, the Russians were the greatest proponents of sights remaining on the gun at all times, possibly because their snipers, as a group (generally excepting the women), were uneducated;

the Germans were the most likely constantly to mount and remove their sights, with unfortunate effects on zeroing. Experienced snipers would always fire warm-up and zero-check shots when they could, but mounting a telescope sight under pressure, to fire a quick shot, has always been risky.

Even in today's environment, with rapid advances in technology and ever-increasing public scrutiny, effective sniper-training is essential. Though many of the basic techniques are centuries-old, the modern sniper has to master a far wider range of equipment than his forebears. This is particularly true of sights (see Chapter Six), from the complicated range-gauging graticles of many telescopes to computer-assisted image intensifiers, designators, and even the cameras that can broadcast sight-pictures to controllers who may be many miles away. Though some have seen this as disadvantageous, reducing the value of an individual's decision-making capabilities, accountability is increasingly paramount in an age when one man's decision can be another's war crime.

Training must also ensure that an individual sniper can operate effectively in environments ranging from a devastated city to the vastness of open desert or Arctic wastes. And it must also consider matching individual tasks with individual weapons.

One underlying change in sniping in the last fifty years has been the gradual increase in engagement ranges. Where once a few hundred yards would have been regarded as normal and 750–800 yards as practicable limits, these figures have doubled as sniper rifles have increased in power. Hits at ranges in excess of 2,000 yards are regularly reported (though there is still no certainty of strike at such ranges, even with guns such as the .338 L115A3). The advent of the AMR, the anti-matériel rifle chambered for rounds such as .50 Browning or the 14.5mm Soviet type, has broadened the role of the sniper to include destruction of vehicles, radar installations, helicopters and even aircraft.

Snipers at War considers only the conventional anti-personnel role in detail, but acknowledges the effectiveness of large-calibre sniper rifles against delicate equipment. This is little more than an extension of the use of big-game rifles in the First World War, and later, especially by the Red Army, of obsolescent anti-tank rifles as effective anti-matériel weapons. *An illustrated Manual of Sniper Skills* (2016) by Mark Spicer, a sniper and instructor of impeccable pedigree, is a marvellous source of additional information. There the value of an AMR against a warship, even a submarine, is graphically illustrated.

Concealment

It is vitally important in a trench, or similar defensive structure, to deny enemy snipers an easy shot. An idea of the way in which an environment can be obscured can be gained from Hesketh-Prichard's remarks on loopholes in trench parapets:

Spot the sniper! The time spent on camouflaging a sniper's position is rarely wasted, but the weapons must also be obscured. This photograph shows how difficult it can be to distinguish a well-hidden marksman even at close range. *Accuracy International*

The efforts to camouflage our loopholes were extraordinarily primitive – indeed, concealment was nearly impossible in the form of parapet then in use. Many of our units took an actual pride in having an absolutely flat and even parapet, which gave the Germans every opportunity of spotting the smallest movement. The parapets were made of sandbags beaten down with spades, and it is not too much to say that along many of them a mouse could not move without being observed by the most moderate-sighted German sniper . . . At a later date a trial was instituted, and proved that in spotting and shooting at a dummy head exposed for two to four seconds over a flat parapet, the number of hits was three to one, as compared with the same exposure when made over an imitation German parapet . . .

The German trenches were deeper, with much more wire in front, and from our point of view looked like the course of a giant mole which had flung up uneven heaps of earth. Here and there, a huge piece of corrugated iron would be flung upon the parapet and pinned there with a stake . . . Here and there lay great piles of sandbags, black, red, green, striped, blue, dazzling our eyes . . .[13]

In addition, from the earliest days of their craft, successful snipers have always made use of personal camouflage and cover. This can range from leaves and twigs woven into scrim stretched over a helmet, and rags tied round a rifle to break its silhouette, to purpose-designed clothing such as the *Tarnjacke*; that descendant of a 'ghillie suit', the Denison Smock; and the variety of pre-printed camouflage clothing that can now be obtained. The paleness of the face and hands can be hidden with a veil or dulled with ash, charcoal or mud, and a man's outline can be disrupted by loose garments or padding. Properly attired, as many a battlefield report testifies, the cunning sniper can lie, undetected and undetectable, no more than yards from his enemy.

Observation

Keen observation and good range-estimation are vital to efficient sniping, and even the earliest training programmes laid emphasis on the problem. Good observation paid vital dividends. As Vasiliy Zaytsev reflected:

> We lay in silence, observing. The rays of the setting sun were lighting up the hill. They illuminated all the dark recesses and bathed every outcrop in bright relief. A bit further down the hill some spent German artillery shell cases lay scattered. I had nothing to do, so I counted them. There were twenty-three.
>
> Then I did a double-take. One was missing a bottom! Through a shell case like that, just like through a telescope, someone could see a long way into the distance. I raised myself a little.
>
> Suddenly there in the casing – it was like a flint struck a spark! An explosive bullet ripped into the embankment behind me.[14]

Other factors also had to be taken into account. For example, Hesketh-Prichard observed that the British Army snipers of the First World War held an advantage over their opponents:

> In the very crack Jäger regiments, such ... as were, I suppose, recruited from Rominten or Hubertusstock districts, where the great [hunting] preserves of the Kaiser lay, and in which were a large percentage of Forest Guards, this was very noticeable. But for long-distance work, and the higher art of observation, the Germans had nothing to touch our Lovat Scouts. This is natural enough when one comes to consider the dark forests in which the German Forest Guards live ... Mostly game is seen within fifty or seventy yards, or even closer, in these sombre shades, and then it is only the twitching of an ear or the movement of an antler lifted in the gloaming. Compare the open Scottish hills. It was the telescope against the field-glass, and the telescope won every time.[15]

The approach

In many ways, the approach to a potential target away from the confines of a static or defensive position, such as a trench, is one of the most important aspects of 'offensive sniping'. The sniper has to decide what to take. The rifle, optical sight and spotting telescope are essential, together with a water bottle and a few basic rations. The right ammunition has to be chosen for the task; this will usually be ball, but armour-piercing, incendiary and occasionally even tracer cartridges can be needed. Each round has to be checked visually, cleaned and, if possible, cycled 'dry' through the rifle-action to reduce the likelihood of jamming.

The best ammunition that can be obtained is essential (now often 'match grade'), ideally taken from the same manufacturing batch to ensure consistency of performance. Notice has to be taken of bullet weights, as ball and tracer, especially, generate differing muzzle velocity and so take different flight paths; this problem can be of very real significance should the target be a thousand yards away.

Most snipers have carried self-defence weapons: handguns, fighting knives or submachine-guns. The US M1 Carbine was once a particular favourite, as it was light and handy, fired a cartridge that was much more powerful than a pistol, and could strike accurately at 200 yards. The M4A1, a short-barrelled derivative of the M16, is a modern favourite even though it cannot sustain fire without 'cooking off'. Pistols with high-capacity magazines have also proved to be useful at close range. France's anti-terrorist GIGN have even used a long-barrelled .357 Magnum MR73 revolver, fitted with a bipod and an optical sight, as an urban sniper weapon.

Two or more grenades were once deemed essential, as they could repel a threat, cover a hasty retreat or, in extreme circumstances, be a way of avoiding capture. Soviet women snipers were particularly vulnerable to terrible repercussions, and there are several well-documented cases of self-sacrifice..

Atmospheric conditions can be important: the angle and position of the sun; the strength and direction of the wind; lingering morning mist over a lake or river; a chill which could highlight a tell-tale wisp of propellant smoke; changes of temperature if observation is to be prolonged from day into night, or if the approach is to be made at night to strike the next day.

Terrain has to be analysed, particularly if cover is minimal or natural features inhibit progress. The approach has to be mapped in the minds of a sniper and his observer to reduce the chances of being spotted. It is obviously far easier to move forward in scrubland or wooded terrain than, for example, was so often the case in the deserts of Iraq or the barren hills of Afghanistan. But it is also much more difficult to identify possible enemy positions in woodland.

Identifying a shooting site before starting the approach, if it can be done, is beneficial. And, most importantly, the locations of enemy pits, posts and trenches, threats of discovery or retaliation, have to be considered. Only when the team is confident of reaching its goal should an advance be made slowly, steadily, constantly keeping watch, and taking detours to avoid detection.

Once 'on site', the sniper team has to ensure that it is sufficiently well dug-in below the enemy's sightline, helped if possible by vegetation and foliage. The men have to be aware that the muzzle blast that accompanies each shot can kick up dust or snow that can immediately (and possibly fatally) betray their presence.

There are many reports of snipers being killed when reflections from their rifles, equipment or, especially, telescope-sight lenses flashed a give-away in the sun. This was a particular problem during both world wars, simply because the metalwork of the guns was customarily blued; only when sand-blasting, Parkerising and similar non-reflective finishes appeared did the problem recede.

Varnished woodwork caused similar problems, though many snipers, using whatever was to hand, applied rudimentary camouflage to disrupt the outlines. The temperature-stable camouflage-finish synthetic stocks of today, which can be acquired in a wide range of styles, are a great advance in concealment. They can be used virtually straight from the box if the pattern-design has been accurately matched to theatre, though even the best camouflaged stock is worthless if no attention has been paid to obscuring the rest of the rifle.

There are many testimonies to the stalking skills of individual snipers, but few can match the solo mission undertaken in 1967 by Carlos Hathcock to eliminate an NVA general headquartered some distance behind the lines.[16]

Target selection

Once the sniper team is in position, unless time or circumstances dictate otherwise, another period of observation begins. A scouting telescope, with better light-gathering qualities than the rifle telescope and generally a wider field of view, is an excellent tool.

If the position of the enemy is unknown, the greater the value of observation becomes. It has not been unusual for snipers to remain motionless for long periods – if they have taken position at night, for example, and are waiting for sunrise to blind the enemy.

Identifying the best target is crucial. Snipers have often been instructed to eliminate high status targets in the hope that killing a senior officer will destabilise command structure. They are taught how to recognise regimental badges, rank identifiers, and the medals and decorations that are often worn by the arrogant or the unwary. And there are the little things that gave a clue to rank when no insignia can be seen: giving commands, carrying binoculars, or a handgun instead of a rifle.

German snipers killed many British officers in the First World War simply because they wore jodhpurs or knee-high puttees, or carried swagger-sticks. One marksman claimed to target soldiers 'with thin legs and moustaches', as they were almost certain to be officers. Russian snipers killed many German officers during the Great Patriotic War simply because rank distinctions and Iron Crosses were still to be seen.

In an extreme case, a single well-placed shot can engender panic and turn the tide of battle. The death of General Simon Fraser at the Battle of Bemis Heights, during the American Revolutionary War, cost the British dearly; and a decision taken in Vietnam in 1967 by US Marine Sergeant Carlos Hathcock and his partner John Burke, first to eliminate the officer and senior NCO of a platoon of inexperienced NVA soldiers, caused such panic that the snipers were able to kill all but a handful of the eighty men over five days.[17]

Missions were often aborted, even after prolonged observation, simply because the sniper's target was impossible to define with certainty. It was far better to try another day than take a chance. The unwise man fired and risked retaliation; the experienced man could accept defeat. Or he could ensure that something more effective dealt with the threat:

> 'See to the left of that brush pile where you can barely see the grass looks a little different?', I asked the platoon leader on the hill, pointing.
>
> 'Yes, sir.'
>
> 'Bore-sight a 106 directly on that spot. We have no way of catching the [NVA] sniper until he returns and fires a shot, but when he does – fire the recoilless and see what happens.'
>
> The Marine swivelled the long 106 millimetre recoilless anti-tank weapon and deflected its barrel until its open snout pointed directly at the sniper's hide. The gunner slammed in an HE round and closed the breech. All he had to do was slap the push-button trigger . . .[18]

A few days later, 'Charlie One Shot' tried his luck one last time; he missed and, seconds later, the 106 mm HE shell blew him to pieces.

Captain Jim Land and Sergeant Don Reincke of the USMC had devised the ideal answer to an otherwise risky mission; unlike their quarry, both lived to fight another day.

There have been many precedents: in the First World War, British snipers used large-bore elephant rifles to penetrate the iron shields that supposedly conferred immunity on German snipers. In the later conflict, the Russians often used their 14.5 mm PTRD and PTRS anti-tank rifles to neutralise snipers' posts which were either out of range or too well protected to attack with the 7.62 mm *snayperskiy vintovkiy*.

Above: a US Special Forces marksman trains with a Russian SVD (Dragunov) rifle. Familiarity with the weapons and equipment of a potential enemy can prove to be a life-saver in battle. *US Army photograph*

Once snipers are in position, there is the range to the target to consider. But many factors can influence range-gauging. Under-estimates can be due to clear weather, or if the sun is behind the firer; if the terrain is rough or undulating; when shots are taken uphill; when the lower part of the target is obscured; if the shot is taken across water or snow; or if the target contrasts too sharply with the surroundings. Conversely, over-estimates can occur if it is rainy or cloudy; at dawn, dusk or in shade; when the shot has to be taken against the sun; or when shooting downhill. Shots will fly high into a headwind, hit low in a tailwind. And it is notoriously difficult to make accurate assessments when prone.

The addition of laser rangefinders to some of the latest sights minimises guesswork, of course, but the snipers of the world wars could rely on nothing but their own observational skills (even though attempts were occasionally made to provide ranging stadia and similar marks on the reticles of telescope sights).

Success depended on the skill with which they could make allowances for range, wind and light. Some men zeroed their sights at a specific distance; others adjusted their sights for each change in distance. Neither method was infallible, though the flat-shooting capabilities of rounds such as the US

.30–06 was a very real advantage. Charles Sasser and Craig Roberts relate how Marine Lance-Corporal Jim Miller, serving in Vietnam in 1968, had zeroed his Remington M700 rifle at 1,000 yards. Miller said that all he had to do at 500 yards, was 'aim at his balls in order to get him in the chest'; at 1,200 yards, he simply aimed at the head to hit the heart.[19]

Such a method in the hands of an experienced sniper could save the vital seconds taken to adjust the sights – a life saver, or giving the near-instant second shot that would otherwise have been impossible.

When the range increased beyond that at which a successful hit could be guaranteed by a single rifleman, wise snipers sometimes fired as a group, setting their sights to bracket the distance estimated to the target. This greatly increased the chance of a hit, even though some sights would be set too low while others were too high. Clifford Shore wrote about:

> A whispered conversation between the four snipers followed as to the range to the German observer [in a tree-top]; they could not agree. Two of them said they estimated the range to be between 250 and 300 yards; the others said it was nearer 350 yards. The matter was settled in a rather novel manner. Three of the snipers set their sights at 250, 300 and 350 whilst the fourth sniper took the binoculars and kept them riveted on the prospective target. When quite satisfied ... he coolly gave the three men a fire order; the three rifles 'spoke as one' and the Boche came somersaulting to the ground.[20]

The shot

When it comes time to fire, the sniper has to take control of his breathing and be aware almost of his heartbeat before gently squeezing the trigger. He also has to heed the moment. It is vital to fire at the instant that a target is exposed to its greatest extent, but also to fire before an opponent recognises any threat.

Though not always employed on anti-personnel duties, particularly when machine guns have to be neutralised (striking the gun could be as effective as hitting crewmen), snipers have often favoured particular types of hit. Effective sniping saps enemy morale, and this effect can be raised by specific targeting. Lyudmila Pavlichenko habitually shot the second man in a column; done often enough, this makes the 'second man' in any line fearful. In extreme circumstances, no one will 'go second' and confusion may reign until officers and NCOs give the orders that can result in near-mutiny. Other well-known snipers deliberately shot targets in the chest or stomach, particularly if they were part of a group. The rationale was simple: colleagues would rush to the injured soldier's aid, exposing themselves to another shot, and medical staff

French Special Forces soldiers demonstrate camouflaging skills, though their efforts are probably not ideally suited to this particular terrain. One man shoulders a 7.62 mm bolt-action FRF-1 and his companion has a 12.7 mm PGM Précision Hecate II anti-matériel rifle. *Author's collection*

would be needed to deal with the wounded in a way a 'clean kill' could never require.

Shooting the last man in a column, especially if strung-out or detached from the group, could pass unnoticed and allow another shot. Firing by precise time, 'death at eight o' clock', could instil ever-growing fear as the clock ticked down (though never to be attempted without constantly changing shooting position). Observing supply tracks, shooting messengers or those attempting to reinforce forward positions could also be profitable.

There were countless ways in which an enemy's routine could be disrupted – hitting metal loop-hole plates in the dead of night to ruin sleep, or shooting holes in water containers to highlight thirst. One of the good sniper's greatest assets has always been the ability to improvise.

Aftermath

A shot is only really successful if the sniper, observer and any associated personnel can withdraw safely to their own lines. Consequently, not only is it essential to plan the approach but it is equally important to plot the escape in such a way that the retreating team cannot be seen. This can be time-consuming, but time has no significance if safety is to be ensured. Wise snipers sometimes plan several routes, selecting the best route as circumstances dictate.

Kills have usually had to be validated by officers to add to an individual's score. But intelligence gathered by the sniper team, which can be in a position to monitor the enemy for hours at a time, has often been more useful. Alterations in the enemy's lines, changes in personnel or the presence of reinforcements, sometimes the prelude to an attack, can influence tactics. Virtually all instructors and sniping textbooks emphasise the importance of reporting observations accurately. This is, therefore, inevitably a keystone of training.

Notes to Epilogue

1. Charles W. Sasser & Craig Roberts, *One Shot – One Kill*, p. 3.
2. Hesketh Vernon Hesketh-Prichard, *Sniping in France*, pp. 33–5.
3. Shore, *With British Snipers to the Reich*, pp. 128–9.
4. Sasser & Roberts, *One Shot – One Kill*, pp. 166–7.
5. *The Sniper Anthology*, telling the story of Bert Kemp.
6. N. Kalinichenko, 'How Soviets View Aiming Problems', *The American Rifleman*, September 1970, pp. 40–3.
7. Iosif Pilyushin, *Red Sniper on the Eastern Front*, p. 140.
8. Sasser & Roberts, *One Shot – One Kill*, pp. 15–16.
9. Shore, *With British Snipers to the Reich*, p. 247.
10. The 'Hesketh Head' – sometimes sporting a dummy cigarette attached to a long rubber tube through which smoke could be blown – could slide vertically on a plank or similar support. It was exposed long enough to be shot by a German sniper then lowered by the precise height of a trench periscope. The operator looked through the bullet holes, from the rear, and into the ocular lens of the periscope. As the lens was now aligned directly with the path of the bullet, the operator could see exactly where the shot had come from. This promoted very effective counter-sniping; a trial showed that of seventy-one 'heads shot', sixty-seven revealed the position of the German sniper.
11. Shore, *With British Snipers to the Reich*, p. 283.
12. Hesketh-Prichard, *Sniping in France*, pp. 31–2.
13. Ibid., pp. 33–4.
14. Zaitsev, *Notes of a Russian Sniper*, pp. 97–8.
15. Hesketh-Prichard, *Sniping in France*, pp. 105–6.
16. Sasser & Roberts, *One Shot – One Kill*, p. 202ff.
17. Ibid., pp. 139–53.
18. Ibid., pp. 182–3.
19. Ibid., p. 158.
20. Shore, *With British Snipers to the Reich*, p. 72.

Sources and Further Reading

Snipers at War has drawn information from websites, on-line publications and conventional printed material. Particularly useful were details extracted from the splendid websites of the US Government Patent Office (www.uspto.gov), the Deutsches Patent- und Markenamt (www.dpma.de) and the Espacenet intellectual-property site (worldwide.espacenet.com). Unfortunately, there are gaps in coverage which, though they are steadily closing, still inhibit research: digitised British patent records currently go back only to the mid-1890s (though English records to 1852 are complete), and French nineteenth-century patents are accessible only into the 1870s.

Canadian military records proved to be extremely helpful. Though digitisation is still far from complete, the service careers of most of the CEF snipers of the First World War were retrieved from the Library and Archives Canada website (www.bac-lac.gc.ca) and some of those yet to be processed were found through genealogical sites.

Among printed works, *The Sniper Anthology* is a particularly useful overview of the subject and John L. Plaster's all-embracing *The History of Sniping and Sharpshooting* remains a benchmark to which we all aspire. Mark Spicer's *An Illustrated Manual of Sniper Skills* is essential reading for anyone involved in the practical aspects of our subject.

I particularly enjoyed Hesketh-Prichard's *Sniping in the Great War* and McBride's *A Rifleman Went to War*: both are not only fascinating to read, but also had a huge effect on the development of sniping techniques which are as relevant today as they were a century ago. Frederick Crum's *A Rifleman Scout* provided details going back to the war in South Africa. *Gallipoli Sniper* by John Hamilton is an impeccably documented story of Billy Sing, and the meticulous research underpinning Martin Pegler's *Sniping in the Great War* allowed me to cut more than a few corners.

Of the Second World War memoirs – which are not, perhaps, always incontestably truthful – I enjoyed reading *Red Army Sniper* by Yevgeni Nikolaev and *Red Sniper on the Eastern Front* by Iosif Pilyushin, who lost an eye to a shell fragment and then taught himself to shoot left-handed. *Sniper Ace* by Bruno Sutkus, though something of a 'catalogue of kills', gives a valuable first-hand German perspective. More recent stories, related not only by Chris Kyle of *American Sniper* fame but also by lesser-known snipers such as Craig Harrison (*The Longest Kill*) and James Cartwright (*Sniper in Helmand*), also merit attention.

Individual kill-totals accessible on many websites are widely contested. The absence of confirmation from Russian sources is often taken to prove that the highest claims are propaganda (see my remarks on pages 199–202) even though the claim of Simo Häyhä to have accumulated 505 kills in less than a hundred days generally passes unchallenged. Similarly, the ranges at which kills have been achieved – growing greatly since the widespread issue of large-calibre Long Range Rifle Systems – have never been subjected to the statistical analysis that would show if 'one shot, one kill' *is* achievable . . . and, if so, under what conditions.

The selective list that follows contains brief details of important and accessible printed sources, though additional information can be found in the chapter notes. Where there has been more than one edition of a book, the date and publisher stated are those of the newest edition.

Alla Begunova: *Angely Smerti: Zhenshchni-snaĭpery, 1941–1945* ('Angels of Death: Women Snipers, 1941–1945'). Veche, 2014

James Cartwright: *Sniper in Helmand: Six Months on the Frontline* (Foreword by Andy McNab). Pen & Sword, 2014

N. A. and R. F. Chandler: *Carlos Hathcock, White Feather*. Iron Brigade Armory, 1997

Philippe Contamine: *War in the Middle Ages*. Basil Blackwell, 1984

Frederick M. Crum: *Memoirs of a Rifleman Scout*. Frontline Books, 2014

Denis Edwards: *The Devil's Own Luck*. Leo Cooper, 1999

Adrian Gilbert: *Sniper: One-on-One*. Sidgwick & Jackson, 1994

——: *Stalk and Kill: The Sniper Experience*. Sidgwick & Jackson, 1997

—— with Tom C. McKenney, Dan Mills, Roger Moorhouse, Tim Newark, Martin Pegler, Charles W. Sasser, Mark Spicer, Leroy Thompson and John B. Tonkin: *The Sniper Anthology: Snipers of the Second World War* (Introduction by John L. Plaster). Pen & Sword, 2012

John Hamilton: *Gallipoli Sniper: The Remarkable Life of Billy Sing*. Frontline Books, 2015

Craig Harrison: *The Longest Kill: The story of Maverick 41, one of the world's greatest snipers*. Sidgwick & Jackson, 2015

Robert Harvey: *Mavericks: The Maverick Genius of Great Military Leaders* (Constable, 2008)

Michael E. Haskew: *The Sniper at War: From the American Revolutionary War to the Present Day*. Thomas Dunne, 2005

Hesketh Vernon Hesketh-Prichard: *Sniping in France, With Notes on the Scientific Training of Scouts, Observers, and Snipers*. Hutchinson & Co., 1920 Roland Kaltenegger: *Eastern Front Sniper: The Life of Matthäus Hetzenauer*. Greenhill Books, 2017

Grigory Krivosheev (editor): *Soviet Casualties and Combat Losses in the Twentieth Century*. Greenhill Books, 1997

Chris Kyle, with Jim DeFelice and Scott McEwen: *American Sniper: The Autobiography of the Most Lethal Sniper in U.S. History*. William Morrow, 2015

Richard D. Law, *Sniper Variations of the German K98k Rifle*. Collector Grade Publications, 1996

Herbert W. McBride: *A Rifleman Went To War: How One Man Revolutionised Sniping in World War One*. Endeavour Press, 2016

Tom. C. McKenney: *Battlefield Sniper: Over 100 Civil War Kills*. Pen & Sword, 2016

George Markham: *Guns of the Elite*. Cassells, 1987

Jean Martin, *Armes à Feu de l'Armée Française 1860 à 1940*. Crepin-Leblond, 1974

Dan Mills: *Sniper One: The Blistering True Story of a British Battle Group under Siege*. Penguin Group, 2007

Roger Moorhouse: *Killing Hitler: The Third Reich and the Plots against the Führer*. Vintage Publishing, 2007

Leigh Neville: *Guns of the Special Forces, 2001–2015*. Pen & Sword, 2015

——: *Modern Snipers*. Bloomsbury, 2016

Yevgeni Nikolaev: *Russian Sniper: A Memoir of the Eastern Front in World War II*. Greenhill Books, 2017

Lyudmila Pavlichenko: *Ya-snaïper* ('Me: the sniper'). Moscow, 2014. (To be published in English in 2018 by Greenhill Books, as *Lady Death: The Memoirs of Stalin's Sniper*)

Martin Pegler: *Out of Nowhere: A History of the Military Sniper*. Osprey Publishing, 2004

——: *Sharpshooting Rifles of the American Civil War*. Osprey Publishing, 2017

——: *Sniping in the Great War*. Pen & Sword, 2008

Josef Pilyushin: *Red Sniper on the Eastern Front*. Pen & Sword, 2010

John L. Plaster: *Sharpshooting in the Civil War*. Paladin Press, 2009

——: *The History of Sniping and Sharpshooting*. Paladin Press, 2008

———: *The Ultimate Sniper.* Paladin Press, 2006

Charles W. Sasser & Craig Roberts: *One Shot—One Kill.* Pocket Star Books, 1990

Peter R. Senich: *The German Sniper.* Paladin Press, 1996

Clifford Shore: *With British Snipers to the Reich.* Frontline Books, 2012

Ian Skennerton: *The British Sniper.* Skennerton Publishing, 1983

———: *The Lee-Enfield Story: The Lee-Metford, Lee-Enfield, S.M.L.E. and No. 4 Series Rifles and Carbines, 1880 to the Present.* Greenhill Books, 1993

Mark Spicer: *An Illustrated Manual of Sniper Skills.* Pen & Sword, 2016

———: *Sniper.* Salamander Books, 2001

Bruno Sutkus: *Sniper Ace From the Eastern Front to Siberia: The Autobiography of a Wehmacht Sniper* (Introduction by David L. Robbins). Frontline Books, 2009

Lyuba Vinogradova: *Avenging Angels: Soviet Woman Snipers on the Eastern Front [1941–5]* (Introduction by Anna Reid). Maclehose Press, 2017

Albrecht Wacker: *Sniper on the Eastern Front: The Memoirs of Sepp Allerberger Knight's Cross.* Pen & Sword, 2016

John Walter: *Central Powers Small Arms of World War One.* Crowood Press, 1999

———: *German Military Letter Codes 1939–1945: A concise identification guide to the manufacturers' marks of the Third Reich: by code, by name and by location.* Tharston Press, 2005

———: *Guns of the Elite Forces.* Frontline Books, 2016

———: *Guns of the Reich: The Small Arms of Hitler's Armed Forces, 1933–1945.* The History Press, 2016

———: *Rifles of the World.* Krause Publications, fully revised 3rd edition, 2006

———: *The World's Elite Forces. Small Arms and Accessories.* Greenhill Books, 2002

Vassili Zaitsev: *Notes of a Russian Sniper: Vassili Zaitsev and the Battle for Stalingrad.* Frontline Books, 2015

Index